EASTERN COLLEGE
L
ST. DAVI

D0564132

Atomic and Molecular Structure: THE DEVELOPMENT OF OUR CONCEPTS

ATOMIC AND MOLECULAR STRUCTURE: The Development of Our Concepts

WALTER J. LEHMANN

John Wiley & Sons, Inc. New York · London · Sydney · Toronto

QD
461
.L5

1-15-77

Copyright © 1972, by John Wiley & Sons, Inc.

All rights reserved. Published simultaneously in Canada.

No part of this book may be reproduced by any means, nor
transmitted, nor translated into a machine language without the written permission of the publisher.

Library of Congress Cataloging in Publication Data

Lehmann, Walter J.   1926-
    Atomic and molecular structure.

    Includes bibliographies.
    1. Atomic theory.   2. Molecular theory.

I. Title.

QD461.L5   541'.22   70-37434

ISBN 0-471-52440-9

PREFACE

This book is based on a course given to liberal arts students and other nonscience majors at the University of Massachusetts at Boston. It emphasizes how scientists think rather than what their current theories are. *How* does the scientist develop his theories; how does he prove them? How is his work affected by developments in other fields, by the general religious and political climate, by economic factors, and by available means of communication?

To explore these questions, we trace the development of our concepts of atomic and molecular structure. And what a wonderful, logical development it frequently is! It has all the elements of a good detective novel: an elusive "culprit"; a few star detectives, supported by many bit actors; good hunches and bad hunches; painstaking follow-up of clues; magnificent deductions of which a Sherlock Holmes could be proud; pleasant surprises; severe disappointments; and a solution of the mystery through sheer perseverence.

The material is presented in five parts. Part One begins with the work of Boyle and Lavoisier and culminates in Mendeleev's development of the periodic table in 1869. Part Two traces the evolution of the nuclear atom—a dense nucleus with orbiting electrons. Because the evidence that led in 1911 to this atom model was mostly electrical in nature, the discussion is introduced by a study of simple electrical phenomena and some basic concepts of physics. Part One establishes the atomicity of matter; Part Two establishes the atomicity of electricity.

Part Three examines the outer structure of the atom and ends with Bohr's 1913 atom model, which placed the outer electrons in well-defined orbits. Since this model was based mostly on spectroscopic evidence, Part Three examines the nature of light, beginning with Newton's early experiments with prisms.

Part Four explores the origin and philosophy of quantum mechanics, which gave birth to the present model of the atom. This model no longer has circling electrons in neat orbits, but instead has a nucleus surrounded by fuzzy charge clouds. Effort is made to develop feeling for quantum mechanics, one of the most important fundamental concepts since the time of Newton. The philosophical implications of quantum mechanics—involving questions of causality, indeterminacy, free will, and the nature of reality—are of great interest to the nonscientist as well.

The structure of molecules is examined in Part Five: the nature of the chemical bond; the shapes of molecules, and the relation of molecular geometry to properties of substances; how relatively few different atoms form such a great variety of substances and how we can understand the behavior of these substances by understanding molecular structure. The discussion focuses not so much on our understanding of molecular structure as on how scientists arrived at this understanding. Accordingly, the last chapter of the book is devoted to a brief discussion of some of the tools and techniques used by the modern structural chemist.

Throughout, the book is concerned with the unity of science and the interrelation of science and the humanities. It also shows that most developments in science result from collective efforts of many scientists, often in spite of separation by national boundaries. We cite the work of scientists who lived three hundred years ago and the work of scientists of more recent times. These examples all illustrate the approach of scientists today and point up the empirical nature of science: *Theory guides, experiment decides.* The scientist's laws and theories must conform to reality and are subject to constant critical reexamination. (In principle, then, scientists are very open-minded people!)

Footnotes point out a few sections that could be omitted without loss of continuity in courses that are hard-pressed for time. For abbreviated courses, Part Five could be omitted and conceivably even Part Four (although that would be regrettable). On the other hand, the reading lists at the end of each part suggest references for amplification and additional reading.

To the Student

When studying a particular chapter, first read the introduction and rapidly scan the rest of the chapter to gain a general orientation. Then return to the beginning and study each section carefully.

Problems interspersed within the text are usually essential to understanding the material and should be solved before proceeding to the next section. Additional problems at the end of some chapters often introduce new ideas of secondary importance or are designed as self-test exercises. All problems are intended to illustrate or enlarge upon material in the text, rather than to develop skill in solving numerical problems. Answers to selected problems can be found at the end of some chapters. Approximate answers usually suffice.

When studying for comprehensive examinations, review especially the introductions and summaries or conclusions of chapters and parts and refer frequently to the detailed Table of Contents to gain overall perspective.

You may find some of the problems difficult, but no previous courses in science or mathematics are required to enable you to solve them. Appendix 1 briefly discusses exponential notation, significant figures, approximate calculations, and the use of units. Appendix 2 introduces the units and history of the metric system. A glossary of important terms also is appended.

Above all, don't memorize definitions or laws—try to understand their meaning. Don't memorize dates either. They are merely included to indicate the sequence of the birth of ideas, not necessarily to assign priority of invention and discovery. We are interested in the logical development of ideas rather than the tracing of their history.

Dates may refer to the first conception of an idea, its preliminary communication (perhaps in a letter or at a meeting), its submission to a scientific journal, or its appearance in print in such a journal; for this reason different authors may cite slightly different dates.

Frequently, several scientists came up with similar ideas independently and simultaneously. To avoid confusion, I intentionally do not always mention co-discoverers.

I hope that this book will give you some insight into the "scientific mind," some feeling for the way scientists approach a problem, analyze it, try to solve it, and attempt to verify conclusions. You may, at some future time, be called upon—as a member of a committee or as a voter— to pass judgment on some project that involves spending public funds or energy in such fields as environmental pollution, defense, disarmament, public health, etc. Perhaps the analytical method you encounter here will help you to appraise the pros and cons and determine what other information might be needed to help you reach a conclusion.

I wish to thank Professor Lowell M. Schwartz for his critical review of the manuscript and his many helpful suggestions. I also wish to thank my wife for her help in improving the readability of the text, and my daughter Monica for her assistance with the index. Finally, I am grateful to Professor H. Siebert and the University of Heidelberg for their material assistance and forebearance during the production stage of this volume.

WALTER J. LEHMANN

CONTENTS

xiii

Atomic and Molecular Structure: THE DEVELOPMENT OF OUR CONCEPTS

PART ONE

Developments Leading to the Periodic Table

Part One (Chapters 1–5) traces the chain of events that led to the discovery that certain chemical elements have predictable similarities. This discovery was one of the key clues to the mystery of atomic structure.

The sequence was as follows. First came the realization that matter is made up of particles and that the particles of a given element are alike but are unlike those of other elements. Then came the realization that elements have well-defined properties and that the properties of certain elements are similar. Finally it was discovered that these similarities appear at regular ("periodic") intervals if the elements are arranged in the order of their atomic weights. These similarities and their so-called periodicity were expressed concisely in 1869 in a schematic arrangement called the periodic table.

The periodic table not only enabled chemists to systematize a vast amount of knowledge about the properties of elements and their compounds but also led to new inquiries, which eventually clarified the structures of atoms and of the matter in our universe.

Chapter 1

THE BEGINNINGS OF CHEMISTRY

EARLY HISTORY

Chemistry as a science is only two or three hundred years old. It is true that the peoples of early civilizations in China, India, Mesopotamia, and Egypt knew how to extract metals from ores, how to prepare dyes and medicines, how to manufacture and glaze ceramics, how to brew beer, and how to embalm their dead. But apparently these peoples made no attempt to understand the properties of matter that made all these chemical processes possible.

On the other hand, the Greeks, beginning about the sixth century B.C., did speculate about the nature of matter. The philosopher Heraclitus (540–475 B.C.) believed that all substances were composed of a single element—fire. Later, Empedocles (490–430 B.C.), and still later, Aristotle (384–322 B.C.) suggested that all matter is composed of four basic elements—earth, fire, air, and water. They believed that variations in the proportions of these elements and of the four basic properties—cold, warmth, dryness, and wetness—accounted for the differences between substances (Fig. 1-1). (The doctrine of four basic elements was subsequently disproved, but the concept of basic building blocks of nature has remained fruitful to this day.) Such generalizations about natural phenomena were typical of the Greek philosophers. However, they did not normally try to verify their ideas through experimentation, which they considered undignified. Most of their "natural philosophy" remained pure speculation.

Chemistry as a practical art continued to flourish under the Greeks and later under the Arabs. After conquering Spain in A.D. 711–713, the Arabs in turn spread their knowledge into western Europe.

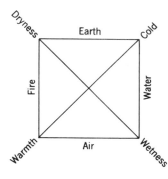

Fig. 1-1. The four basic elements of Empedocles and Aristotle.

At that point, chemistry took an unfortunate turn. One group of chemists, the alchemists, wanted to convert ordinary metals into silver and gold. Such transformations would have been reasonable if it were true that all matter is composed of only four elements. Another group, the iatrochemists (from the Greek *iatros,* physician), wanted to discover an "elixir" of eternal life. Unfortunately, the search for these elusive goals not only was unsuccessful but also fostered secrecy, quackery, and fraudulence. On the other hand, these workers did accumulate considerable knowledge about the chemical behavior of specific substances. In the process of pursuing their objectives they discovered many new substances, including phosphorus, various metals, mineral acids, opium, and other medicines. They also developed and refined useful laboratory techniques, including distillation, filtration, and crystallization. But they came up with only few theories of general value.

SCIENTIFIC CHEMISTRY

We usually think of "scientific" chemistry as beginning with Robert Boyle (Anglo-Irish, 1627–1691). Before Boyle, investigators were usually men of independent means, mostly English, who pursued science as a hobby and whose independence and leisure allowed them to develop new theories about natural phenomena. These investigators abandoned the secretive ways of the alchemists. They organized scientific societies, through whose meetings and journals they exchanged findings and ideas. Such societies were founded in Rome (1602), in London (the Royal Society, 1662), and in Paris (1666). This open exchange of information in part explains why distantly separated scientists thought along similar lines and almost simultaneously came up with identical but seemingly independent discoveries.

Boyle often is called the founder of scientific chemistry for at least three reasons: (1) He was among the first to study chemistry as an intellectual pursuit, not merely as an aid to medicine or metallurgy; (2) he introduced a rigorous experimental method into chemistry; and (3) he defined (or, rather, redefined) the term "element."

THE DEFINITION OF "ELEMENT"

Perhaps Boyle's most far-reaching contribution was his redefinition of the word "element." According to Boyle, *an element is a substance that is not made up of other substances;* nonelemental substances are composed of two or more elements and can be decomposed into these elements.

Boyle went on to show that the four "elements" of the Greeks (earth, fire, air, and water) were not elements in the sense of his definition. He did this by demonstrating that these substances could be converted into each other; therefore, they were not "elemental." For example, he converted water into "earth" (in the form of plants), and he generated "air" (gas) from "earth" (iron) by the action of acid. Thus Boyle could use his definition to demonstrate experimentally that a given substance is *not* an element; but his definition did not explicitly provide a positive test for telling whether a given substance *is* an element.

To be useful, a definition should prescribe some operation for detecting, and perhaps measuring in the laboratory, that which is being defined. Let us therefore redefine an element as

a substance that cannot be broken down by chemical means into two or more other substances.

This definition provides a definite operation for telling whether a given substance is an element: Can we or can we not break it down into other substances? We call this an *operational definition.*

Our refinement of Boyle's definition still has a major shortcoming. Identification of an element is only provisional, for how can we be sure that a substance tentatively identified as an element will not be decomposed by a clever chemist at some future time? In fact, Boyle himself defined as elements some substances that were later shown not to be elements after all. Among these were quicklime—a compound of calcium and oxygen—and light and heat—not material substances, but forms of energy (*see* Chapter 6).

Our definition has another, more subtle shortcoming. Today we can convert mercury into gold. Does this mean that gold is not an element? No, such a transmutation requires the use of a cyclotron or similar equipment and is not considered to have been accomplished by "chemical means." We restrict "chemical means" to the relatively simple methods used in chemical laboratories. Such an argument brings us dangerously close to a vicious circle.

These shortcomings of the definition of "element" persisted until 1913. In that year H. G. J. Moseley's discovery of atomic numbers provided a nonprovisional, unambiguous definition of "element" (*see* Chapter 11). Meanwhile, Boyle's ideas enabled chemists to be more systematic in their quest for new substances. By 1800, some twenty-five or thirty elements were identified; today we recognize more than a hundred elements (*see* endpaper).

THE INTRODUCTION OF QUANTITATIVE METHODS

Some people start their chronology of scientific chemistry a hundred years after Boyle, with the Frenchman Antoine Lavoisier (1743–1794). Lavoisier's work is characterized by systematic application of quantitative methods and conscientious use of the chemical balance in experiments.

In 1774 Lavoisier heated tin with air in a closed vessel. He demonstrated through careful weighings on the balance that there was no change in the weight of the system even though the tin had combined with the oxygen of the enclosed air to yield a white powder (tin oxide). Lavoisier generalized this observation as the *Law of Conservation of Mass:*

> *The total weight of the reagents before a reaction takes place is equal to the total weight of the products after completion of the reaction.*

"Matter can be neither created nor destroyed" is another way of saying it.[1]

We can demonstrate the Law of Conservation of Mass, in a manner similar to Lavoisier's, by weighing a photoflash bulb before and after a flash. After the flash, the metal filament has disappeared and a white powder has appeared in its place. Obviously, a chemical reaction has taken place in the bulb during the flash: The metal magnesium has combined with the oxygen in the bulb to yield white magnesium oxide. But the weight of the bulb remains the same (Fig. 1-2).

---

[1] Since Einstein's introduction in 1905 of his famous equation, $E = mc^2$, we realize that energy can be converted into matter. Nonetheless, the Law of Conservation of Mass is still applicable to chemical reactions, because any mass gains or losses during chemical reactions are much too small to be measured in practice.

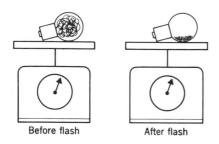

Fig. 1-2. Conservation of mass: The weight of the flash bulb remains unchanged.

Before flash   After flash

A NEW THEORY OF COMBUSTION

Lavoisier made another major contribution to chemistry by his revolutionary explanation of the combustion process. According to the then widely accepted "phlogiston theory," burning was the escape of the "phlogiston" (fire principle) trapped in a substance. What remained after burning was dephlogistinated ash, called calx.

$$\text{combustible substance} \rightarrow \text{calx} + \text{phlogiston}$$

Such substances as coal and sulfur were thought to consist almost entirely of phlogiston, because little or no ash remained after they were burned. But why did the burning process require air? According to the phlogiston theory, air was required to "soak up" the escaping phlogiston.

However, the phlogiston theory did not adequately explain the combustion of metals. When metals are heated in air, they lose their metallic character and are transformed into other substances—for instance, iron into iron rust. Lavoisier, using his balance, discovered that rust weighed *more* than the original metal. The defenders of the phlogiston theory explained this finding by ascribing *negative* weight to phlogiston!

Lavoisier considered that explanation implausible. In 1775 he suggested that combustion is not a process in which something (phlogiston) is released, but a process in which something is added—namely, the gas oxygen, a constituent of air:

$$\text{metal} + \text{oxygen} \rightarrow \text{metal oxide}$$

(Oxygen had been discovered only the year before by the Englishman Joseph Priestley [1733–1804] and, separately, by the Swede Carl Wilhelm Scheele [1742–1786].)

Lavoisier verified his hypothesis through the following experiment. Over a charcoal fire he heated mercury metal in a retort whose neck opened into a bell jar containing a measured volume of air (Fig. 1-3).

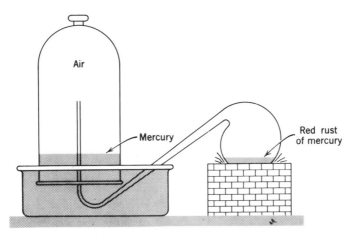

Fig. 1-3. Lavoisier's arrangement for heating mercury in a confined volume of air.

After a while he noticed the formation of a red powder on the surface of the heated mercury. After twelve days the amount of red powder no longer increased, and Lavoisier allowed the apparatus to cool. He now observed that the volume of the air in the retort and bell jar had lost one-fifth of its original volume (50 cubic inches had shrunk to 42). The remaining "air" no longer supported burning, and a mouse was quickly suffocated in it. Lavoisier called this "air" *azote*; we call it *nitrogen* today. Lavoisier collected the red powder and found it to weigh 45 grains. The powder when reheated hotter than before gave up a gas and was reconverted into 41½ grains of mercury. This gas behaved exactly like Priestley's oxygen; it "was more capable of supporting respiration and combustion than ordinary air." When it was added to the previous nitrogen, ordinary air was formed.

Thus Lavoisier proved that air consists of two parts: one-fifth oxygen (the active ingredient) and four-fifths nitrogen (the inactive ingredient). He also showed that oxygen combines with mercury (and other elements) to produce an oxide:

$$\text{mercury} + \text{oxygen} \rightarrow \text{mercuric oxide}$$

Lavoisier's weighings indicated mercuric oxide to contain about 92% mercury (41.5/45) and therefore 8% oxygen. Modern weighings would indicate 92.6% mercury.

Through this and similar experiments, Lavoisier converted most of his contemporaries to his theory of combustion—with the ironic exception of

Priestley, one of the discoverers of oxygen! Today no one questions that combustion is an *oxidation* process.

SUMMARY

In this chapter we have outlined the beginnings of chemistry from prehistoric times until the latter part of the eighteenth century. As we shall do throughout the book, we have selected only those matters germane to our central theme—the quest for understanding atomic structure.

Early civilizations in the Far East and Near East used chemical processes for practical ends, such as metal extraction, dyeing, brewing, and embalming. The Greeks, on the other hand, speculated about the nature of matter but failed to follow through with experimentation. Experimental techniques were developed in the Middle Ages by alchemists and so-called medicinal chemists, who unfortunately employed them toward futile ends.

Scientific chemistry, as it is known today, began in the latter half of the seventeenth century with Boyle, to whom it was an intellectual pursuit, not just a means to an end. He insisted that theories of chemistry must be based on experimental evidence. To Boyle we also owe the definition that an *element is a substance that cannot be broken down by chemical means into two or more other substances.*

A hundred years later, Lavoisier introduced the chemical balance and quantitative methods. He formulated the *Law of Conservation of Mass.* His experiments proved the old phlogiston theory untenable: Combustion is not a release of phlogiston but an oxidation process. Lavoisier's rigorous quantitative methods prepared the way for the first concrete steps toward a comprehensive atomic theory.

Problem 1-1. Oxygen is an element. When electric sparks are passed through pure oxygen, ozone is generated. Is ozone an element (a) according to Boyle's original definition; (b) according to our operational redefinition? *Note:* Many elements exist in more than one form, such as sulfur (rhombic and monoclinic) and carbon (graphite and diamond).

Chapter 2

THE LAWS OF CHEMICAL COMPOSITION AND THE ATOMIC THEORY

In order to establish a firm foundation for his theory of atoms, John Dalton (1766–1844) needed two more fundamental laws of chemistry. The first was the Law of Constant Composition (1799), usually credited to the French chemist Joseph Proust (1754–1826). The other, the Law of Multiple Proportions, was formulated by Dalton himself and others, after the original announcement of his atomic theory in 1803. In part, discovery of this law was a consequence of the predictions of the early version of the atomic theory, but in turn it led to a revision in that theory in 1808.

THE LAW OF CONSTANT COMPOSITION

The *Law of Constant Composition,* or, as it is sometimes called, the Law of Definite Proportions, asserts that

> *a given compound always contains the same elements chemically united in the same proportions by weight.*

For example, when 9 grams (g) of pure water is broken down into its elements, it will always yield 8 g of oxygen and 1 g of hydrogen.[1] The weight-percentage composition of water, therefore, is as follows:

$$\text{Oxygen:} \quad \frac{8 \text{ g}}{(8 + 1) \text{ g}} \, 100\% = \frac{800}{9} \, \% = 88.9\%$$

[1] The metric system is discussed in Appendix 2.

$$\text{Hydrogen}: \frac{1 \text{ g}}{(8 + 1) \text{ g}} \, 100\% = \frac{100}{9} \% = 11.1\%$$

$$\text{Total} = 100\%$$

$$\text{Weight-ratio of oxygen to hydrogen} = \frac{88.9 \text{ g}}{11.1 \text{ g}} = 8:1$$

The composition of water is constant, as long as it is pure, regardless of its source or its mode of preparation. This is true for rainwater, for water purified from human perspiration, for recondensed water of crystallization, or for water made by the explosive union of hydrogen and oxygen gases.

The reverse takes place in the formation of compounds. Thus 1 g of hydrogen will always combine with 8 g of oxygen to yield 9 g of water. If a mixture of 1 g of hydrogen and 11 g of oxygen is exploded, only 9 g of water will be formed; 3 g of oxygen will remain uncombined. If a mixture of 1 g of hydrogen and only 2 g of oxygen is exploded, 0.75 g of hydrogen will be left over; the combining ratio is still 0.25 : 2, or 1:8.

Problem 2-1.[2]   If 5 g of hydrogen are exploded with 50 g of oxygen, how much water will be formed? How much of which gas, if any, will be left over?

Problem 2-2.   (a) How many pounds of hydrogen and of oxygen can be obtained from 27 lb of water?
(b) What is the weight-percentage composition of this water?

Problem 2-3.   In an experiment, 63.5 g of finely granulated copper were placed into a glass tube and, while the tube and its contents were heated over a flame, an airstream was passed over the reddish copper until it turned back (Fig. 2-1). The resulting product weighed 74.4 g. From these data, determine the percentage composition of the product.

Problem 2-4.   The foregoing experiment was continued. The material and apparatus used in Problem 2-3 were reassembled, and more air was passed over the previous product for several minutes. Afterward the product, still black, weighed 77.2 g. What do you deduce from this result?

[2] Problems interspersed throughout the text are essential to understanding the material. It is advisable to work these problems before going on, especially since they usually require little time. Some chapters have additional problems at the end. These occasionally introduce new ideas—interesting, but of secondary importance—or else they are self-testing exercises. Answers to selected problems can be found at the end of each chapter. Approximate answers usually suffice (*see* discussion of significant figures in Appendix 1.)

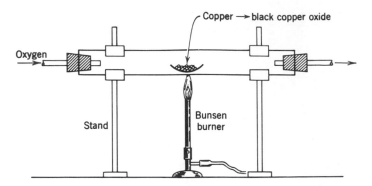

Fig. 2-1. Reaction of oxygen with copper to form copper oxide.

Problem 2-5. When the previous steps were repeated twice more, the products weighed, respectively, 79.6 g and 79.4 g. What do you conclude now? Calculate the percentage composition on the basis of these two results.

The Law of Constant Composition can be demonstrated even more dramatically through volume measurements of gases. Thus a specific volume of hydrogen will combine completely with a definite volume of oxygen to form water. This relationship will be constant no matter what the reaction conditions are, so long as the volumes are measured at identical temperatures and pressures. Any excess of either gas will remain unreacted (Fig. 2-2).

COMPOUNDS VERSUS MIXTURES

The Law of Constant Composition applies only to *compounds,* not to mixtures, whose components may be mixed in practically any proportion. Common examples of mixtures are rocks, dirt, seawater, and air. The components of most mixtures, unlike the constituents of compounds, can be separated easily. In some mixtures we can actually see the various components. Rocks and dirt, for example, contain colored granular material, which can be sorted manually. Seawater, on the other hand, is quite uniform, or homogeneous, in its makeup; we call these uniform mixtures *solutions.* Still, we can separate their components rather easily: Freezing the seawater will cause pure ice (water) to crystallize; or, boiling off the water will leave a residue of salt. Water often helps to separate other mixtures; for example, by dissolving the salt in a salt-sand mixture. Water is

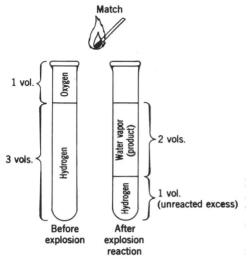

Fig. 2-2. Schematic representation of the Law of Constant Composition: Two volumes of hydrogen *always* react with one volume of oxygen to form water; any excess of either gas remains unreacted.

also used in panning for gold: The dirt is washed away, and gold nuggets remain in the pan. Magnets can separate a mixture of iron filings and almost anything else. But the components of air, a solution, are not easily separated, and it is no wonder that air was long thought to be an elemental substance.

Compounds may also be distinguished from mixtures by their properties. The properties of compounds generally differ markedly from those of their constituents. For example, sodium is a shiny metal, and chlorine is a greenish poisonous gas; the compound sodium chloride, common table salt, is a white, edible crystal. Hydrogen and oxygen are gases; but they react chemically to form water. By contrast, mixtures and solutions, often retain the properties of their components. For instance, seawater or mixtures of sand and salt still taste salty, with an intensity related to the salt concentration. Air, a mixture of nitrogen and oxygen, still retains the life-giving properties of oxygen.

The most characteristic property of compounds is that they obey the Law of Constant Composition. For most purposes the following definition suffices.

Compounds are substances of characteristically constant composition that can be broken down into two or more simpler substances.

(Substances that cannot be broken down into two or more substances have been defined previously as elements.)

The critical student will note a circular argument here: We define a compound as a substance having constant composition and then we observe, as a law of nature, that compounds have constant composition! Sound absurd? Actually, the observation of constant composition was made first, and on that basis a special group of substances—namely, compounds—was recognized. Fully aware of the logical circle in which we may have placed ourselves, let us nevertheless make heuristic use of this operational definition of "compound" and see where it leads us. The important point is that there seems to be a special category of nonelemental substances that have definite, constant composition.[3]

THE ATOMIC THEORY

The Englishman John Dalton first recognized clearly the implications of the Law of Constant Composition. The observation that one element will combine with only a definite amount of another element to form a specific compound indicated to him that matter is not a continuous fluid but comes in discrete particles, which he called *atoms*. Dalton described the properties of these particles as follows: They cannot be divided (because they are nature's *basic* building blocks); and they cannot be destroyed or created (because of the Law of Conservation of Mass). All atoms of a given element are identical and differ from the atoms of all other elements, particularly in weight. Atoms preserve their identity in chemical reactions; atoms of different elements combine in definite, simple ratios to form the *molecules* of compounds.

The concept of atoms was not original with Dalton. Almost 2500 years earlier the Greek philosophers Leucippus and Democritus had proposed the concept of atoms (from *atomos*, indivisible). Their ideas were mere speculation, unsupported by experimental evidence, even contradicted by common sense, and were all but forgotten until Dalton's time. Several developments had paved the way for a revival of the idea of atoms; among them, increasing disenchantment with the contention of alchemists that different kinds of matter could be converted into one another; clarification of the concept of elements by Boyle in 1660; Lavoisier's formulation of the Law of Conservation of Mass in 1774; and Proust's Law of Constant Composition in 1799. Undoubtedly, Newton's hundred-year-old

[3] The word *heuristic* is closely related to Archimedes' legendary exclamation "Heureka!" or "Eureka!" It means: "helping to discover or learn; guiding or furthering investigation," or, "Let's try it and see where it leads us."

corpuscular theory of light, essentially an atomistic theory (*see* Chapter 12), also influenced Dalton. In retrospect, it seems that by 1800 the formulation of an atomic theory was almost inescapable.

There were others who proposed atomic theories almost simultaneously. Dalton's theory, however, had two important new features: (1) introduction of a quantitative aspect by ascribing different weights to atoms of different elements; and (2) application of the principle of simplicity— i.e., if there is no information to the contrary, make the simplest possible assumption and pursue its consequence. Thus Dalton pictured the formation of molecules by atoms combining in ratios of small whole numbers, such as 1:1 or 1:2, or perhaps 2:3, rather than 19:43. In accordance with this principle,[4] Dalton originally pictured the formation of a water molecule as the union of one hydrogen atom with one oxygen atom. In today's symbols this would be

$$H + O \rightarrow HO$$

Let us take another look at some experimental evidence in the light of Dalton's atomic theory. It has been demonstrated that water can be decomposed into hydrogen and oxygen in the volume ratio of 2:1; conversely, hydrogen and oxygen combine to form water *only* in the volume ratio of 2:1. If combining weights rather than volumes are determined, the hydrogen:oxygen ratio is found to be *always* 1:8 weight or mass units (grams, pounds, kilograms, ounces, etc.). Finding that these facts are always so, we can deduce that:

1. Water is not an element, because it can be decomposed into two different substances.

2. Water is a compound, not a mixture, because it is characterized by constant composition.

3. Hydrogen and oxygen may be elements or compounds. We list them as elements today because we have never been able to decompose them into simpler substances.[5]

4. The observations are compatible with the atomic theory (but they do not prove it). If there *are* atoms and if formation of compounds involves interaction of these atoms, it is reasonable that all molecules of a

[4] Today we realize that there are many exceptions to this simple rule. The formula of morphine, for instance, is $C_{17}H_{19}NO_3$; 17:19 is not really a "simple" ratio. (The subscripts indicate the number of atoms of each element contained in one molecule of the compound.) Nevertheless, Dalton's heuristic assumption of simplicity enabled him to develop his atomic theory.

[5] Although both ordinary (molecular) hydrogen and oxygen can be decomposed into atoms, they cannot be decomposed into "two or more substances;" therefore they meet the criterion for elements.

given compound contain the same number of atoms; and, if all atoms of a given element have the same weight, a compound must have a definite composition by weight.

Problem 2-6. (a) *If,* as Dalton mistakenly believed, a water molecule were composed simply of one oxygen atom and one hydrogen atom (HO), what would be the ratio of the weight of the oxygen atom to that of the hydrogen atom? (*Hint:* Look at the combining weights.)

(b) If the formula instead is H_2O, what is the weight ratio of the oxygen atom to the hydrogen atom? (Note that you need to know the formula, or atomic composition, of water in order to calculate atomic weights.)

(c) Does the observed volume ratio of 2:1 prove that water contains two hydrogen atoms for each oxygen atom? (Remember, we have not yet shown that equal volumes contain equal numbers of atoms.)

THE LAW OF MULTIPLE PROPORTIONS

Let us now look more critically at Dalton's application of the rule of simplicity. Even if atoms combine in simple ratios, two elements A and B could still combine in more than one simple ratio (as in fact they do)— say, 1:1 (AB) as well as 2:1 (A_2B). Each compound would still have a definite composition.

Take the case of copper and oxygen, for example. We found (*see* Problem 2-5) that 63.5 g of copper combined with 16 g of oxygen to form 79.5 g of black copper oxide. There is another compound of these two elements, a red oxide, containing 127 g of copper per 16 g of oxygen. The ratio of the two quantities of copper that combine with the *same* amount of oxygen (16 g) is 63.5:127, or 1:2. (Today we know that the formulas are, respectively, CuO and Cu_2O.

The then unknown existence of these two copper-oxide compounds caused a famous dispute in 1799–1808, between C. L. Berthollet (1748–1822), a colleague of Lavoisier, and J. L. Proust. Berthollet observed that copper could combine with *any* amount of oxygen, up to a fixed limit, and cited this fact as evidence *against* the validity of the Law of Constant Composition. Proust, however, demonstrated in his laboratory that Berthollet's "compound," which appeared to be an arbitrary combination of elements, was just a variable mixture of two compounds, each with a quite definite composition—namely, black and red copper oxides (CuO and Cu_2O). Because Proust confirmed the Law of Constant Composition through this work, he is usually regarded as its discoverer.

The controversy had some additional results. It established a clear definition of a chemical compound, as distinct from a mixture; and it directed the attention of chemists to the fact that two elements frequently combine in more than one proportion. This observation is stated in the *Law of Multiple Proportions:*

> *If two elements A and B form more than one compound, the weights of B in all compounds that unite with a fixed weight of A are in the ratio of small whole numbers.*

This law is, of course, in harmony with the atomic theory.

Nitrogen and oxygen also form more than one compound. There are, in fact, three nitrogen-oxygen gases (nitrous oxide, nitric oxide, and nitrogen dioxide), each with different properties (e.g., color, boiling temperature, and support of combustion). The oxygen content of each can be determined experimentally by passing the gas over a weighed quantity of hot copper wire, which then forms black copper oxide (CuO). Each experiment is continued until a given volume of nitrogen has been formed and collected. The increased weight of the copper oxide over the original copper wire represents the amount of oxygen given up by each gas. The data below show the amount of oxygen that combines with a fixed amount of nitrogen.

	Compound A	Compound B	Compound C
1. Final mass of tube containing copper oxide	122.75 g	211.80 g	195.85 g
2. Initial mass of copper-filled tube	121.95 g	210.25 g	192.60 g
3. Weight gain (1−2) = mass of oxygen	0.80 g	1.55 g	3.25 g

The weights of oxygen that reacted with a fixed amount of nitrogen to form the three compounds are in the ratio of 0.80:1.55:3.25, or 1:1.94: 4.06; this equals 1:2:4 within the experimental error (i.e., give or take 0.05 g).[6]

Problem 2-7. (a) Write a possible set of three molecular formulas for the three nitrogen oxides.

(b) Write a different set of three, also compatible with the data.

[6] See discussion of significant figures in Appendix 1.

The Law of Multiple Proportions was a logical outgrowth of the atomic theory and its rule of simplicity. Dalton believed that atoms combine in simple ratios to form compounds. Proust, in resolving his controversy with Berthollet, had shown that two elements might combine in more than one weight proportion. Dalton, in turn, in his Law of Multiple Proportions showed that these proportions were not arbitrary but were related to each other in a simple way.

EXCEPTIONS TO THE LAW OF CONSTANT COMPOSITION

Isotopes

According to Dalton's atomic theory, each element is made up of identical atoms, characterized especially by identical weights. About a hundred years later, scientists discovered that some elements are made up of atoms that are nearly identical but have slightly different weights. Such related forms of elements or atoms are called *isotopes* (*see* Chapter 11).

For example, the element boron as found in nature is a mixture of light and heavy boron atoms; these isotopes differ in weight by about 10%. Also, a heavy hydrogen atom, called deuterium, weighs twice as much as the ordinary hydrogen atom. Consequently there can be two types of water: ordinary water, with an oxygen-hydrogen weight ratio of 8:1; and "heavy water" with an oxygen-hydrogen (deuterium) weight ratio of 8:2.

Strictly speaking, the existence of two forms of a compound with two different weight ratios is a violation of the Law of Constant Composition. Fortunately, for the sake of the atomic theory, these discrepancies were not known in Dalton's day.

Nonstoichiometric Compounds

There is another, more serious, violation of the Law of Constant Composition. Gaseous compounds consist of individual molecules. Only *whole* atoms—one, two, or more—can unite to form molecules. Thus nitrogen-oxide molecules can be NO, N_2O, NO_2, etc. (each with quite distinct properties and identifiable as a separate compound); but there can be no $N_{1.7}O$ molecule. In some solids, however, we *can* have such nonintegral ratios. For example, titanium oxide, a solid crystalline compound, has no individual TiO molecules; the crystal is, rather, a regular array of titanium atoms alternating with oxygen atoms. In such a crystal there may be vacant spots in the crystal lattice ("lattice defects"), due to missing titanium or oxygen atoms, depending on the method of preparation (Fig. 2-3). The average composition therefore may vary continuously from

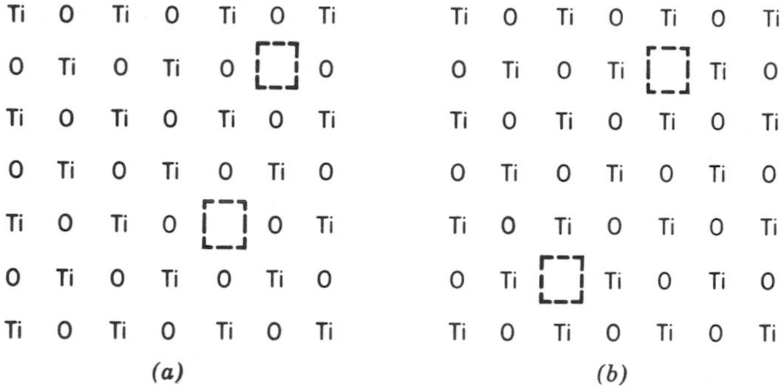

Fig. 2-3. Lattice defects in a titanium oxide crystal. (a) Missing titanium atoms. (b) Missing oxygen atoms.

$Ti_{0.7}O$ to $TiO_{0.7}$. These formulas represent *average* compositions of solid crystals and not the formulas of single molecules.

In contrast to the nitrogen-oxygen compounds, the variation in the composition of titanium oxide does not greatly affect its chemical properties; nor does it affect the basic crystalline structure. The variations do, however, affect the electrical and optical properties of titanium oxide. Thus these so-called *nonstoichiometric compounds* conform in almost all respects to our ideas of a compound, except in their failure to have constant composition.[7]

Interestingly, some of the compounds used to "prove" the Law of Constant Composition were in fact nonstoichiometric compounds, but the variations fell within the allowable experimental error of the chemical analyses of that time. Thus this "law," so crucial to the development of atomic theory, is only an approximation, originally "proved" by data inadequate to reveal its failures.

CONCLUSION

The existence of nonstoichiometric compounds and of isotopes points up the fact that the validity of our "laws" of nature depends on the accuracy of the experiments from which they are derived and on the number and variety of cases investigated. As more accurate experiments are performed, in more varied situations, some laws may have to be refined or even discarded. Other laws, such as the Law of Constant Composition, remain useful if we keep their limitations in mind.

Note that the three laws we discussed—Conservation of Mass, Constant

[7] Stoichiometry deals with measuring the combining quantities of elements.

Composition, and Multiple Proportions—do not prove the atomic theory. These laws are compatible with the atomic theory, even derivable from it, but by themselves do not prove the correctness of the atomic theory, nor even the existence of atoms. However, as more and more data were gathered and were correlated with the help of the atomic theory, scientists became increasingly confident of the general validity of that theory.

ANSWERS TO SELECTED PROBLEMS

2-3. % oxygen $= 100\%$ $(74.4-63.5)/74.4 = 14.7\%$.
        % copper $= 85.3\%$.

2-6. (a) 8:1.

2-5. % oxygen $= 20.1\%$ (average).

Chapter 3

GASES: THE BEGINNING OF ATOMIC WEIGHTS AND MOLECULAR FORMULAS

Once it had been established that atoms exist, that the atoms of each element have different weights, and that atoms combine to form molecules of compounds, it was logical to ask, "What are the relative values of these different atomic weights?" and "What are the formulas of various compounds; that is, in what ratios do atoms combine to form particular molecules?" These two questions are closely related.

Problem 3-1.[1]   (a) Given only the weight ratio of hydrogen to oxygen in water as 1:8 and no information about atomic weights, write down several possible molecular formulas for water.
(b) What is the simplest possible formula?
(c) On the basis of this formula, what would the atomic weight of oxygen be, taking hydrogen as equal to one unit?

We have noted that the principle of simplicity led Dalton to a false conclusion about the formula of water and the atomic weight of oxygen (*see* Problem 2-6). It is also evident from Problem 3-1a that combining weights alone, without atomic weights, will not yield a molecular formula for a product. Nor can atomic weights be deduced from the weight ratios of the reactants unless the formula of the product is known (*see* Problem 2-6). However, we will see in this chapter how knowledge about atoms

---

[1] In solving the problems of this chapter, do not apply any knowledge about chemical formulas that you may have acquired in a previous course, because such knowledge was not available at the time of Dalton.

was advanced by deductions based on *volume* measurements of combining gases and their products.

We will also note how the same experimental evidence was variously interpreted by different investigators. "Science" is not absolute, and "scientific conclusions" are often influenced by the experience and prejudices of an investigator. This was as true in 1800 as it is today. Usually time and further experimentation reveal who is "right" and who is "wrong." However, it is the author's intent throughout this book not so much to belabor the theories themselves as to show how scientists arrived at theories.

THE LAW OF COMBINING VOLUMES

We have indicated that the Law of Constant Composition can be demonstrated by determining combining weights or by measuring combining gas volumes. In 1808 the French chemist J. L. Gay-Lussac (1778–1850) showed that in the formation of compounds the volumes of combining gases not only have *fixed* ratios (the Law of Constant Composition) but also are in *the ratios of small whole numbers.* What is more, *the products, if gaseous, also bear simple whole-number ratios to the reactants,* when measured under the same conditions of temperature and pressure. For example:

> *One* volume of carbon monoxide combines with *one* volume of chlorine to form *one* volume of phosgene;[2]

and

> *two* volumes of ethane combine with *seven* volumes of oxygen to form *four* volumes of carbon dioxide and *six* volumes of water vapor.

This so-called Law of Combining Volumes further bolsters the atomic theory. If matter were a continuous fluid instead of atomistic, the foregoing observations would be completely unintelligible. Why would volumes of fluids combine only in small-integer ratios? If, on the other hand, we think of each volume as containing tiny particles (atoms) that combine with one another, we can come up with a reasonable explanation for the observed phenomenon, as we shall do in the next section.

"EVEN"

Let us for the moment assume that a given volume of any gas contains the same number of particles as an identical volume of any other gas. (Why this might be so is another question, which we shall investigate later.) With this assumption we can readily explain the observation noted

---

[2] There is no small-integer relationship for combining *weights*: 28 g of carbon monoxide combine with 71 g of chlorine to form 99 g of phosgene.

earlier, that one volume of carbon monoxide reacts *completely* with one volume of gaseous chlorine to yield one volume of the gas phosgene:

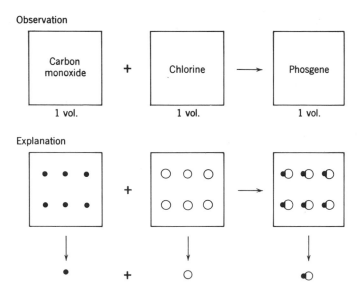

In our illustration we have taken one particle out of each box to form one molecule of phosgene. Since by our assumption the two boxes contain the same number of particles, they will be empty at the same time; i.e., the two gas volumes will react *completely*. If, on the other hand, the boxes contained unequal numbers of particles—say, 6 and 7, respectively, or 190 and 200—some particles would remain in one box and the two gases would not react completely. It was considerations such as these that led several scientists during Dalton's time to theorize that, under the same conditions of temperature and pressure,

Equal Volumes of gases contain Equal Numbers of particles (EVEN).

Problem 3-2. Assuming the equal-volumes-equal-numbers (EVEN) principle, draw the *simplest* picture (analogous to the foregoing illustration) that is compatible with the reported volume ratios for each of the following reactions of gases.

(a) *Three* volumes of hydrogen combine with *one* volume of nitrogen to form ammonia.

(b) *One* volume of ammonia combines with *one* volume of hydrogen chloride to form solid ammonium chloride.

(c) *Two* volumes of hydrogen combine with *one* volume of oxygen to form water.

(d) *Two* volumes of nitric oxide combine with *one* volume of oxygen to form nitrogen oxide.

The Law of Combining Volumes by no means proves the EVEN principle. The observations on combining gas volumes would also be compatible with the idea that the numbers of particles in equal volumes of gases are integral *multiples* of one another. For example, ammonia is formed by the reaction of three volumes of hydrogen with one volume of nitrogen (Problem 3-2b). We might explain this by saying that nitrogen simply contains three times as many atoms as an equal volume of hydrogen, illustrated as follows.

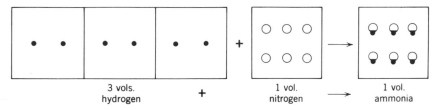

| 3 vols. hydrogen | + | 1 vol. nitrogen | → | 1 vol. ammonia |

Even when the reaction ratio is 1:1, the explanation is not obvious. For example, the combination of carbon monoxide (A) and chlorine (B) in a 1:1 ratio to form one volume of phosgene could be explained equally well by these various alternatives:

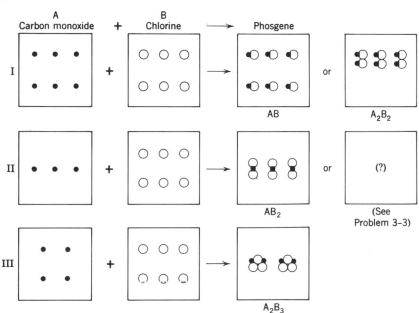

In alternative I, the reactants contain equal numbers of particles per unit volume. If we assume that the product also contains an equal number of particles, its formula would be ⦵; i.e., AB. (This is the "explanation" offered on page 23.) But if the product contains only half as many particles in the same volume, the formula would be ⦵; i.e., A_2B_2, or $(AB)_2$.

In alternative II, chlorine has twice as many particles per unit volume as carbon monoxide. The product formula AB_2 is compatible with this supposition.

Problem 3-3. In the empty box for alternative II draw in another possible product. What would be its formula?

In alternative III, the ratio of reactant particles is 4:6 (i.e., 2:3). A possible product is A_2B_3.

We can draw several conclusions from an analysis of these alternatives:

1. Volume measurements of gaseous reactions and the Law of Combining Volumes do not by themselves prove the equal-volumes-equal-numbers principle.

2. Without the EVEN principle we cannot deduce the formula of the product merely from data on reactant volumes.

3. If the contents of the boxes are to react completely to form small product-molecules, the numbers of atoms in the reactant boxes must be in the ratio of small whole numbers; e.g., 6:6 $(= 1:1)$; 3:6 $(= 1:2)$; 4:6 $(= 2:3)$. If the ratio of the reactant particles in two equal volumes instead were, say, 147:125, they could not react completely to form small molecules; the simplest formula of the product would be $A_{147}B_{125}$.

Problem 3-4. (a) Without assuming EVEN, draw at least two conceivable pictures for the reaction of two volumes of hydrogen with one volume of oxygen to form water.
(b) What is the relationship of the number of hydrogen and oxygen particles per unit volume in each of your pictures?

Problem 3-5. If in the reaction of one volume of hydrogen (A) with one volume of chlorine (B) there are 100 particles and 60 particles per unit volume, respectively, what is the simplest possible formula of the product produced by complete reaction?

The third conclusion that the numbers of particles in equal volumes of gases are integrally related warrants further examination. Dalton's atomic theory says that atoms of various elements are different. Then why would a gas volume always contain *exactly* 2, 3, 4 or 2/3, 3/4, etc., times as many atoms as an equal volume of some other gas? Why not 17, 5.63, or 0.3214 times as many? Why are there always such small-integer relationships? If the atoms of hydrogen occupy more space than those of oxygen, why would they occupy *exactly* twice the space? And why would exactly the same relationship hold for other pairs, such as nitric oxide and oxygen? There seems to be no reasonable answer. But we can make a case for the special integral relationship of 1:1, that is, for the equal-volumes-equal-numbers principle. We can devise models in which the *effective* volumes of all particles are *identical*.

There are analogous situations where effective volumes are independent of actual volumes. For example, the number of racing cars that can be accommodated on a race track depends more on the speeds than on the actual sizes of the individual cars. Likewise, the number of people that can be accommodated in an auditorium generally does not depend on the sizes or weights of the individuals (Fig. 3-1).

Our model of gases is based on the kinetic theory of gases. According to that theory the molecules of a gas are in very rapid random motion, so that the *effective* volume occupied by a gas molecule is much larger than its actual volume, as in our racing-car analogy. We can get an idea of the relation of these two volumes by contrasting the volume occupied by steam and the volume that the steam occupies when it is condensed into liquid water, a ratio of about 1000:1. In liquids the particles presumably are touching each other, so that the observed volumes correspond approximately to the actual total volumes of the particles. The observed volumes

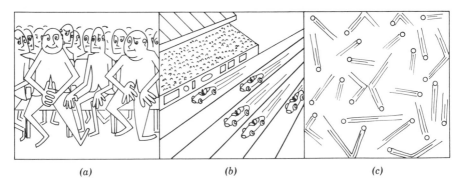

(a)   *(b)*   *(c)*

Fig. 3-1. The volumes occupied by bodies often are independent of their actual volumes. (*a*) Fat and slender people in auditorium seats. (*b*) Racing cars on a race track. (*c*) Molecules bouncing around in a closed volume.

of gases, on the other hand, depend on the speeds of the particles and are almost independent of the actual particle volumes.

The kinetic theory of gases thus explains why equal volumes of gases contain equal numbers of particles, independent of the nature of the particles. This explanation was not available to most chemists in 1810; it was developed in its present form only in the latter half of the eighteenth century. Nevertheless, even though scientists had no plausible explanation in 1810, they used the equal-volumes-equal-numbers principle as a working hypothesis to fit the observations embodied in Gay-Lussac's Law of Combining Volumes.

The EVEN principle brought a bonus with it: an easy method of determining the relative molecular weights of gases. If two equal volumes contain the same number of particles, the individual particle weights will be in the same ratio as the total weights. For example, if the weights of two equal volumes of oxygen and hydrogen are in the ratio of 16:1, we can conclude that the oxygen particle weighs sixteen times as much as the hydrogen particle.

Problem 3-6. A given volume of nitrogen weighs 14 times as much as an equal volume of hydrogen. What is the particle weight of nitrogen, on the basis of hydrogen = 1?

Problem 3-7. What is the particle weight (on the basis of hydrogen = 1) of a gas whose density (i.e., weight per unit volume) is 1.5 times that of nitrogen?

THE CONTROVERSY

Ironically, the EVEN principle was not accepted by Dalton, the "father of the atomic theory." To his mind, two types of experimental observations were incompatible with EVEN. One involved the relationship of reactant and product volumes; the other involved gas densities.

The EVEN principle enabled scientists to explain the reaction of one volume each of carbon monoxide and chlorine to form one volume of phosgene as pictured on page 23: One particle of each reactant combines with one particle of the other reactant to form one (combined) particle, or molecule, of product. In Problem 3-2c you were asked to picture in an analogous simple manner the reaction of two volumes of hydrogen with one volume of oxygen to form a product called water (or steam, in the gaseous phase). Your picture should have looked something like this:

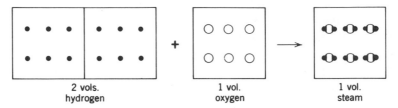

2 vols.
hydrogen

1 vol.
oxygen

1 vol.
steam

The trouble is—and we intentionally did not mention this earlier, to avoid complications—that this reaction yields not one but *two* volumes of steam:

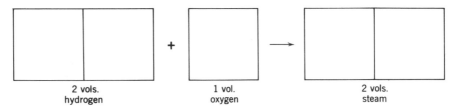

2 vols.
hydrogen

1 vol.
oxygen

2 vols.
steam

Assuming EVEN for the *reactant* atoms, we would have to picture the reaction as follows (remember, we cannot create atoms!):

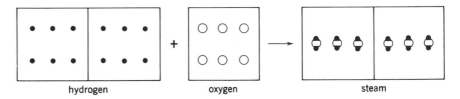

hydrogen

oxygen

steam

Note that, under the given assumptions, EVEN could not apply to the product.

Problem 3-8. Assuming EVEN for both reactants and product, picture the reaction of hydrogen and chlorine in a 1:1 volume ratio to form hydrogen chloride.

Problem 3-9. Now picture the same reaction given the added fact that one volume of hydrogen plus one volume of chlorine yields *two* volumes of hydrogen chloride. Assume EVEN only for the reactants. What is the relationship of the number of particles in each product volume (i.e., the particle density) to that in the reactant volumes? Can you picture this reaction so that EVEN applies to reactants and products alike?

Problem 3-10. Now redraw the pictures for reactions (a) and (d) in Problem 3-2, given the added facts that three volumes of hydrogen combine with one of nitrogen to form *two* volumes of ammonia; and that two volumes of nitric oxide combine with one of oxygen to form *two* volumes of nitrogen oxide.

Such observations of volume relationships convinced Dalton that the equal-volumes-equal-numbers principle was not valid. He did believe that equal volumes apparently had integral (i.e., whole-number) relationships other than 1:1; for example, oxygen and hydrogen both had twice as many particles per unit volume as water. But it seems that he just shrugged off this apparent relationship as a curiosity of nature. He even suggested that more precise measurements might reveal that these ratios, as well as the ratios of combining volumes, were not exactly integral.

Dalton was strengthened in his skepticism about EVEN by other observations, involving gas densities—i.e., the weights of unit volumes of gases. Modern gas-density measurements reveal that oxygen weighs sixteen times as much as hydrogen, and steam weighs nine times as much as hydrogen, under the same conditions of temperature and pressure. In other words, the weight relationships of equal volumes of hydrogen, oxygen, and steam are in the ratios of 1:16:9. Dalton saw no way of both satisfying the EVEN principle for these reactants and products *and* being compatible with the volume and density observations. He thought the reaction was adequately represented by a diagram like the one preceding Problem 3-8, reproduced here for convenience.

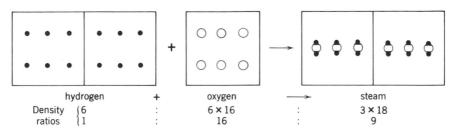

	hydrogen	+	oxygen	⟶	steam
Density ratios	$\begin{cases} 6 \\ 1 \end{cases}$	$\begin{matrix} : \\ : \end{matrix}$	$\begin{matrix} 6 \times 16 \\ 16 \end{matrix}$	$\begin{matrix} : \\ : \end{matrix}$	$\begin{matrix} 3 \times 18 \\ 9 \end{matrix}$

Note that the product in this picture violates the EVEN principle. According to this diagram we explain the observed densities as follows. If we use the weight of the hydrogen atom as the standard weight (hydrogen = 1), and assign 16 to the oxygen atom, then each box of hydrogen as pictured, weighs 6 units; the oxygen box weighs $6 \times 16 = 96$ units; and the steam, which we picture as H_2O, weighs $3 \times (2 + 16) = 3 \times 18 = 54$ units. The weight, or density, ratios are 6:96:54, which equals 1:16:9, as observed.

Such reasoning persuaded Dalton that the EVEN principle was invalid. But once again, the fact that Dalton's picture was *compatible* with measured densities does not *prove* the correctness of his picture or its underlying assumptions. As a matter of fact, Dalton suggested a further simplification. If the EVEN principle is not valid for gaseous *products*, why assume it for the reactants? Dalton applied the rule of greatest simplicity, which led him to the formula HO instead of H_2O for water. He erroneously pictured the reaction of hydrogen and oxygen to form water simply as a 1:1 addition of atoms:

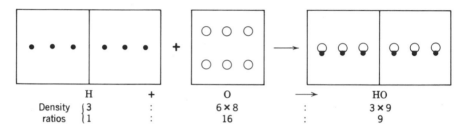

	H	+	O	→	HO
Density ratios	3	:	6 × 8	:	3 × 9
	1	:	16	:	9

He assumed that each oxygen atom weighs 8 times as much as each hydrogen atom (instead of 16 times as much, as in our previous picture). Thus each hydrogen box would weigh 3 units; each oxygen box would weigh $6 \times 8 = 48$ units; and each steam box would weigh $3(8 + 1) = 27$ units. The ratios would be 3:48:27, or 1:16:9, as observed. Thus this diagram explains the observed volume and density data just as well as the previous diagram and has the added advantage of greater simplicity. Dalton felt that in order to explain the experimental evidence he had to sacrifice the beauty and simplicity of the equal-volumes-equal-numbers principle. As we shall see in the next section, Dalton should instead have questioned the validity of his almost-hidden assumption that the hydrogen, nitrogen, and oxygen particles are *atoms*.

Problem 3-11. Ammonia, formed by adding hydrogen to nitrogen, is less dense than nitrogen itself. Explain this on the basis of your answer to Problem 3-10. The density ratio of nitrogen to hydrogen is 14:1. What is the density of your predicted product? How does this compare to the actually observed density of 8.5? Does the picture you drew in answering Problem 3-2 fit the observed density ratios of 1:14:8.5 (or, 2:28:17)?

AVOGADRO'S HYPOTHESIS

The Italian physicist Amedeo Avogadro (1776–1856) rescued the equal-volumes-equal-numbers principle in 1811 by coupling it with the novel suggestion that the particles of gaseous elements could consist of more than one atom—for example, H_2, N_2, O_2. Avogadro felt that saving the simple and logical EVEN principle was worth sacrificing the simple but unnecessary idea of monatomic (single-atom) elements.

We showed in the previous section that, for the reaction of hydrogen and oxygen producing water and for the reaction of hydrogen and chlorine producing hydrogen chloride, the assumption of monatomic elements precluded the EVEN principle (Problem 3-9). Conversely, the assumption of EVEN for reactants and products in these reactions *demands* the existence of polyatomic molecules of these elements. Assuming EVEN, if *one* volume of oxygen containing n particles is transformed into *two* volumes of water, the water contains a total of $2n$ molecules and therefore at least $2n$ oxygen atoms. Hence, each of the original oxygen particles must have contained at least two atoms. The hydrogen-oxygen reaction thus tells us that oxygen molecules are diatomic (O_2), because *one* oxygen volume is transformed into *two* water volumes. But by itself this reaction tells us nothing about the hydrogen particles, because two volumes of hydrogen are simply transformed into two volumes of water.

Other reactions, however, do tell us about the composition of hydrogen. For example, the reaction of *one* volume of hydrogen with *one* of chlorine producing *two* volumes of hydrogen chloride tells us that both hydrogen and chlorine particles are polyatomic. If we accept the EVEN hypothesis and assume that hydrogen and chlorine molecules are diatomic (two atoms per molecule: H_2 and Cl_2), we picture the formation of hydrogen chloride as follows.

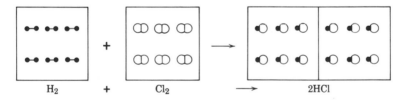

$$H_2 \quad + \quad Cl_2 \quad \longrightarrow \quad 2HCl$$

Similarly the formation of water is pictured as follows.

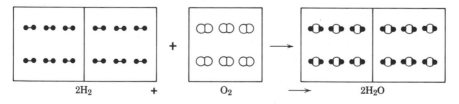

$$2H_2 \quad + \quad O_2 \quad \longrightarrow \quad 2H_2O$$

In this case we used the information obtained from the hydrogen-chlorine reaction to tell us that hydrogen is diatomic. The water reaction alone does not tell us this. Had we not known that hydrogen particles contain two atoms, we might have pictured the formation of water more simply, but wrongly, as follows.

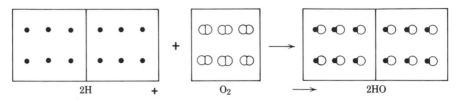

2H   + O_2 $\longrightarrow$ 2HO

CHEMICAL SYMBOLS, FORMULAS, AND EQUATIONS

The symbols we use merit a word of explanation. In 1811 the Swedish chemist Jöns Berzelius (1779–1848) replaced Dalton's cumbersome picture symbols for the elements with a new set of letter symbols, which are essentially those in use today. The *symbol* is merely the first letter of the element's name, followed by a second letter when it is needed to avoid ambiguity. Sometimes the symbol is derived from the Latin name for the element, especially if the element was known to the ancient Romans. Some examples follow. A complete list is given in the endpaper.

C–Carbon	**S**–Sulfur	**Hg**–Mercury (*Hydrargyrum*)
Ca–Calcium	**B**–Boron	
Cl–Chlorine	**Ba**–Barium	**Cu**–Copper (*Cuprum*)
H–Hydrogen	**Au**–Gold (*Aurum*)	**K**–Potassium (*Kalium*)
N–Nitrogen	**Pb**–Lead (*Plumbum*)	**Na**–Sodium (*Natrium*)
P–Phosphorus	**Ag**–Silver (*Argentum*)	**Fe**–Iron (*Ferrum*)

A molecule of any substance can be represented symbolically by a *formula,* if we know its composition. The number of atoms of each element in one molecule of an element or compound is indicated by a subscript; thus the formula O_2 tells us that oxygen in its normal form is diatomic, having two atoms in each molecule. Water, containing two hydrogen atoms combined with one oxygen atom, has the formula H_2O; carbon dioxide, as the name implies, is CO_2.

A coefficient preceding the formula—e.g., $2H_2O$— indicates the number of molecules (or volumes in gaseous reactions) of that element or compound participating in a chemical reaction. Thus the *equation* $2H_2 + O_2 \rightarrow 2H_2O$ tells us that two molecules (or volumes) of diatomic hydrogen combine with one molecule (or volume) of diatomic oxygen to form two molecules of water (or two volumes of steam). In a properly "balanced" *chemical equation* the numbers of each type of atom in the reactants must equal those in the products. (In chemical reactions atoms are neither created nor destroyed but merely rearranged.) The foregoing equation indicates the combination of four hydrogen atoms (2×2) with two oxygen atoms to yield two molecules of water, containing four hydrogen atoms and two oxygen atoms. (A product, of course, may be more than a single compound, as in $2C_2H_6 + 7O_2 \rightarrow 4CO_2 + 6H_2O$.)

Problem 3-12. Balance the following equations:
   (a) $...C + ...O_2 \qquad \rightarrow ...CO$
   (b) $...CO + ...O_2 \qquad \rightarrow ...CO_2$
   (c) $...C + ...O_2 \qquad \rightarrow ...CO_2$
   (d) $...Na + ...Cl_2 \qquad \rightarrow ...NaCl$
   (e) $...C_2H_6O + ...O_2 \rightarrow ...CO_2 + ...H_2O$

MOLECULAR AND ATOMIC WEIGHTS

Relative atomic weights can sometimes be determined by proper interpretation of gaseous reactions. Previously we pointed out a bonus of the EVEN principle: Measurements of relative gas densities directly yield relative molecular weights. For example, the observation that a box of oxygen is sixteen times as heavy as an equal-size box of hydrogen tells us that the *molecular weight* of oxygen is sixteen times that of hydrogen. Analysis of two chemical reactions ($H_2 + Cl_2 \rightarrow 2HCl$; $2H_2 + O_2 \rightarrow 2H_2O$) told us that both hydrogen and oxygen molecules are diatomic—i.e., contain two atoms; therefore, the *atomic-weight* ratios also are 16:1. Thus our values of atomic weights may depend very much on our understanding reactions and formulas of compounds. This is an important point to which we shall return in the next chapter's discussion of Cannizzaro's method of determining atomic weights.

Problem 3-13. From the foregoing picture of the formation of water, $2H_2 + O_2 \rightarrow 2H_2O$, calculate the density of steam on the basis of hydrogen atom = 1 (hydrogen molecule = 2) and oxygen atom = 16. Compare the ratios of hydrogen to oxygen to steam densities calculated in this manner with the observed ratios cited in "The Controversy," p. 29.

Problem 3-14. Assume that both hydrogen and nitrogen are diatomic. Draw a picture for the reaction of three volumes of hydrogen with one volume of nitrogen producing two volumes of ammonia. What, then, is the formula of ammonia? Write an equation for the reaction, making sure that the numbers of each kind of atoms are balanced on both sides of the arrow. What is the molecular weight of ammonia on the basis of hydrogen atom = 1, nitrogen atom = 14? Calculate the density ratios of reactants and products and compare them to the observed ratios cited in Problem 3-11.

Problem 3-15. If the formula of nitric oxide is NO and that of oxygen is O_2, draw the correct picture and write an equation for the reaction of two volumes of NO with one volume of O_2 producing two volumes of nitrogen dioxide.

We should bear in mind that the reaction

tells us that neither hydrogen nor chlorine can be monatomic; but it does not tell us that they must necessarily be diatomic. Either or both elements could just as well be tetratomic:

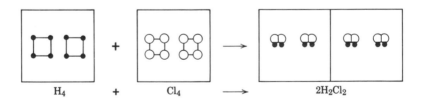

(Note that in this case the formula of hydrogen chloride would be H_2Cl_2.) The observation that one volume of a reactant becomes two volumes of product—in other words, that the atoms in one volume of reactant are divided between two volumes of product—merely requires that the reactant molecules contain an *even* number of atoms. In the absence of any compelling evidence to the contrary, we made the simplest assumption: that these elementary molecules contain the smallest number of atoms compatible with the experimental data—that is, two. No reaction has ever been found in which one volume of oxygen or hydrogen produces four or six volumes of product; this vindicates our assumption.

Problem 3-16. Consider the following observations.
  I.   2 vols of element A + 1 vol of element B → 2 vols of product.
  II. 1 vol of A + 1 vol of C → 2 vols of product.
  III. 1 vol of A + 1 vol of B → 1 vol of product.
(a) Accept Avogadro's hypothesis (EVEN) and write all the chemical formulas that are compatible with observation I (only); simply assume that gaseous molecules contain no more than four atoms. Can you make any general statements about the composition of molecules A and B on the basis of observation I?
(b) Which of the formulas listed in (a) are also compatible with observation II?
(c) If in addition you consider observation III and retain the assumption in (a), what can you say?
(d) Write the one correct version for each of the reactions I, II, and III.
(e) Do the three observations and your reasoning suffice to prove the composition of molecules A and B? Explain.

CONCEPTUAL MODELS

Scientists frequently devise some sort of "model" to help explain experimental observations. Such models may be merely schematic diagrams (e.g., boxes filled with dots and circles) or more elaborate physical pictures (e.g., particles in a box, cars on a racetrack to suggest electrons, and bouncing billiard balls to suggest molecules of gas). Models may also consist of chemical equations (e.g., $2H_2 + O_2 \rightarrow 2H_2O$) or even mathematical expressions (e.g., $PV = nRT$). To be useful, a model should allow us to make predictions that we can check by further observation or experimentation. If a model fails this test, it must be altered or replaced. On the other hand, if a model helps us predict correctly, it is a good one—at least for the purpose at hand.

But correct predictions alone do not prove the "correctness" of the model. We say that Dalton's gas model yielded correct predictions of densities. Thus it was good for predicting density. Later, however, it was shown that this model yielded incorrect molecular formulas and had to be discarded. Similarly, Avogadro's model of equal volumes containing equal numbers of particles yielded correct volume and formula predictions. Therefore it was a good model by that criterion of usefulness; yet more recent measurements have revealed that Avogadro's model yields only approximately correct values.

We should not label any model as "correct" or "true." (After all, molecules are *not* billiard balls.) Models are merely analogies and are "good" if they are useful in helping us interpret observations and in predicting the results of experiments. Fortunately, models and scientific theories are generally based not on one but on many interrelated observations, which help us avoid or eliminate false hypotheses and poor models.

The role of conceptual models is a recurring theme throughout this book.

CONCLUSION

Avogadro's hypothesis answered Dalton's objections to the equal-volumes-equal-numbers principle by introducing a new idea: the existence of polyatomic molecules of elements. Today we know that Avogadro was right, but during his time he was not believed. For one thing, Avogadro could not compete against Dalton's prestige. For another, the idea of polyatomic gaseous molecules was difficult to accept in the absence of a good theory of bonding. Why would two atoms of the *same* element attract each other and unite? (A theory of electrostatic attraction might explain this behavior in the case of *unlike* atoms.)[3] If like atoms can get together, why does this bonding involve only a relatively small number? Why don't more atoms unite, so that the element ends up as a liquid or solid rather than a diatomic gas?

Another problem, revealed in 1827, stemmed from the logical conclusion that if Avogadro's hypothesis is correct, most gaseous elements would be diatomic (H_2, N_2, O_2, F_2, Cl_2, etc.). In that year the French chemist Jean-Baptiste Dumas (1800–1884) announced that mercury vapor is monatomic (Hg) and sulfur vapor is hexatomic (S_6).[4] Such divergences from "standard" behavior seemed unbelievable at the time. This problem and the questions related to the bonding of like atoms remained unanswered for more than a century—until the application of quantum mechanics finally produced a satisfactory theory of chemical bonding (*see* Chapter 25). In view of these doubts and unsolved problems, it is no

[3] Berzelius believed that chemical bonding is due to electrostatic attraction between positive and negative atoms. He published an early version of his theory in the same year (1811) and in the same journal in which Avogadro's proposal appeared. Berzelius' theory drew on discoveries made in 1800 that electricity can be produced chemically and that it can decompose chemical compounds (*see* Chapter 7).

[4] We know today that sulfur vapor consists of S_8, S_6, S_4, and S_2 molecules in relative amounts that depend on temperature.

wonder that Avogadro's hypothesis found slow acceptance. In fact, it was largely forgotten until its revival by Cannizzaro almost a half century later, as we shall see in the next chapter.

ANSWERS TO SELECTED PROBLEMS

3-1. (a) *Any* formula will do: e.g., HO, H_2O, HO_2, H_2O_3, etc.
(b) HO.
(c) 8. (From the formula H_2O, the observed weight ratio 1:8 leads to an atomic weight of 16 for oxygen—2:16 = 1:8. And from the formula HO_2 we would deduce an atomic weight of 4—1:[2 × 4] = 1:8).

3-2. (a)

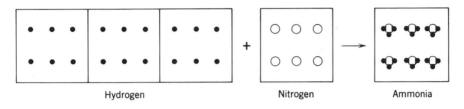

| Hydrogen | Nitrogen | Ammonia |

(b) Analogous to example (a).

3-3. For example, A_3B_6. (Also possible: A_2B_4—easily seen if box A contains an even number of particles.)

3-5. A_5B_3.

3-6. 14.

3-7. 21.

3-8. Analogous to the pictured carbon monoxide-chlorine reaction (p. 23).

3-9. Relationship: one half.

3-14. $3H_2 + 2N_2 \rightarrow 2NH_3$.

3-16. (a)

$$2A + B_2 \rightarrow 2AB \qquad (1)$$
$$2A_2 + B_2 \rightarrow 2A_2B \qquad (2)$$
$$2A_3 + B_2 \rightarrow 2A_3B \qquad (3)$$
$$2A + B_4 \rightarrow 2AB_2 \qquad (4)$$
$$2A_2 + B_4 \rightarrow 2A_2B_2 \qquad (5)$$

etc.

Element B must be polyatomic with an *even* number of atoms. (1 vol → 2 vols)

(b) Element A now must be even-atomic; therefore, only (2) and (5) are compatible.

(c) Both elements must be even-atomic. If element B is B_2, then $A_2 + B_2 \rightarrow A_2B_2$. If the element B is B_4, then $A_2 + B_4 \rightarrow A_2B_4$; but this is contrary to our assumption of no more than four atoms per molecule. Therefore element B is B_2 and element A is A_2.

(d) I. $2A_2 + B_2 \rightarrow 2A_2B$

    II. $A_2 + C_2 \rightarrow 2AC$

    III. $A_2 + B_2 \rightarrow A_2B_2$

(e) No! Our reasoning is based on the assumption of no more than four atoms per molecule; but *this has not been proved*. (Dalton made a similar mistake when he implicitly assumed that elementary particles consist of no more than one atom.)

Chapter 4

ATOMIC-WEIGHT TABLES

THE PERIOD OF CONFUSION

In the years between 1811 and 1858, the problem of determining atomic weights became more and more vexing because of the uncertainty about formulas of polyatomic elements. Solutions were proposed, only to be abandoned when they failed to account for the growing body of experimental facts. In 1814 to 1826 Berzelius published tables of relative atomic weights, which on the whole agree closely with today's values. But his contemporaries distrusted his results, and some of them produced their own tables. Eventually, it seemed that there were as many atomic-weight tables as there were investigators, and some scientists concluded that it was impossible ever to determine atomic weights and molecular formulas.

There were, however, a few milestones along the way to the ultimate determination of atomic weights, particularly the logical deductions of Cannizzaro (1858), the work of Dulong and Petit on the specific heats of solids (1819), and Prout's proposal of hydrogen as the basic building block of all elements (1815).

CANNIZZARO

The Italian chemist Stanislao Cannizzaro (1826–1910) brought an end to the confusion in 1858 through his new method of determining atomic weights. He reasoned that a molecule must naturally contain one or more whole atoms of each of its constituent elements, because atoms by definition are indivisible.[1] A molecule of water, for example, contains 2 hydro-

---

[1] Atoms *are* indivisible in ordinary chemical reactions, but we know today that they may be split by high-energy collisions, as in a cyclotron.

gen atoms, not 1.5 atoms or 0.37 atoms; similarly, it contains one whole oxygen atom. It is clear then that in one molecular weight of a compound there must be one atomic weight of each element or some integral multiple of this weight. Accordingly, Cannizzaro chemically analyzed a series of compounds containing a particular element and determined the weight of this element contained in one molecular weight of each compound. All weights obtained were integral multiples of the atomic weight; in other words, the greatest common factor would be the atomic weight or a multiple thereof. In a large number of compounds, at least one compound would probably contain only one atom of the element per molecule.

To use this method, Cannizzaro first had to find the molecular weights of the various compounds. For this, he revived Avogadro's almost forgotten EVEN principle and used the method described on p. 27: Because equal volumes of gases contain equal numbers of molecules, the weights of these volumes must stand in the same ratio as the weights of their molecules. Thus Cannizzaro could determine relative molecular weights. He then arbitrarily defined the molecular weight of the lightest known gas, hydrogen, to be 2, and fixed the values of all other gases on that basis. He chose the value 2 rather than 1 because he realized that the hydrogen molecule is diatomic.

Let us illustrate Cannizzaro's procedure by outlining the determination of the atomic weight of oxygen. First, we determine the molecular weights of a number of gaseous compounds of oxygen by measuring their densities relative to hydrogen gas at the same temperature and pressure. For example, a given volume of nitric oxide weighs fifteen times as much as an equal volume of hydrogen; i.e., its density is fifteen times that of hydrogen. Therefore the molecular weight of nitric oxide is 30 on a scale on which the molecular weight of hydrogen gas is 2. Similar measurements of other oxygen-containing gases provide the molecular-weight data given in Table 4-1.

The second step is the determination of the weight of oxygen contained in 1 gram-molecular weight of each compound. (A *gram-molecular weight* is the molecular weight expressed in grams. Thus the gram-molecular weight of hydrogen is 2 g, and the gram-molecular weight of nitric oxide is 30 g. Any other convenient unit could be used instead, such as a pound-molecular weight). To obtain the oxygen content of a compound, we must perform a chemical analysis. When 30 g (1 gram-molecular weight) of nitric oxide are analyzed, 16 g of oxygen are obtained. Repetition of this procedure with the other compounds yields the weight of oxygen contained in 1 gram-molecular weight of each of the gases listed in the table.

Table 4-1. The Molecular Weights of Various Compounds Containing Oxygen

Compound	Molecular Weight (relative to $H_2 = 2$)	Grams of Oxygen in 1 Gram-molecular Weight
Water	18	16
Nitric oxide	30	16
Nitrous oxide	44	16
Nitrogen dioxide	46	32
Sulfur dioxide	64	32
Carbon dioxide	44	32
Oxygen	32	32
Ozone	48	48
Sulfur trioxide	80	48

Examination of these data immediately discloses that the smallest weight of oxygen found in 1 gram-molecular weight of any of the compounds is 16 g; and that all other oxygen weights are integral (whole-number) multiples of 16 g—$2 \times 16 = 32$, $3 \times 16 = 48$. We may conclude that the atomic weight of oxygen is 16.

Problem 4-1. Assuming the atomic weight of oxygen to be 16 and using a modern atomic-weight table for the other elements (*see* endpaper), write the formula for each compound in Table 4-1.

Problem 4-2. Wouldn't an atomic weight of 8 for oxygen also fit the numerical data of the table? What would be the formula of each compound in that case? Do you notice anything *odd* about the numbers of oxygen atoms in the latter formulas? Is this peculiarity enough reason to reject an atomic weight of 8? If not, how would you proceed to prove or disprove an atomic weight of 8 for oxygen?

Problem 4-3. If the first three compounds had not been included in the table, would you have concluded that the atomic weight of oxygen is 32 instead of 16? In that case, what would be the molecular formula of ozone?

Problem 4-4. On the basis of your thinking in Problem 4-3, formulate a general rule for assigning atomic weights from data such as those in Table 4-1. Will this rule always lead to the correct result (*see* Problem 4-2)? If not, can you say anything about the type of error that would result?

On the basis of the foregoing discussion we can offer two distinct definitions of *molecular weight*. A *theoretical* definition is that molecular weight is the weight of a molecule relative to that of the hydrogen atom. An *operational* definition is that molecular weight is twice the ratio of the weights of equal volumes of a gas and hydrogen. The latter definition implicitly tells us what operations to perform in order to determine molecular weight: First, determine the weight of hydrogen in a vessel; next, determine the weight of the unknown gas in a similar vessel; then divide twice the latter by the former (twice, because we are using hydrogen = 2 as the basis of our scale).

OTHER GUIDES TO ATOMIC WEIGHTS

Cannizzaro's method of determining atomic weights was generally accepted and had great impact on the development of chemistry, in part because it came at the right time. In the half century between Avogadro's original formulation of his equal-volumes-equal-numbers principle of gases, coupled with the idea of polyatomic elements, and Cannizzaro's revival and extension of the theory, various other methods of determining atomic weights had been developed. Some of these yielded conflicting answers—especially if the EVEN principle was ignored, as was often the case. With Cannizzaro's scheme, the other methods suddenly fell into place. When the work of previous investigators was reinterpreted, many of the discrepancies disappeared, and the conclusions of the other methods were seen to support each other instead.

We shall discuss two methods that predated Cannizzaro's work: the method of Dulong and Petit, which relied on measuring the specific heat of an element; and the use of equivalent weights, in which the combining weights of elements in a compound are determined by chemical analysis. We shall also mention the use of the mass spectrometer, a modern instrument that was invented about a half century after Cannizzaro introduced his method.

Specific Heats

In 1819 two Frenchmen, the chemist Pierre Dulong (1785–1838) and the physicist Alexis Petit (1791–1820), jointly discovered a relationship between the specific heat of a solid element and its atomic weight. The *specific heat* of any substance is the amount of heat in calories required to raise the temperature of one gram of the substance by one degree Celsius (centigrade). (The specific heat of water, then, is 1.00 cal/g°C; *see* the definitions of Celsius and calorie in Appendix II.) Dulong and

Petit found that the higher the atomic weight of an element, the lower its specific heat; and that the product of specific heat and atomic weight (which we call *heat capacity*) is approximately constant:[2]

$$\text{Specific heat} \times \text{atomic weight} \approx 6$$

Table 4-2 lists some modern heat capacities. The variation of the values in the last column (the products of the atomic weights and specific heats) gives some idea of the approximate nature of the Rule of Dulong and Petit. Three serious exceptions—there are others—are listed at the bottom of the table; note that these are among the lighter elements.

Table 4-2. Heat Capacities of Some Solid Elements

Element	Atomic Weight	Specific Heat (cal/g °C)	Heat Capacity (cal/g °C-atom)
Sodium	23.0	0.30	6.9
Aluminum	27.0	0.21	5.7
Phosphorus	31.0	0.19	5.9
Sulfur	32.0	0.18	5.8
Iron	55.8	0.11	6.1
Silver	107.9	0.056	6.0
Iodine	126.9	0.052	6.6
Gold	197.2	0.031	6.1
Lead	207.2	0.031	6.4
Beryllium	9.0	0.39	3.5
Boron	10.8	0.3	3
Carbon (Diamond)	12.0	0.12	1.4

At the time of its discovery, the relationship between specific heat and atomic weight was an unexplained empirical relationship. If it were a universal law—i. e., if there were no exceptions—it could of course be used to calculate unknown atomic weights from experimentally determined specific heats:

$$\text{atomic weight} \approx 6/\text{specific heat}$$

Example 4-1.   Find the approximate atomic weight of iron.

Solution

The specific heat of iron is determined experimentally by dropping a weighed quantity of hot iron into a measured quantity of cool water

[2] We use the symbol $\approx$ to mean "is approximately equal to."

and observing the temperature changes. We shall start with 100 cc (i. e., 100 g) of water in a Thermos bottle at room temperature (20°C), and with 200 g of iron at 100°C (heated in boiling water). After the iron is dropped into the Thermos bottle, the water-and-iron system is found to attain a constant temperature of 35°C. The water thus has been warmed 15° (from 20°C to 35°C) and the amount of heat absorbed is:

weight of H_2O × specific heat of H_2O × temperature rise
= 100 g × 1 cal/g°C × 15°C
= 1500 calories

The heat lost by the iron in cooling from 100°C to 35°C is:

wt. of Fe × sp. ht. of Fe × temp. drop
= 200 g × sp. ht. × 65°C = 13,000 × sp. ht.

The heat lost by the iron equals the heat absorbed by the water:

13,000 × sp. ht. = 1,500 calories

Therefore, the specific heat of Fe = $\dfrac{1,500}{13,000}$ = 0.11 cal/g°C.

Applying the Rule of Dulong and Petit, we obtain the approximate atomic weight of iron:

AW ≈ 6/sp. ht. = 6/0.11 = 57

The actual atomic weight is 55.8.

Dulong and Petit did not have a reliable table of atomic weights; thus there were too few checkpoints for their rule. This in turn made many scientists unwilling to believe wholly in the atomic weights obtained with that rule, especially because other methods yielded conflicting results.

Today we recognize the utility of the Rule of Dulong and Petit. In spite of its approximate nature, it can help us determine accurate atomic weights when it is used in conjunction with other techniques of atomic-weight determination, such as Cannizzaro's method. Suppose, for example, that we wish to determine the atomic weight of a metallic substance that we have just "discovered." We proceed by the following steps.

1. After considerable experimentation, we succeed in preparing a compound that contains this element and is gaseous at a reasonable temperature—slightly above the boiling point of water, in this instance.

2. Careful weighings tell us that 0.5419 g of the metal were used to prepare 0.5589 g of the gas. The fraction of the "new" material in the gas

is therefore $0.5419/0.5589 = 0.9696$, or 96.96%. (The other component must be very light!)

In applying this method we are not limited to binary (two-element) compounds, nor must we know the other component(s).

3. Our next step is a molecular-weight determination. We weigh the gas and compare its weight to that of an equal volume of hydrogen. We find that our gas is 111.4 times as heavy as hydrogen (allowing for the weights of the containers). Using the value 2.016 as the molecular weight of hydrogen (see endpaper), we calculate that the gas has a molecular weight of $111.4 \times 2.016 = 224.6$ atomic-weight units.

4. Of these 224.6 units, 96.96% are due to the presence of the "unknown" element—that is, $224.6 \times 0.9696 = 217.8$ atomic-weight units. These 217.8 atomic-weight units in each molecule represent one or more atoms of our unknown.

5. Several possibilities are now open to us: (a) As one alternative, we can follow the rest of Cannizzaro's procedure by laboriously preparing several other gaseous compounds of our unknown, repeating steps 2 to 4, and looking for the greatest common factor. If, for example, the next compound contains 145 atomic weight units of the unknown per molecule, we could see that approximately 72.5 is the greatest common factor and may therefore represent the atomic weight. This would mean that the first compound contains three atoms ($3 \times 72.5 = 217.5$) and the second contains two atoms ($2 \times 72.5 = 145$) per molecule.

But the atomic weight could also be half of 72.5, meaning six and four atoms, respectively, per molecule. To see whether any molecule contains only 36.3 atomic-weight units of unknown per molecule, we would be faced with the task of making many more compounds. And even then we would never be absolutely certain of the atomic weight.

5.(b) The Rule of Dulong and Petit comes to our rescue. In fact, with luck, we can use their method with only one of the gaseous compounds and need not prepare others. We determine the specific heat of the original "unknown" metal by measuring the number of calories necessary to raise its temperature one degree Celsius. Let us assume that we find a specific heat of 0.074 cal/g°C.

6. Using the Rule of Dulong and Petit, we calculate the approximate atomic weight:

$$\text{AW} \approx \frac{6}{0.074} = 81$$

7. Comparison of this figure, 81, with the value 217 (the weight of metal per molecular weight, obtained in step 4) indicates that our gas molecule contains *three* atoms of "unknown" metal (217.8/81 = 2.7).

8. The exact atomic weight of the metal, therefore, is

$$\frac{217.8}{3} = 72.6$$

Problem 4-5. If 72.6 represents the correct atomic weight, what is the metal? According to an atomic weight table (*see* endpaper) what are the only two or three possible formulas of this compound? (By subtraction determine how many atomic weight units per molecule belong to the other component(s), then THINK!)

The Rule of Dulong and Petit and the method of Cannizzaro complement each other very nicely. Cannizzaro's method depends on the availability of a number of gaseous compounds of the element in question. This limits it mostly to light nonmetallic elements, because compounds of heavy elements and of metals are more likely to be solids. The Rule of Dulong and Petit, on the other hand, is applicable only to solid elements, especially heavy metallic elements. (Table 4-2 indicated some failures of the rule for light elements.)

The Rule of Dulong and Petit by itself results at best in a crude approximation of atomic weights, but the accuracy of Cannizzaro's method is also limited, because it involves the determination of molecular weights by the gas-density method or by comparison of equal volumes of gases. The validity of that procedure depends on the validity of the EVEN principle: Direct comparison makes sense only if equal volumes do, in fact, contain equal numbers of molecules. But alas, they do not. The EVEN principle is only an approximation, although not a bad one. The volume occupied by a gas is not completely independent of the nature of its molecules. Fortunately, another classical method, the method of combining, or equivalent, weights, involves no approximation and is limited only by the purity of the compounds and the accuracy of our weighings.

Combining Weights

In principle, finding atomic weights from combining weights is quite simple. Through chemical analysis, we determine the weight of the new element that will combine with a gram-atomic weight of an element of known atomic weight. The atomic weight of the new element is an integral multiple or submultiple of that combining weight. This factor can be determined by referring to approximate atomic weights, obtained through the Rule of Dulong and Petit.

For example, chemical analysis shows that 51.795 g of unknown element Y combine with one gram-atomic weight (35.453 g) of chlorine. The formula of this compound could be YCl, or YCl_2, or Y_2Cl, etc.—in which cases the atomic weight of Y would be 1, 2, or ½ times 51.795, respectively. The specific heat of the element is determined as 0.0303 cal/g°C., which, by the Rule of Dulong and Petit, yields an approximate atomic weight as follows.

$$ AW \approx \frac{6}{0.03} = 200 $$

This atomic weight is approximately four times the combining weight (51.795). The *exact* atomic weight accordingly is found to be

$$ 51.795 \times 4 = 207.18 $$

The literature value for lead is 207.19. The compound presumably is $PbCl_4$.

These calculations are quite similar to those in our previous example. The results of the combining-weight method, however, do not depend on approximative assumptions, such as the EVEN principle.

Problem 4-6. Suppose that the specific heat in the foregoing example had been crudely determined as 0.0280 cal/g°C instead of 0.0303 (i.e., an error of −0.023, or 8%). How would this have affected the calculation of the precise atomic weight?

Problem 4-7. Suppose that the combining weight had been determined as 52.000 instead of 51.795 (i.e., an error of 0.205, or less than 0.5%). How would this have affected the calculated results?

The application of the combining-weight method is not restricted to binary compounds, but it does require knowledge of the atomic weight of one constituent of the compound in question. It may be necessary to obtain this knowledge by analyzing another compound. For example, let us assume that YZ is the only available compound of a new element, Y, whose exact atomic weight we wish to determine, but the atomic weight of Z has not yet been determined. Therefore, we must first determine the atomic weight of Z by means of another compound, ZW, in which W represents a well-known element. If the atomic weight of W is also unknown, we can continue the process until *some* element is set up as an exact reference point. This brings us to the next topic—the establishment of atomic-weight scales.

ATOMIC-WEIGHT SCALES

Hydrogen, discovered in 1766 by the English scientist Henry Cavendish (1731–1810), is the lightest known element. In 1815 another Englishman, the physician William Prout (1785–1850), noted that most atomic weights appear to be integral multiples of the atomic weight of hydrogen. For example, the carbon atom weighs 12 times as much as the hydrogen atom; nitrogen, 14; oxygen, 16; sulfur, 32. Prout believed (correctly) that such a relationship is more than accidental,[3] and (not quite correctly) interpreted this observation to mean that the other "elements" were made up of hydrogen as the basic building block of nature.[4]

Even at the time, these relationships were only approximate, and there were also some quite drastic exceptions. Prout simply ascribed the discrepancies to experimental error. He committed the cardinal scientific sin of making the facts fit the theory.[5]

Prout's hypothesis was dealt the death blow in the 1860s when careful experiments by the Belgian chemist J. S. Stas (1813–1891) revealed that the atomic weights of boron and chlorine are in fact, 10.8 and 35.5 times that of hydrogen, respectively.

Just before World War I the discovery of isotopes (see p. 18) brought about a revival of the whole-number idea. We now explain fractional atomic weights as representing *average* weights of mixtures of isotopes, such as boron-10 and boron-11, or chlorine-35 and chlorine-37; and we explain the nearly integral numbers, not by asserting that hydrogen is the basic element, but by claiming that *all* elements, including hydrogen, are composed of the same basic building blocks—namely, protons, neutrons, and electrons (see Chapter 11).

Regardless of any theory of atomic structure, it is convenient to use the weight of the lightest element, hydrogen, as the unit of an empirically determined atomic-weight scale. On that basis the atomic weight of oxygen is *almost* 16 (in relative units).[6]

However, there were two principal reasons for revising the scale slightly by setting oxygen equal to *exactly* 16. First, the oxygen scale is more convenient to use, and the slight upward revision resulted in more

[3] In 1897 the Swedish spectroscopist J. R. Rydberg (1854–1919), whom we shall meet again in Chapter 15, calculated a random probability of one in a billion that the first 22 elements would be as close to whole numbers as observed.

[4] Why is "elements" in quotation marks here?

[5] Akin to the flustered young lawyer who said: "And these, ladies and gentlemen of the jury, are the conclusions upon which I base my evidence."

[6] Although convenient, the selection of hydrogen as the unit for the scale is quite arbitrary. For instance, Berzelius in 1813 used a scale on which oxygen was 100 and hydrogen was 6.64.

nearly integral numbers. Second, most atomic weights were determined by the method of combining weights. In this method we must determine the weight of an unknown element which combines with a certain weight of a known element. The combining ratio of an element to oxygen, and therefore its atomic weight relative to oxygen, can often be determined directly, because most elements form compounds with oxygen; but the relationship of an unknown element to hydrogen would often have to be determined indirectly and thus would involve more experimental error.

In 1961 international agreement was reached on a still more convenient scale. One of the most precise methods of determining atomic weights is by means of the mass spectrometer, developed in 1919 by the American physicist Francis Aston (1877–1945). The mass spectrometer essentially separates the various "nuclides" (isotopes) and determines the mass of each individually; boron-10 and boron-11, for example, are "weighed" separately in a mass spectrometer.[7] With the aid of the mass spectrometer, scientists found that ordinary oxygen is a mixture of several isotopes: oxygen-16 (99.76%), oxygen-17 (0.04%), and oxygen-18 (0.20%). Thereupon, physicists began to use an atomic-weight scale based on oxygen-16 equals exactly 16; while chemists continued to use the scale based on ordinary oxygen equals exactly 16. In order to bring the two groups together and to relate atomic weights directly to mass-spectrometric determinations, scientists agreed upon another nuclide as the reference standard. The standard of the new unified scale is carbon-12 equals exactly 12. (Ordinary carbon is composed of two isotopes, 98.89% of car-

Table 4-3. Atomic Weights on Three Scales

Species[a]	Old Chemical Scale	Old Physical Scale	New Unified Scale
O	16 exactly	16.0044	15.9994
^{16}O	15.99560	16 exactly	15.99491
^{12}O	12.00052	12.00382	12 exactly
C	12.011	12.015	12.011
H	1.0080	1.00827	1.0080

[a] We use superscripts to indicate the weight (or mass number) of the species. In some works the superscript appears on the right, e.g., C^{12}.

[7] See Chapter 11 for a discussion of the historical aspects of mass spectrometry and Chapter 27 for its modern applications.

bon-12 and 1.11% of carbon-13.) Table 4-3 compares some values on the three scales. The atomic weights on the unified scale of all known elements are listed in the endpaper.

It should be emphasized that the atomic-weight scale is a *relative* scale: The weights of the atoms of the various elements are compared to the weight of the carbon-12 atom. In other words, the unit of mass is 1/12 the atomic mass of the carbon-12 species. Because atomic weights are relative numbers, merely expressing the ratios of two masses, there are no dimensions or units attached, such as grams; the atomic weight of boron, for example, is just 10.811. Sometimes we will call this **10.811** "atomic mass units" (amu).[8]

THE MOLE CONCEPT

The *absolute* weight or mass of an atom (in pounds or grams or other units) can be obtained only if we know how many atoms it takes to make up a gram or some other chosen weight. It is convenient to use the number of carbon-12 atoms that it takes to make up 12 grams of carbon. This number—call it N—is identical to the number of oxygen-16 atoms in 16 grams of oxygen, the number of oxygen *molecules* in 32 grams of oxygen (remember that oxygen is diatomic), and the number of water molecules in 18 grams of water. We call this number of atoms or molecules a *mole*.

A mole is the number of atoms in a gram-atomic weight of a substance, or *the number of molecules in a gram-molecular weight of a substance.*

It should be emphasized that a mole is a *number* (N) of particles, not a weight (although, of course, N particles possess a certain weight). We can talk about a mole of atoms, or a mole of molecules, or even a mole of people (although the earth does not contain that many people.) A mole is simply a convenient "packaging unit," like a dozen or a gross—only it is much larger.

The actual value of N—that is, the number of atoms or molecules in a mole—cannot be derived theoretically but must be determined by experiment (see next section). It was first obtained in 1865 by the Austrian scientist Joseph Loschmidt (1821–1895) and therefore is sometimes called

---

[8] Purists insist that this relative scale should be called the atomic *mass* scale, since we are comparing masses rather than weights. They claim, in fact, that throughout this book we should be referring to atomic *masses*, not weights. Strictly speaking, this is correct, but the term "atomic weight" is so conventional that we shall continue to use it in this book, interchangeably with "atomic mass." (*See* "Mass Versus Weight" in Chapter 6.)

Loschmidt's number. More commonly, though, it is known as *Avogadro's number*, in honor of Avogadro's original contribution in 1811. (Unfortunately, Avogadro did not live to see the revival and vindication of his hypothesis through the work of Cannizzaro and Loschmidt.) Since atoms are very small and light, the number of atoms in a mole is very large indeed. The modern value of Avogadro's number, N, is about[9]

$$6.02 \times 10^{23}$$

This is the number of atoms in a gram-atomic weight, or the number of molecules in a gram-molecular weight of a substance. For example, a gram-molecular weight of hydrogen (2g) contains 6×10^{23} molecules; we call this a mole of hydrogen molecules. A gram-atomic weight of hydrogen (1g) contains 6×10^{23} *atoms;* this is a mole of hydrogen atoms. A gram-molecular weight of water (18 g) contains 6×10^{23} molecules; we speak of a mole of water. And so on.

Problem 4-8.   Calculate the weight in grams of a single hydrogen atom.

A quantity that often comes up is the volume occupied by Avogadro's number of gas molecules. Remember EVEN: Equal volumes of gases (under identical conditions of temperature and pressure) contain equal numbers of molecules; hence, N molecules of *any* gas occupy the *same* volume as N molecules of any other gas. Again, this quantity must be determined experimentally. At 0° Celsius and sea-level pressure—the arbitrarily agreed-upon *s*tandard conditions of *t*emperature and *p*ressure (STP)—one mole of any gas (e.g., 2 g of H_2, 32 g of O_2, 16 g of CH_4, 4 g of He) occupies approximately 22.4 liters, or almost 6 gallons. This is known as the *molar volume.*

The mole is a most helpful concept to chemists. We do not usually observe reactions of individual atoms and molecules. Rather, we observe the aggregate effect of atomic and molecular interactions in bulk samples containing very large numbers of particles. We often think in terms of Avogadro's number of particles; that is, we think of chemical quantities in terms of moles or fractions of moles. The equation

$$2H_2 + O_2 \rightarrow 2H_2O$$

thus has many meanings to the chemist:

(a) Two hydrogen molecules combine with one oxygen molecule to produce two molecules of water.

[9] *See* Appendix 1-A for a discussion of exponential notation, a technique used to facilitate the expression and manipulation of very large and very small numbers.

(b) $2 \times 6 \times 10^{23}$ molecules of hydrogen combine with 6×10^{23} molecules of oxygen to produce $2 \times 6 \times 10^{23}$ molecules of water.

(c) 2 moles of hydrogen combine with one mole of oxygen to produce 2 moles of water.

(d) 4 g of hydrogen combine with 32 g of oxygen to produce 36 g of water. This is equivalent to: 1 g of hydrogen combines with 8 g of oxygen to produce 9 g of water.

(e) Two times 22.4 liters of hydrogen combine with 22.4 liters of oxygen to produce 2×22.4 liters of water vapor (at STP). Or, more generally: Two volumes of hydrogen combine with one volume of oxygen to produce two volumes of water vapor.

Note that the formulas in the equation (H_2, O_2, and H_2O) describe the chemical species involved: diatomic hydrogen and oxygen molecules, and water molecules containing two hydrogen atoms and one oxygen atom. The coefficients preceding the formulas in a chemical equation indicate the quantities involved—i.e., the number of molecules, moles, or volumes of gases (*see* the section "Chemical Symbols, Formulas, and Equations" in Chapter 3).

The following examples illustrate use of the mole concept in chemical calculations.

Example 4-2. Calculate the number of (a) moles, (b) molecules, and (c) atoms in 8 g of O_2. (d) What is the volume at STP?

Solution.

(a) $\text{moles} = \dfrac{\text{grams}}{\text{g-molecular weight}} = \dfrac{80 \text{ g}}{32 \text{ g}} = 0.25$

(b) $\text{molecules} = (\text{moles})(\text{Avogadro's no.}) =$
$(0.25)(6.02 \times 10^{23}) = 1.5 \times 10^{23}$

(c) atoms: Since there are two oxygen atoms in each oxygen molecule, the number of oxygen atoms in 8 g of oxygen is twice the number of molecules:

$$2(1.5 \times 10^{23}) = 3.0 \times 10^{23}$$

(d) $\text{volume} = (\text{moles})(\text{volume per mole}) =$
$(0.25)(22.4 \text{ l}) = 5.5 \text{ l}$

Example 4-3. How many grams of H_2O can be prepared from 0.40 g of H_2 and 8 g of O_2? How much O_2 will be left?

Solution.

In problems of this type, *always* first calculate the number of moles of each component of the reaction:

$$\text{moles of } H_2 = \frac{0.40}{2.0} = 0.20$$

$$\text{moles of } O_2 = \frac{8.0}{32} = 0.25$$

According to the balanced equation, we need 2 moles of H_2 for each mole of O_2. From the number of moles of each component, we see that there is not enough H_2 to react with all of the O_2. Thus, H_2 is the *limiting reagent*. We need only half as much O_2 and H_2. Therefore, if we use all the H_2 (0.20 mole), we will need 0.10 mole of O_2. Hence, 0.15 mole of O_2 will be left after the reaction.

The balanced equation indicates that 2 moles of H_2 produce 2 moles of H_2O. Therefore, 0.20 mole of H_2 produces 0.20 moles of H_2O. Since the molecular weight of H_2O is 18, 0.20 mole equals 0.020 × 18 grams, or 3.6 g of H_2O.

EXPERIMENTAL DETERMINATION OF AVOGADRO'S NUMBER

One method of determining Avogadro's number—a method frequently used by students in introductory chemistry courses—involves comparing the volume of a liquid with the area it occupies when it is spread out in a monomolecular layer on a water surface. Let us illustrate by an analogy. Suppose a bunch of little cubes is stacked in an orderly lattice and found to occupy 512 cc. When we place these little cubes only one layer high onto a flat surface so that they form a square or rectangular area, the total area they occupy will depend upon the size of the little cubes. Consider three cases.

1. If each cube is 4 cm on a side, having a volume of $4 \times 4 \times 4 = 64$ cc, the 512-cc volume would contain only 512/64 or 8 cubes; that is, a cubic stack would have only 2 cubes along each edge (Fig. 4-1*a*). When these eight cubes are spread out onto a rectangular area, there would be 4 cubes along one side and 2 along the other (Fig. 4-1*b*). Since each cube is 4 cm long, the total area would be: $(4 \times 4) \times (2 \times 4) = 16 \times 8 = 128 \text{ cm}^2$.

2. If each cube is 2 cm on a side, having a volume of $2 \times 2 \times 2 = 8$ cc, the 512-cc volume would contain 512/8, or 64 cubes; that is, a cubic stack would have 4 cubes along each edge, which is $4 \times 2 = 8$ cm long (Fig. 4-2*a*). When these 64 cubes are spread out onto, say, a square area,

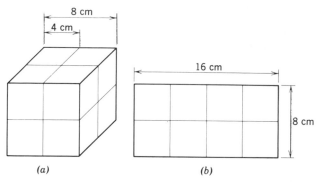

Fig. 4-1. (*a*) A 512-cc stack of 8 cubes, each 4 cm long. (*b*) Unstacked, these 8 cubes occupy a $16 \times 8 = 128$-cm² area.

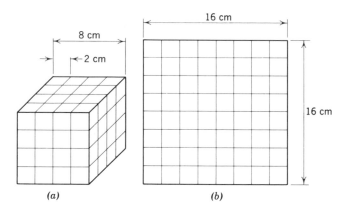

Fig. 4-2. (*a*) A 512-cc stack of 64 cubes, each 2 cm long. (*b*) Unstacked, these 64 cubes occupy a $16 \times 16 = 256$-cm² area.

there would be 8 cubes along each side (Fig. 4-2*b*). Since each cube is 2 cm long, the total area would be: $(8 \times 2) \times (8 \times 2) = 16 \times 16 = 256$ cm². Note that the smaller cubes spread out into an area twice as large as in the prior example.

3. If each cube is only 1 cm on an edge, the 512-cc volume would contain 512 cubes. These could be arranged into a rectangular area of, say, $32 \times 16 = 512$ cm², still larger than either previous area. It is easy to see that still smaller cubes—say, of molecular dimensions—would yield a much larger area.

It should now be evident that comparison of the volume and area in each example would have yielded the dimensions and total numbers of the little cubes, as follows. The volume of the original cubic stack equals its length cubed:

$$V = L^3$$

$$[512 \text{ cc} = (8 \text{ cm})^3]$$

The number, N, of little cubes along a side of the stack, equals the length, L, of the stack divided by the length, l, of a little cube:

$$N = \frac{L}{l} \quad \text{or} \quad l = \frac{L}{N}$$

The total number of cubes is N^3, and the square area, A, occupied by these N^3 cubes equals $N^3 l^2$. We can now solve for the total number, N^3, in terms of the measured area and the length, L, of the stack. The ratio of the area to volume is

$$\frac{A}{V} = \frac{N^3 l^2}{L^3} = \frac{N^3}{L^3}\left(\frac{L}{N}\right)^2 = \frac{N}{L}$$

Then,

$$N = \frac{AL}{V} = \frac{AL}{L^3} = \frac{A}{L^2}$$

Therefore the total number of little cubes is

$$N^3 = \left(\frac{A}{L^2}\right)^3$$

In the second example this would be:

$$N^3 = \left[\frac{256 \text{ cm}^2}{(8 \text{ cm})^2}\right]^3 = 4^3 = 64$$

In a typical demonstration of this method, we measure the volume of a small amount of liquid, such as oleic acid. We then allow the liquid to spread out as a thin film on water; this film will be only one molecule thick, forming a "monomolecular layer." Preferably we have placed the water in a rectangular tray and spread a little talcum powder on the water surface before adding the fatty acid. The spreading fatty acid will push aside the powder, making its outer limits clearly visible. To make it easier to measure its area, we can push it with a thin wire into a rectangular area. From the measured area and volume we can then calculate the number of molecules. From that number we can, in turn, determine the number of molecules in a gram-molecular weight of this acid—that is, Avogadro's number. Note that the accuracy of this method depends on our knowing the shape of the molecules. (They of course are not little cubes.)

Three other methods of determining Avogadro's number are based on principles discussed in later chapters. One of these methods involves electrolysis. In 1833 Faraday measured the amount of electric charge required to electrolyze, or discharge, a mole of atoms (Chapter 7). In 1910 Millikan determined, in effect, the charge on one of these atoms (Chapter 8). The ratio of these two charges represents the number of atoms in a mole—Avogadro's number.

Another method, described in detail in Chapter 9 (p. 140 and Problem 9-2), was used by Rutherford in 1911. Certain radioactive materials emit alpha particles, which can be observed through flashes produced on a fluorescent screen, or counted with a Geiger counter. Each alpha particle then is converted into an atom of helium gas. By counting alpha particles and then measuring the volume of collected helium, Rutherford could calculate the number of helium atoms in a molar volume (22.4 l) of helium gas.

One of the most precise methods for determining Avogadro's number involves X-ray diffraction, discovered by Max von Laue in 1912 (Chapters 13 and 27). With this technique, X-ray crystallographers can measure very precisely the spacing between atoms in a crystal lattice. From the spacing we can compute the number of atoms in a given quantity of solid—which in turn yields Avogadro's number. (*Also see* Problem 13-8.)

SUMMARY

Dalton's atomic theory, formulated between 1803 and 1808, stated that elements are composed of indivisible particles called atoms; that atoms of a given element are identical and differ from the atoms of other elements, particularly in weight; and that atoms combine in a variety of simple whole-number ratios to form compounds.

Dalton's "model" was supported by the Law of Constant Composition, discovered in 1799, and the Law of Multiple Proportions, discovered in 1808.

In 1808 Gay-Lussac observed that the volumes of combining gases are in the ratios of small whole numbers. This observation suggested that the numbers of particles contained in equal volumes of different gases are also in the ratios of small whole numbers.

In 1811 Avogadro suggested that the numbers of these particles are *equal;* that is, that equal volumes of gases contain equal numbers of particles. He further suggested that the individual particles (molecules) of elements may be composed of several identical atoms.

In 1858 this assumption became the basis of Cannizzaro's method for finding atomic weights. Cannizzaro's method involved determin tion of the molecular weights of several gaseous compounds by the gas-density method, followed by chemical analyses of the compounds.

We discussed two other methods for finding atomic weights. Dulong and Petit's method was developed in 1819 and involved determination of the specific heat of a solid element. The method of combining weights involved chemical analysis of a compound composed of the "unknown" element and an element of known atomic weight.

Modern atomic weights are expressed on a relative scale, on which the atomic weight of the carbon-12 isotope is set at exactly 12. On this scale the atomic weight of hydrogen is 1.0080.

In 1865 Loschmidt devised a method for "counting" the number of atoms in a given weight of substance. The modern value for the number of atoms in 12 grams of carbon-12 (or 1.0080 g of hydrogen) is 6.02×10^{23} and is known as Avogadro's number. Avogadro's number of particles is called a mole (analogous to calling 144 items a gross).

The weight in grams of a mole of atoms of an element is its gram-atomic weight; it is numerically equal to its atomic weight. The weight in grams of a mole of molecules of a compound is its gram-molecular weight and is numerically equal to its molecular weight.

A mole of any gas—that is, Avogadro's number of molecules—occupies a volume of about 22.4 liters. This is known as the molar volume.

✿   ✿   ✿

The stage is now set for the next development in the quest for understanding the atom: With the availability of relative atomic weights scientists could study the elements more systematically and thus discovered new relationships.

ADDITIONAL PROBLEMS

Problem 4-9. Calculate the approximate diameter of the water molecule. [*Hint:* Compute the number of molecules in one gram of water. By definition, one gram of water occupies a volume of one cubic centimeter (*see* Appendix II). Now determine the approximate volume of one molecule. From that value calculate the approximate diameter, thinking of the molecule as a little cube.]

Problem 4-10. If 5.9 gallons of a diatomic elementary gas at standard conditions are liquefied (through cooling and applying pressure), the weight of the liquid is found to be 71 grams. Referring to an atomic-weight table, identify the element.

Problem 4-11. In a vessel filled with water vapor what approximate percentage of the space is occupied by the actual molecules, and what percentage is empty space? (*Hint:* Compare the volume occupied by 18 g of water vapor with that occupied by 18 g of liquid water.)

Problem 4-12. Chapter 3 showed that one volume of carbon monoxide (CO) combines with one volume of chlorine (Cl_2) to produce one volume of phosgene.
(a) What is the formula of phosgene?
(b) Write a balanced equation for the reaction.
(c) How much phosgene is produced by the reaction of 28 grams of carbon monoxide?

Problem 4-13. Sulfur dioxide and hydrogen sulfide react in accordance with the following equation.

$$SO_2 + 2H_2S \rightarrow 3S + 2H_2O$$

The atomic weight of sulfur is 32.
(a) What weight of hydrogen sulfide will react with 64 g of sulfur dioxide?
(b) How much sulfur and how much water will be produced? (Watch conservation of mass!)
(c) What is the *liquid* volume of the water?
(d) If only 1 gram of hydrogen sulfide is used, what quantity of sulfur dioxide will be consumed, and how much sulfur will be produced?
(e) The reactants are gases. What volume of sulfur dioxide is required to react completely with one liter of hydrogen sulfide?
(f) Approximately how many pounds of each reactant are required to produce 100 pounds of sulfur?

Problem 4-14. An unknown gas weighs 1.34 grams per liter (at standard conditions of temperature and pressure).
(a) What is the molecular weight of this gas?
(b) This substance is found to be composed of only carbon and hydrogen. Write all possible formulas of this compound.
(c) Crude analysis shows that more than half the weight of this substance is due to carbon. Is this information enough to tell you its formula?
(d) Subsequent more careful analysis reveals a weight composition of 80% carbon, 20% hydrogen. What is the formula?
(e) If one mole of this gas is burned to produce only carbon dioxide

(CO_2) and water, how many moles of each will be formed? Write an equation for the reaction, balancing the carbon and hydrogen atoms.

(f) How many grams each of carbon dioxide and water will be formed by the combustion of one mole of the gas? By the combustion of one gram?

(g) How many grams of oxygen are required to burn up one mole of the gas? Determine this by direct calculation from the equation in part (e) and also by application of the Law of Conservation of Mass to the results of part (f).

Problem 4-15. The formula of the main constituent of a popular beverage is C_2H_5OH.

(a) What is its molecular weight?

(b) How many atoms of hydrogen are there in each molecule?

(c) How many gram-atomic weights of hydrogen are there in each gram-molecular weight—i.e., in 1 mole of hydrogen gas?

(d) How many hydrogen atoms are there in each mole?

(e) How many grams of hydrogen are there in each mole?

(f) What is the weight in grams of 1 mole of this substance (a liquid)?

(g) What is this weight in ounces?

(h) What, very approximately, is the volume of a mole of this liquid, in cubic centimeters and ounces? (Assume a density similar to that of water.)

(i) In the human body, this substance is eventually burned up according to the following equation:

$$C_2H_5OH + 3O_2 \rightarrow 2CO_2 + 3H_2O$$

The heat liberated by this combustion process is 330 kilocalories (or large calories") per mole of C_2H_5OH. Approximately how many large calories are liberated by the combustion of 1 ounce of this substance?

(j) By now you probably realize that C_2H_5OH is alcohol. A calorie chart lists 80 calories for a 1-ounce shot of bourbon whiskey. Is your last answer compatible with this figure? Account for the apparent discrepancy.

ANSWERS TO SELECTED PROBLEMS

4-8. N atoms of hydrogen weigh 1 gram. Therefore, one atom weighs $1/N$ gram, or 1.7×10^{-24} g.

4-9.   1 g of H_2O = 1/18 mole; no. of molecules in 1 g = $1/18 N$; vol. of

$$1 \text{ g} = 1 \text{ cc}; \text{ vol. of 1 molecule} = \frac{\text{vol. of 1 g}}{\text{no. of molecules/g}} = \frac{1 \text{ cc}}{1/18 N} \approx$$

3×10^{-23} cm^3; diameter = $\sqrt[3]{\text{volume}}$ = $\sqrt[3]{30 \times 10^{-24}}$ cm^3 $\approx$ 3×10^{-8} cm (A very small number indeed!)

4-10. 5.9 gal $\approx$ 22.4 l $\therefore$ Molec. Wt. $\approx$ 71; AW $= \dfrac{71}{2} = 35.5$

The gas is chlorine.

4-11. % space of molecules $= \dfrac{\text{liq. vol.}}{\text{gas vol.}} \times 100 = \dfrac{18 \text{ cc}}{22{,}400 \text{ cc}} \times 100 = \dfrac{18}{224}$

$= 0.08\%$ (less than 1/10%) $\therefore$ empty space is 99.92%.

4-12. (a) $COCl_2$.

4-13. (a) 2×34 g $= 68$ g.

(b) 3×32 g; 2×18 g (total $= 132$ g).

(c) 36 cc.

(d) $1/68 \times 64$ g; $1/68 \times 96$ g.

(e) 1/2 liters.

(f) $100/96 \times 64 \approx 67$ lb; $100/96 \times 68 \approx 71$ lb.

4-14. (a) 1.34 g/liter $\times$ 22.4 liter/mole $= 30$ g/mole; *MW = 30*.

(b) (I) CH_{18}; (II) C_2H_6.

(c) I is 12/30 carbon; II is 24/30 carbon $\therefore$ formula is C_2H_6.

(d) $24/6 = 4/1 = 80/20$ (C_2H_6).

(e) $C_2H_6 \rightarrow 2CO_2 + 3H_2O$.

           $\uparrow$ $\uparrow$

(f) Part I: 2×44 g; 3×18 g.

Part II: 1/30 of foregoing quantities.

(g) To balance the equation, we need $(2 \times 2) + (3 \times 1) = 7$ oxygen atoms, or, 7×16 g $= 112$ g.

4-15. (c) 6.

(d) $6N \approx 4 \times 10^{24}$.

(e) 6 g.

(f) 46 g.

(g) $\sim$1.6 oz.

(h) vol. of 46 g $\approx$ 46 cc (assume a similar density as water) $\therefore$ $\sim$1.6 oz.

(i) $330/1.6 \approx 200$ calories ($=200$ kcal).

(j) Yes. 80 calories is for "80 proof" alcohol, meaning a solution of which alcohol comprises only 40%. Therefore, the calories in 1 ounce of pure (100%) bourbon would be $\dfrac{100}{40} \times 80 = 200$ calories.

Chapter 5

THE PERIODIC TABLE

In Chapter 1 we noted that early chemical investigators accumulated a large body of information, even though they did not understand why certain chemical reactions took place. As time went on, more and more elements were discovered and their chemical behavior was studied. In 1789 Lavoisier recognized about twenty-six elements; in 1869 sixty were known, and today we recognize one-hundred-and-four elements—a far cry from the four "elements" of the ancient Greeks!

Scientists soon realized that certain groups of elements have similar properties, and they tried to find a scheme relating these similarities. Several schemes were tried, but it was not until Cannizzaro's reliable table of atomic weights had become available in 1858 that a workable overall scheme was possible.

EARLY CLASSIFICATION SCHEMES

The German scientist Johann Döbereiner (1780–1849) was one of the first to discover a relationship among elements. He noticed in 1817 that there are several groups of three elements having related properties. Chlorine, bromine, and iodine make up such a group of three, called a *triad*. Döbereiner noticed that the atomic weight of bromine, 80, is about halfway between chlorine's 35 and iodine's 127: $\frac{1}{2}(35 + 127) = 81$. Other triads are calcium (40), strontium (88), and barium (137); and sulfur (32), selenium (79), and tellurium (128). In these triads other properties of the middle member seem to be halfway between those of the two other members. For example, considering the properties color and state, chlorine is a greenish-yellow gas, bromine is a brownish liquid, and iodine is a violet-black solid. The chemical reactivity of bromine also is intermediate between those of chlorine and iodine (*see* p. 73).

Döbereiner's scheme left much to be desired because it merely grouped the elements in threes without relating the properties of *all* the elements. Nonetheless, it was important because it represented one of the first attempts to classify or group elements according to their properties.

In the following years several other classification schemes were proposed, notably J. A. R. Newlands' (English, 1837–1898) law of octaves in 1864. This scheme arranged the elements in order of increasing atomic weights and recognized that "the eighth element, starting from a given one, is a kind of repetition of the first, like the eighth note in an octave of music." The first three of Newlands' octaves are shown here:[1]

H	Li	Be	B	C	N	O
F	Na	Mg	Al	Si	P	S
Cl	K	Ca	Cr	Ti	Mn	Fe

Elements in the vertical columns have similar properties. The scheme unfortunately breaks down beyond calcium (Ca, in the third row); but it paved the way for the more elaborate proposals of Mendeleev and Meyer. It should be noted that Newlands' scheme worked only when the new atomic weights of Cannizzaro (1858) were used.

THE MENDELEEV TABLE

In 1869 an arrangement of the elements was proposed that is essentially the one in use today. Two chemists, one a German, Lothar Meyer (1830–1895), and the other a Russian, Dmitri Mendeleev (1834–1907), did their work independently and published their results in the same year. Although their periodic tables were quite similar, Meyer and Mendeleev arrived at them in very different ways. Meyer based his arrangement on the physical properties of the elements, whereas Mendeleev used chemical properties as his criteria. Mendeleev is generally given more credit because he went quite a bit further than Meyer. For example, he predicted the chemical properties of elements unknown at that time and left space for them in his arrangement.

Mendeleev, as we have pointed out, used chemical properties of the elements as the basis of his scheme. Essentially, he arranged the elements from left to right according to increasing atomic weight, as Newlands had

---

[1] See the endpaper for a complete listing of the elements, their symbols, and their atomic weights.

Table 5-1. Periodic Table of the Elements

	Representative Elements		Transition Elements									Representative Elements					Noble Gas Elements
1	1 **H** 1.0080																2 **He** 4.00260
2	3 **Li** 6.941	4 **Be** 9.01218										5 **B** 10.81	6 **C** 12.011	7 **N** 14.0067	8 **O** 15.9994	9 **F** 18.9984	10 **Ne** 20.179
3	11 **Na** 22.9898	12 **Mg** 24.305										13 **Al** 26.9815	14 **Si** 28.086	15 **P** 30.9738	16 **S** 32.06	17 **Cl** 35.453	18 **Ar** 39.948
4	19 **K** 39.102	20 **Ca** 40.08	21 **Sc** 44.9559	22 **Ti** 47.90	23 **V** 50.941	24 **Cr** 51.996	25 **Mn** 54.9380	26 **Fe** 55.847	27 **Co** 58.9332	28 **Ni** 58.71	29 **Cu** 63.546	30 **Zn** 65.37	31 **Ga** 69.72	32 **Ge** 72.59	33 **As** 74.9216	34 **Se** 78.96	35 **Br** 79.904	36 **Kr** 83.80
5	37 **Rb** 85.4678	38 **Sr** 87.62	39 **Y** 88.9059	40 **Zr** 91.22	41 **Nb** 92.9064	42 **Mo** 95.94	43 **Tc** 99.9062	44 **Ru** 101.07	45 **Rh** 102.9055	46 **Pd** 106.4	47 **Ag** 107.868	48 **Cd** 112.40	49 **In** 114.82	50 **Sn** 118.69	51 **Sb** 121.75	52 **Te** 127.60	53 **I** 126.9045	54 **Xe** 131.30
6	55 **Cs** 132.9055	56 **Ba** 137.34	57–71 ★	72 **Hf** 178.49	73 **Ta** 180.947	74 **W** 183.85	75 **Re** 186.2	76 **Os** 190.2	77 **Ir** 192.2	78 **Pt** 195.09	79 **Au** 196.967	80 **Hg** 200.59	81 **Tl** 204.37	82 **Pb** 207.2	83 **Bi** 208.980	84 **Po** 209	85 **At** 210	86 **Rn** 222
7	87 **Fr** 223	88 **Ra** 226.025	89–103 ★	104														

★Lanthanide Series	57 **La** 138.9055	58 **Ce** 140.12	59 **Pr** 140.9077	60 **Nd** 144.24	61 **Pm** 145	62 **Sm** 150.4	63 **Eu** 151.96	64 **Gd** 157.25	65 **Tb** 158.9254	66 **Dy** 162.50	67 **Ho** 164.9303	68 **Er** 167.26	69 **Tm** 168.9342	70 **Yb** 173.04	71 **Lu** 174.97
★Actinide Series	89 **Ac** 227	90 **Th** 232.0381	91 **Pa** 231.0359	92 **U** 238.029	93 **Np** 237.0482	94 **Pu** 244	95 **Am** 243	96 **Cm** 247	97 **Bk** 247	98 **Cf** 251	99 **Es** 254	100 **Fm** 257	101 **Md** 258	102 **No** 255	103 **Lr** 256

TABLE OF ATOMIC WEIGHTS

Element	Symbol	Atomic Number	Atomic Weight	Element	Symbol	Atomic Number	Atomic Weight	Element	Symbol	Atomic Number	Atomic Weight
Actinium	Ac	89	227	Hafnium	Hf	72	178.49	Promethium	Pm	61	145
Aluminum	Al	13	26.9815	Helium	He	2	4.00260	Protactinium	Pa	91	231.0359
Americium	Am	95	243	Holmium	Ho	67	164.9303	Radium	Ra	88	226.0254
Antimony	Sb	51	121.75	Hydrogen	H	1	1.0080	Radon	Rn	86	222
Argon	Ar	18	39.948	Indium	In	49	114.82	Rhenium	Re	75	186.2
Arsenic	As	33	74.9216	Iodine	I	53	126.90455	Rhodium	Rh	45	102.9055
Astatine	At	85	210	Iridium	Ir	77	192.22	Rubidium	Rb	37	85.467
Barium	Ba	56	137.34	Iron	Fe	26	55.847	Ruthenium	Ru	44	101.07
Berkelium	Bk	97	247	Krypton	Kr	36	83.80	Samarium	Sm	62	150.4
Beryllium	Be	4	9.01218	Lanthanum	La	57	138.9055	Scandium	Sc	21	44.9559
Bismuth	Bi	83	208.9806	Lawrencium	Lr	103	256	Selenium	Se	34	78.96
Boron	B	5	10.81	Lead	Pb	82	207.2	Silicon	Si	14	28.086
Bromine	Br	35	79.904	Lithium	Li	3	6.941	Silver	Ag	47	107.868
Cadmium	Cd	48	112.40	Lutetium	Lu	71	174.97	Sodium	Na	11	22.9898
Calcium	Ca	20	40.08	Magnesium	Mg	12	24.305	Strontium	Sr	38	87.62
Californium	Cf	98	251	Manganese	Mn	25	54.9380	Sulfur	S	16	32.06
Carbon	C	6	12.011	Mendelevium	Md	101	258	Tantalum	Ta	73	180.9479
Cerium	Ce	58	140.12	Mercury	Hg	80	200.59	Technetium	Tc	43	98.9062
Cesium	Cs	55	132.9055	Molybdenum	Mo	42	95.94	Tellurium	Te	52	127.60
Chlorine	Cl	17	35.453	Neodymium	Nd	60	144.24	Terbium	Tb	65	158.9254
Chromium	Cr	24	51.996	Neon	Ne	10	20.179	Thallium	Tl	81	204.37
Cobalt	Co	27	58.9332	Neptunium	Np	93	237.0482	Thorium	Th	90	232.0381
Copper	Cu	29	63.546	Nickel	Ni	28	58.71	Thulium	Tm	69	168.9342
Curium	Cm	96	247	Niobium	Nb	41	92.9064	Tin	Sn	50	118.69
Dysprosium	Dy	66	162.50	Nitrogen	N	7	14.0067	Titanium	Ti	22	47.90
Einsteinium	Es	99	254	Nobelium	No	102	255	Tungsten	W	74	183.85
Erbium	Er	68	167.26	Osmium	Os	76	190.2	Uranium	U	92	238.029
Europium	Eu	63	151.96	Oxygen	O	8	15.9994	Vanadium	V	23	50.9414
Fermium	Fm	100	257	Palladium	Pd	46	106.4	Xenon	Xe	54	131.30
Fluorine	F	9	18.9984	Phosphorus	P	15	30.9738	Ytterbium	Yb	70	173.04
Francium	Fr	87	223	Platinum	Pt	78	195.09	Yttrium	Y	39	88.9059
Gadolinium	Gd	64	157.25	Plutonium	Pu	94	244	Zinc	Zn	30	65.37
Gallium	Ga	31	69.72	Polonium	Po	84	209	Zirconium	Zr	40	91.22
Germanium	Ge	32	72.59	Potassium	K	19	39.102				
Gold	Au	79	196.9665	Praseodymium	Pr	59	140.9077				

Based on the 1969 report of the Commission on Atomic Weights of the International Union of Pure and Applied Chemistry.

done. Elements having similar chemical properties are listed under each other, forming vertical groupings called *families* or *groups* (*see* Table 5-1). It is evident that similarities of chemical properties are repeated "periodically," at more or less regular intervals. Hence, we call this tabular arrangement a *periodic table* and call the horizontal rows *periods*.

PROPERTIES OF ELEMENTS

Metals Versus Nonmetals

By *properties* of materials we mean the way they affect our senses or our instruments and how they behave under certain circumstances.

On the basis of their properties, we classify elements such as sodium, magnesium, iron, gold, and mercury as *metals;* and other elements such as oxygen, sulfur, and bromine as *nonmetals*. Some of the properties that distinguish metals are shiny metallic luster, malleability, and the ability to conduct electricity and heat. Nonmetals lack these properties.

In the periodic table metals are located on the left, and nonmetals are on the right. A few elements in the middle of the table exhibit intermediate properties: Carbon in the form of graphite, for example, does not have metallic luster but does conduct electric current; boron, on the other hand, has some metallic luster but does not conduct current. Of the 104 known elements, approximately 76 are distinctly metals, 10 are nonmetals, 12 are of intermediate character, and 6 are inert (noble) gases.

Metals differ from nonmetals in other ways, too. For example, nonmetals do not liberate hydrogen gas from water or from acids; but many metals do. Sodium, for instance, reacts vigorously with water, producing hydrogen and sodium hydroxide:

$$2Na + 2H_2O \rightarrow H_2 + NaOH$$
sodium + water → hydrogen + sodium hydroxide

And zinc reacts with acids:

$$Zn + 2HCl \rightarrow H_2 + ZnCl_2$$
zinc + hydrochloric acid → hydrogen + zinc chloride

The vigor of these reactions decreases from left to right in the periodic table. For instance, gold (Au)—a metal with luster and electrical conductivity—does not liberate hydrogen from either water or acids.

The metals on the left side of the periodic table react also with oxygen, producing oxides. These oxides in turn react with acids to produce water and are called basic oxides, or simply *bases*. The following are typical reactions.

(1a) $\qquad$ $4Na + O_2 \rightarrow 2Na_2O$
sodium + oxygen → sodium oxide

(1b) $\qquad$ $Na_2O + H_2O \rightarrow 2NaOH$
sodium oxide + water → sodium hydroxide (lye)

(2a) $\qquad$ $2Mg + O_2 \rightarrow 2MgO$
magnesium + oxygen → magnesium oxide

(2b) $\qquad$ $MgO + H_2O \rightarrow Mg(OH)_2$
magnesium oxide + water → magnesium hydroxide
(milk of magnesia)

Nonmetals also react with oxygen to produce oxides; *these* oxides can react with bases. The following are typical reactions.

(1a) $\qquad$ $2S + 3O_2 \rightarrow 2SO_3$
sulfur + oxygen → sulfur trioxide

(1b) $\qquad$ $SO_3 + Na_2O \rightarrow Na_2SO_4$
sulfur trioxide + sodium oxide → sodium sulfate

(2a) $\qquad$ $4P + 5O_2 \rightarrow 2P_2O_5$
phosphorus + oxygen → phosphorus pentoxide

(2b) $\qquad$ $P_2O_5 + 3MgO \rightarrow$
phosphorus pentoxide + magnesium oxide →
$Mg_3(PO_4)_2$
magnesium phosphate

There are many other chemical reactions that distinguish metals from nonmetals, but we need not go into those now.

Combining Power

The power of an element to combine with other elements or atoms was the chemical property most useful to Mendeleev in the formulation of his periodic table.[2] In hydrogen fluoride, HF, both elements have equal combining power—one atom, or one gram-atomic weight, of fluorine combines with one atom, or one gram-atomic weight, of hydrogen. On the other hand, oxygen in water, H_2O, has *twice* the combining power of hydrogen—one atom, or one gram-atomic weight, of oxygen combines with *two* atoms, or gram-atomic weights, of hydrogen. If we set the combining power of hydrogen equal to *one*, then that of fluorine also is one, and the combining power of oxygen is two.

[2] In Mendeleev's time the word *valence* was used for combining power.

We can picture the combining power of atoms in terms of imaginary bonding hooks. In such a model, hydrogen and fluorine atoms have one hook each, while oxygen has two:

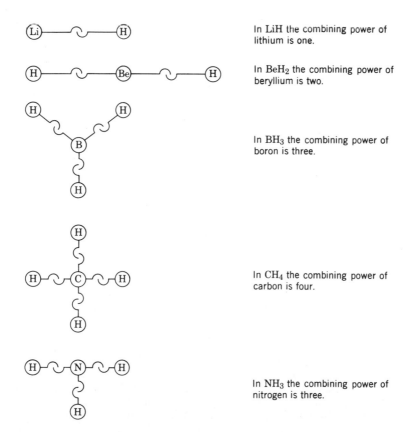

HF   HOH, or H₂O

We would represent hydrogen compounds of other second-row elements as follows.

In LiH the combining power of lithium is one.

In BeH₂ the combining power of beryllium is two.

In BH₃ the combining power of boron is three.

In CH₄ the combining power of carbon is four.

In NH₃ the combining power of nitrogen is three.

Other compounds may be represented similarly; some may involve multiple bonding (double and triple hooks).

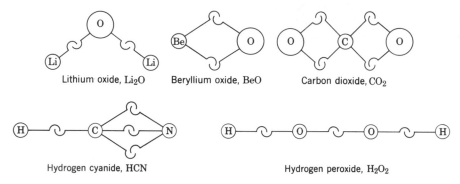

Lithium oxide, Li_2O Beryllium oxide, BeO   Carbon dioxide, CO_2

Hydrogen cyanide, HCN   Hydrogen peroxide, H_2O_2

In these compounds the combining power of Li and H still is one; beryllium and oxygen, two; nitrogen, three; and carbon, four.[3]

We see that, as we go across the periodic table, the combining powers increase from one to four and then drop back to one. The final element in the second row, neon (Ne) seems to be chemically inert—does not combine with other elements—and is said to have zero combining power. This pattern is repeated in the next row: the combining power of sodium (Na) is one; of magnesium (Mg), two; and so on until argon (Ar), which has zero combining power.[4]

Mendeleev gave priority to the combining power of elements in assigning positions in his periodic table. First he arranged the elements in order of increasing atomic weights, then he started and ended each period with elements with combining powers of one. Some of these periods became rather long, with the inclusion of *transition elements*. Occasionally, discrepancies arose. For example, according to atomic weights tellurium (Te) in the fifth row, atomic weight 127.6, should follow iodine (I), atomic weight 126.9. Mendeleev reversed their order in his table because of their combining powers. Similarly, after its discovery, the inert gas argon (Ar), atomic weight 39.9, was placed before the metal potassium (K), atomic weight 39.1.[5]

[3] Some elements exhibit variable combining powers. For example, phosphorus has a combining power of three in PF_3, but five in PF_5. This variability, however, did not seem to confuse Mendeleev.

[4] The so-called *inert gases* in the last column—helium (He) to radon (Rn)—were not known in Mendeleev's time. The interesting story of their discovery and placement into the periodic table is told in Chapter 14. In 1962 some of these "inert" gases were shown to form compounds, after all; we now refer to them as the *noble gases*.

[5] These reversals are discussed in Chapter 11 in connection with atomic numbers. Further discussions of combining power are to be found in Chapter 7 (in the section "Ionic Charges and Chemical Formulas") and in Chapter 19.

PREDICTION OF NEW ELEMENTS

Even more interesting than these reversals were the several unfilled positions in Mendeleev's original table. These appeared when Mendeleev lined up his elements according to their chemical properties and combining powers. With great assurance, Mendeleev concluded that these positions would eventually be filled by then undiscovered elements. He then predicted the properties of these elements by assuming gradation of properties within families, as in Döbereiner's triads.

Three of these unfilled positions were in the fourth row and are now filled by the elements gallium (Ga), scandium (Sc), and germanium (Ge). These were discovered, as one might guess from their names, respectively in France (1875), Sweden (1879), and Germany (1886). In part their discoverers were guided by Mendeleev's predictions. Germanium in particular, illustrates the remarkable correspondence between Mendeleev's predictions and the properties actually found (*see* Table 5-2).

Table 5-2. Predicted and Observed Properties of Germanium

	Mendeleev's Prediction (1871)	Observation
Atomic weight	72	72.59
Specific gravity	5.5	5.47
Atomic volume	13	13.22
Combining power	4	4
Specific heat	0.073	0.076
Color	gray	grayish white
Specific gravity of dioxide	4.7	4.703
Boiling point of tetrachloride	Under 100°	86°
Specific gravity of tetrachloride	1.9	1.887

In 1898 Marie and Pierre Curie were guided by another unfilled position in the periodic table in their search for radium (*see* Chapter 9).

THE MODERN PERIODIC TABLE

Mendeleev's original table has undergone many revisions. His essential ideas, however, remain intact in today's periodic table. There are several forms of the table in current use, but we shall confine ourselves to the form shown in Table 5-1.

The first period contains only hydrogen and the noble gas helium. For convenience of comparison, hydrogen (combining power, one) is listed twice, once in the first column and again in the next-to-last column, because it is chemically similar in some respects to both families. The other periods begin with *alkali metals*—lithium (Li), sodium (Na), potassium (K), rubidium (Rb), cesium (Cs), and francium (Fr)—and end with noble gases—neon (Ne), argon (Ar), krypton (Kr), xenon (Xe), and radon (Rn). The second and third periods each have eight members, like Newlands' octaves. The next two periods each have eighteen members, through the addition of ten *transition metals*. The next period, beginning with cesium, has thirty-two members, through the further addition of fourteen elements following lanthanum (La) in the third column. These fourteen elements are called the *lanthanides,* or *rare earths,* and have been put into a separate row near the bottom of the table to keep it from getting too wide. The last period, beginning with francium, also contains fourteen members of a series, called the *actinides.* They follow actinium (Ac) in the third column and are also placed at the bottom of the table for greater convenience. This last period is incomplete, ending with element 104, discovered between 1964 and 1967.[6]

Each element in the periodic table has been given a serial number, called its *atomic number;* these run from 1 (for hydrogen) to 104 (unnamed). Because only one atomic number (but not atomic weight) is associated with each element, we sometimes refer to elements by their numbers. Element 6, for example, is carbon; element 9 is fluorine, and so on. For the present purpose it suffices to consider atomic numbers as serial numbers in the periodic table; in Chapter 11 we discuss the physical significance of atomic numbers. In Chapters 19 and 24 we shall find out what causes atoms to behave in this "periodic" manner and why some periods have two or eight members while others have eighteen or thirty-two members.

Let us now mention some uses of the periodic table. Two of these have already been discussed: Unfilled positions in the original periodic table served as stimulus and guide in the discovery and isolation of new elements; and an element's position in the table usually tells us its combining power. After all, combining power was Mendeleev's main criterion for placing the elements. Thus combining powers of elements in the first two columns are one and two, respectively; those in the last six columns generally are three, four, three, two, one, and zero, respectively. Combining

---

[6] More elements will probably be "discovered," i.e., produced artificially, in the next few years. An interesting discussion can be found in an article by G. T. Seaborg and J. L. Bloom in *Scientific American,* 220: 57 (April 1969).

powers of the transition elements are not so readily apparent; we shall not discuss them here, to avoid unnecessary confusion.

Beyond those two applications, the periodic table helps us organize and systematize the properties and chemistry of the elements into a manageable subject. Let us briefly look at some properties of a few families to illustrate how the periodic table enables us to view similarities and differences among elements.

The Alkali Metals

The members of the first group starting with lithium are known as the *alkali metals.* They exhibit most strongly the properties we usually associate with metals: high metallic luster, high electrical conductivity, high heat conductivity, and high chemical reactivity with nonmetals. We speak of this group as having strong "metallic character." Because of their high affinity for nonmetals, the alkali metals are found in nature only as compounds, not as uncombined metals.

Metallic character increases as we pass downward in the periodic table from lithium to cesium and francium. Reactivity, for example, increases in that sequence. Other properties also change gradually, as shown in Table 5-3: Atomic weights and densities increase (the first three members

Table 5-3. Some Properties of the Alkali and Alkaline Earth Metals

Alkali Metals[a]	Atomic Weight	Density (g/cc)	Melting Point (°C)	Boiling Point (°C)	Formula of Chloride	Formula of Hydroxide
Lithium	6.9	0.53	179	1317	LiCl	LiOH
Sodium	23.0	0.97	98	892	NaCl	NaOH
Potassium	31.1	0.86	63	774	KCl	KOH
Rubidium	85.5	1.53	39	688	RbCl	RbOH
Cesium	132.9	1.90	28	690	CsCl	CsOH
Alkaline Earth Metals						
Beryllium	9.0	1.86	1280	2970	$BeCl_2$	$Be(OH)_2$
Magnesium	24.3	1.74	651	1107	$MgCl_2$	$Mg(OH)_2$
Calcium	40.1	1.54	845	1487	$CaCl_2$	$Ca(OH)_2$
Strontium	87.6	2.6	769	1384	$SrCl_2$	$Sr(OH)_2$
Barium	137.3	3.5	725	1140	$BaCl_2$	$Ba(OH)_2$
Radium	226.0	5(?)	700	1140	$RaCl_2$	$Ra(OH)_2$
General trends	incr.	incr.	decr.	decr.	similar	similar

[a] Francium, the last member of this family, is not included here; it is very rare and found only in trace amounts.

are light enough to float on water), while melting and boiling points decrease; all are so soft that they can be cut with a knife, and this softness increases as the atomic weight increases.

Another important aspect of the periodic table is that if we understand the chemical reactions of one typical member of a family, we usually have a key to the chemistry of the other members. Sodium, for example, reacts violently with chlorine gas to produce sodium chloride (ordinary table salt).

$$2Na + Cl_2 \rightarrow 2NaCl$$

The other alkali metals react similarly, producing chlorides whose formulas are analogous to sodium chloride: LiCl, NaCl, KCl, RbCl, CsCl, and FrCl. The alkali metals also combine with *other* members of the chlorine family, such as fluorine, and produce analogous compounds: LiF, NaF, KF, Rb, CsF, and FrF.

These formulas are similar because members of the same family have identical combining powers—one for both the alkali metals and the chlorine family. When alkali metals combine with elements whose combining power is two, such as oxygen, the formulas of the products are of the type Li_2O, Na_2O, etc.

Not only do elements of a family have similar properties, but compounds produced by members of two families also have similar properties. For example, the various compounds of alkali metals and members of the fluorine family—the alkali halides—all resemble NaCl. KCl, for instance, is sometimes used as a dietary substitute for ordinary table salt, NaCl.

The Alkaline Earth Metals

The members of the second group (combining power, two), the beryllium family, are known as the *alkaline earth metals*. Their metallic character is less strong than that of the alkali metals; but they are still so reactive that they also are found in nature only as compounds, not as uncombined metals. They are considerably harder than the alkali metals and also have higher densities as well as higher melting and boiling points. Within the family, properties of individual elements change gradually with changing atomic weight, but these changes are not quite so regular as those of the alkali metals (*see* Table 5-3).

The Halogens

From left to right in the periodic table metallic properties diminish. Boron, which follows beryllium, no longer is a typical metal; for example, it is only a poor conductor of electricity. Carbon, the next element, definitely is not a metal, but does have some vestiges of metallic character, such as fair electrical conductivity. In nitrogen and oxygen all metallic characteristics have disappeared.

The members of the fluorine family, on the right side of the periodic table, are called *halogens*. They are the most nonmetallic of the elements, not counting the practically inert gases of the helium family. Their physical state progresses from the gaseous through the liquid to the solid: Fluorine and chlorine are gases at room temperature, bromine is a liquid, and iodine and astatine (extremely rare) are solids. Fluorine is the most reactive member of the group. We see thus that the element with the most nonmetallic character is in the upper right corner of the periodic table, whereas the most metallic elements (Cs and Fr) are in the lower left corner.

SUMMARY, AND CONCLUSION OF PART ONE

It is generally conceded that scientific chemistry had its beginning in the seventeenth century with Robert Boyle's redefinition of the word element. Nearly one-and-a-half centuries later, between 1803 and 1808, John Dalton developed his atomic theory of the elements: Matter is made up of discrete particles; the particles of a given element are like one another and unlike those of other elements. This theory grew out of the following developments: Lavoisier's extensive use of the analytical balance as a scientific tool, his overthrow of the phlogiston theory, and his formulation of the Law of Conservation of Mass (1774); Proust's Law of Constant Composition (1799); and Dalton's own Law of Multiple Proportions (1808).

In 1808 Gay-Lussac discovered the Law of Combining Gas Volumes. This led Avogadro in 1811 to propose the equal-volumes-equal-numbers principle of gases and the existence of polyatomic molecules of elements. His proposal was generally ignored until Cannizzaro revived Avogadro's hypothesis in 1858 and added a reasonable procedure for determining atomic weights and molecular formulas.

Once the elements were arranged in the order of their atomic weights, it became possible to recognize a regular recurrence of similar properties. In 1869 Mendeleev and Lothar Meyer independently revealed this systematic repetition of properties in a clever serial arrangement of the elements, called the periodic table. This scheme not only showed the relationships of the known elements, but also, through conspicuous gaps in the table, pointed to the existence of as yet unknown elements, whose physical and chemical properties Mendeleev predicted with remarkable accuracy.

The periodic table helped chemists systematize what they knew about the physical and chemical properties of elements and their compounds. In the periodic table the elements are listed in order of increasing atomic weight in a series of horizontal rows, known as *periods*. These rows are arranged so that elements with similar properties, especially with similar combining powers, are placed under each other in columns called *groups* or *families*. Within groups, properties of elements generally are similar but change gradually with changing atomic weight. Properties change more drastically as we proceed across the periodic table from the metals on the left to the nonmetals on the right. The most metallic elements are at the lower left, while the most nonmetallic are at the upper right of the periodic table. The far-right column contains the nearly inert noble gases.

A word of caution: Chemists must take the generalizations implicit in the periodic table with a grain of salt, because there are exceptions to the rules. Yet the generalizations are sufficiently valid to serve as a useful reference basis, from which it is relatively easy to note the exceptions.

Correlating properties of the elements in a systematic scheme was indeed an achievement of first magnitude. It was, however, a strictly empirical arrangement, without theoretical basis at the time, and left many unanswered questions:

1. What is the cause of "periodicity"; why are properties of the elements similar at regular intervals, when they are arranged in order of increasing atomic weights?

2. Why must a few elements be arranged out of order, such as the pairs tellurium-iodine and argon-potassium?

3. Why are some elements similar even though they differ widely in their atomic weights—such as lithium (7), potassium (39), and cesium (133)? And why are others with similar atomic weights quite different—such as chlorine (35.5), argon (40), and potassium (39)? Are the properties of an element determined by a factor more fundamental than atomic weight?

4. What structural features of the atom are responsible for an element's properties, such as its physical state (solid, liquid, or gas); its metallic or nonmetallic character; its combining power; its existence as monatomic molecules (like Hg and the inert gases), or as diatomic molecules (like H_2 and O_2), or even as polyatomic molecules (like S_8)?

How scientists found answers to these questions is taken up in the following parts of this book.

ADDITIONAL READING FOR
PART ONE

Asimov, Isaac. *A Short History of Chemistry*. Garden City, N.Y.: Double-day and Co., 1965 (paperback).

Conant, James B. *Science and Common Sense*. New Haven: Yale University Press, 1951 (paperback). A straightforward but sophisticated portrayal of the methods of science through an historical approach.

Farber, Eduard. *Great Chemists*. New York: John Wiley and Sons, 1961. More than a hundred biographies from ancient to modern times.

Kieffer, William F. *The Mole Concept in Chemistry*. New York: Reinhold Publishing Corp., 1965 (paperback).

Leicester, Henry M. *The Historical Background of Chemistry*. New York: John Wiley and Sons, 1965 (paperback).

Nash, Leonard K. *The Atomic-Molecular Theory* (Case 4 of the Harvard Case Histories in Experimental Science, edited by J. B. Conant), Cambridge: Harvard University Press, 1965 (paperback). History of the atom from Dalton to Avogadro, much of it in the words of the original investigators.

Nash, Leonard K. *The Nature of the Natural Sciences*. Boston: Little, Brown and Company, 1963.

Partington, J. R. *A Short History of Chemistry*. 3rd ed. Harper and Row, New York, 1957 (paperback).

Taton, R. *Reason and Chance in Scientific Discovery*. New York: Science Editions, 1962. Emphasizes the psychology of discovery, rather than its logic, in various fields.

Walker, Marshall. *The Nature of Scientific Thought*. Englewood Cliffs, N.J.: Prentice-Hall, 1963 (paperback).

Weeks, Mary E., and Henry M. Leicester. *Discovery of the Elements*. 7th ed. Easton, Pa.: Journal of Chemical Education, 1968. The story of the disclosure, one by one, of the chemical elements.

Young, Louise B., ed. *The Mystery of Matter*. New York: Oxford University Press, 1965. A selection of writings by eminent scientists on the nature of the physical universe and the life it supports. Contributors range from Lucretius to Darwin, Einstein, and Linus Pauling.

PART TWO

Evolution of the Nuclear Atom

By 1870 Dalton's idea of atoms had become firmly established, after an interim period of controversy and doubt. Nearly seventy elements were known. Their atomic weights had been determined, their properties and reactions were being studied, and Mendeleev had managed to systematize some of these properties in his periodic table. Scientists of course wanted to find out the reasons for this peculiar periodicity of properties. Obviously more needed to be known about the structure of these strange little particles called atoms.

In the next few chapters we shall trace the investigations and discoveries that led to the idea that atoms are closely analogous to tiny solar systems, with very dense nuclei, revolving "planetary" electrons, and much empty space between the electrons. But before we turn to the key experiments that confirmed the existence of the nuclear atom, we shall discuss certain basic physical concepts that are essential to an understanding of this development.

Chapter 6

BASIC PHYSICAL CONCEPTS: MOTION AND ENERGY

Centuries ago, the study of nature in all its aspects was the domain of "natural philosophers." This unified approach to science lasted for a long time, as late as the midnineteenth century. It was personified by men like Michael Faraday (1791–1867), whom both chemists and physicists claim as their own.

As our scientific knowledge expanded, it became more and more difficult for a single person to master all of it, and the age of specialization and compartmentalization began. Courses in our universities became more and more specialized, and the various scientists had less and less in common. Even physicists and chemists found it increasingly difficult to communicate with each other—they spoke different scientific languages, often used different symbols, and even had different atomic-weight scales.

Lately, this trend has been reversed; there is more interaction between disciplines. Much of our biological knowledge is now explained on the basis of chemical principles. In turn, much of our chemical understanding is based on physical principles, and many of the chemist's tools are based on physical properties. Today we have university departments of geochemistry, biochemistry, and biophysics; and journals of physical chemistry and chemical physics; and we can attend chemistry colloquia on biophysics. The contemporary chemist must have considerable knowledge in areas that used to be considered the private domain of physicists. We can no longer separate scientific knowledge into packages neatly labeled "physics" and "chemistry."

Many of the following chapters are concerned with the motions of molecules and atoms, the paths of particles in electric and magnetic fields, the energies of circling electrons in atoms, and the energies of chemical bonds. Therefore, this chapter deals with some elementary aspects of motion and energy.

SPEED

The *speed* of a moving object is simply the ratio of the distance traveled to the time required to traverse that distance:

$$\text{speed} = \frac{\text{distance}}{\text{time}}$$

For example, if it takes us half an hour to drive 6 miles to work, then our average speed is 6 mi/0.5 hr, or 12 mph.

It should be clear that the foregoing definition refers to *average speed*. Most likely we did not drive at a constant speed of only 12 mph. There were times when the speedometer registered as much as 40 mph, and often we barely moved at 5 mph. The average speed is simple to calculate, but in a sense it is a fictitious quantity, merely the total distance divided by the total time. *Instantaneous speed* (indicated on a car's speedometer) is speed during a very small time interval. It is the ratio of a very small distance and the corresponding small time interval.

In addition to miles per hour, other common units of speed are kilometers per hour, feet per second, centimeters per second—always distance *per* (i.e., divided by) time. Some conversions (*also see* Appendixes 1-E and 2) are

$$1 \text{ mph} \approx 1.5 \text{ ft/sec} \approx 27 \text{ meters/min}$$

For convenience, we sometimes write ft/sec or cm/sec as ft-sec^{-1} and cm-sec^{-1}. The negative exponent merely indicates that this unit appears in the denominator.

Problem 6-1. Convert 1 mph to cm/sec.

ACCELERATION

When an object changes its speed, it is said to be accelerating. *Acceleration* is the rate at which speed is changing with respect to time; that is, change of speed per unit time:

$$\text{acceleration} = \frac{\text{final speed} - \text{initial speed}[1]}{\text{time interval}}$$

[1] For an expanded definition of acceleration *see* p. 86.

For example, if a car accelerates from 0 mph to 30 mph in 10 seconds, the acceleration is 30 mph/10 sec, or 3 mph each second (conventionally written as 3 mi/hr/sec or 3 mi/hr-sec). If the car reaches 50 mph after another 10 seconds, the acceleration during the second interval is

$$\frac{50 \text{ mph} - 30 \text{ mph}}{10 \text{ seconds}} = \frac{20 \text{ mph}}{10 \text{ sec}}$$

$$= 2 \text{ mph/sec, or } 2 \text{ mi/hr/sec, or } 2 \text{ mi/hr-sec}$$

If the car slows down from 30 mph to a standstill in 10 sec, there has been a *negative acceleration*, or "deceleration," of 3 mph/sec.

Note that the units of acceleration involve distance and *two* time units. (Where these two time units come from should be quite clear in the reader's mind.) We could also have expressed the speed of the car in feet per second:

$$30 \text{ mph} = 44 \text{ ft/sec}$$

Our first acceleration, then, is

$$\frac{44 \text{ ft/sec}}{10 \text{ sec}} = 4.4 \text{ ft/sec/sec, or } 4.4 \text{ ft/sec-sec}$$

We write this in conventional form[2] simply as

$$4.4 \text{ ft/sec}^2, \text{ or } 4.4 \text{ ft-sec}^{-2}$$

Problem 6-2. How many cm/sec^2 equal 1 ft/sec^2?

NEWTON'S FIRST LAW OF MOTION: INERTIA

Isaac Newton (the famous English philosopher and mathematician, 1642–1727) began to formulate his system of mechanics when he was in his early twenties and published it in 1687 in his famous *Mathematical Principles of Natural Philosophy* (called the *Principia*). The story is told that Newton had to be hit on the head by a falling apple to start thinking in the right direction. Actually he was building on the ideas of his predecessors, notably those of the Italian astronomer Galileo Galilei (1564–1642), who had studied accelerated motion.[3]

[2] There is no such thing as a "square second," of course. The notation "sec²" simply means "per second per second." From the given examples it should be evident that there is no inherent need for the two time units to be identical—the first example had miles/hr-sec as the unit of acceleration.

[3] Newton: "If I have seen farther than Descartes [a French philosopher and mathematician of the early seventeenth century, regarded as the founder of analytical geometry], it is by standing on the shoulders of giants."

Newton's *First Law of Motion* states:

Every body continues in its state of rest, or of uniform motion in a straight line, unless it is compelled to change that state by forces acting on it.

We might restate this law as follows:

The speed or direction of motion of a body does not change unless a force acts on the body.

This law means that in the absence of forces acting on a body (a) if at rest, it will remain at rest; and (b) if in motion, it will continue at constant speed and (c) in a straight line. This tendency of material bodies to persist in their state of rest or of uniform motion is called *inertia*.

The First Law of Motion, like any other law of nature, represents the interpretation and generalization of a large amount of observation. Actually, Newton went beyond observation when he stated that a body will continue to do something *indefinitely*. He had no way of knowing whether a body will continue to act in the future as it has in the past. Whether such a law is true or not can be judged in part by drawing conclusions from it—that is, basing predictions on it—and then seeing whether these predictions are supported by experimental evidence.

Newton's laws, and the system of mechanics based on them (Newtonian mechanics), have withstood the test of time. Predictions about the behavior of bodies have been borne out by observation and experiment. However, when speeds approaching the speed of light are involved, Newtonian mechanics is inadequate and Einstein's relativistic mechanics must be used. In a sense, relativistic mechanics is a refinement of Newtonian mechanics.[4]

NEWTON'S SECOND LAW OF MOTION: F = ma

Newton's Second Law of Motion answers the question implicit in the First Law; namely, how will a body behave if there *is* a force acting on it? The answer is that its speed and/or its direction of motion will be changed. More precisely, the change is in the direction of the force. If force is applied on a moving body in the direction of its motion, then the body will speed up, or accelerate. Conversely, if the force opposes the

[4] In the submicroscopic realm of atomic and subatomic particles, Newtonian mechanics is not applicable and must be replaced by quantum mechanics. This subject is taken up in Part Four.

direction of motion, the speed will decrease; that is, negative acceleration (deceleration) will take place. If the body is at rest originally, it will accelerate in the direction of the force.

We use these observations to define *force* as an influence that changes, or tends to change, the state of rest or motion of a body.

Simple experiments show that the magnitude of the acceleration, a, is proportional to the applied force, F ($a \propto F$). Thus doubling the force, or push, acting on a given body will double its acceleration.

We find further that the same force acting on a heavier body will produce less acceleration. A heavier body offers more resistance to change; we say that a heavier body has more *inertia*. Quantitatively, the acceleration is inversely proportional to the mass, m, of the body; in other words, it is proportional to the reciprocal of the mass ($a \propto 1/m$). If we double the mass—for example, by coupling two equal-sized bodies—then we halve the acceleration for a given force; it takes twice the force to produce the same acceleration.

The two relationships ($a \propto F$ and $a \propto 1/m$) can be combined into one:

$$a = \frac{F}{m}, \qquad \text{or} \qquad F = ma$$

The First Law is actually a special case of the Second Law: When $F = 0$, then $a = 0$; no acceleration occurs when there is no force, and the body simply continues at rest or in uniform motion, whichever is the case.

The equation $F = ma$ is the basis for defining units of force. A *dyne* is the force required to accelerate a mass of one gram[5] at the rate of 1 cm/sec². A dyne, then, is equivalent to a g-cm/sec² ($F = ma$). Thus a force of 30 dynes acting on a mass of 6 grams will accelerate that body at the rate of:

$$a = \frac{F}{m} = \frac{30 \text{ dynes}}{6 \text{ g}} = \frac{30 \text{ g-cm/sec}^2}{6 \text{ g}} = 5 \text{ cm/sec}^2$$

MASS VERSUS WEIGHT

We have so far avoided use of the word *weight*. A body possesses weight only by virtue of gravitational attraction. This attraction varies slightly from place to place on the surface of the earth, and thus the same body exhibits slightly different weights at different locations. It would weigh drastically less on the surface of the moon, whose gravitational

---

[5] A *gram* is the mass of one cubic centimeter of water (at a temperature of 4°C). *See* Appendix 2.

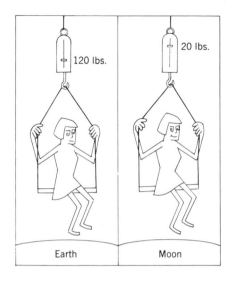

Fig. 6-1. A body's weight depends on the force of gravity. We all would weigh less on the moon.

force is weak; and this same body would weigh nothing in an orbiting satellite.

Yet, regardless of the body's weight at a particular place and time, a given force acting on it will produce a certain acceleration, even if the body has no weight at all, as in a satellite. The magnitude of that acceleration depends on the inertial *mass* of the body, which is inherent in the body and exists even without gravitational forces.

Weight, then, is a *force*—namely, the gravitational force acting on a body. On earth the weight of a body is the force with which the earth attracts it. Measurements on a freely falling 1-g mass will show that its speed increases at the rate of about 980 cm/sec each second; in other words, its acceleration is 980 cm/sec^2. It takes a force of 980 dynes to produce this acceleration in 1-g body.

$$F = ma = (1 \text{ g}) (980 \text{ cm/sec}^2) = 980 \text{ g-cm/sec}^2 = 980 \text{ dynes}$$

That is, a 1-g body is being pulled down by a gravitational force of 980 dynes; we say, therefore, that a 1-g body has a weight of 980 dynes. (A mosquito weighs perhaps 1 dyne.)

But a 2-g body falls at the same rate (disregarding friction); its acceleration also is 980 cm/sec^2. (Galileo showed in his legendary experiment at the Tower of Pisa that light and heavy bodies fall at the same rate.) This means that the force acting on a 2-g mass is 2×980 g-cm/sec^2, or 2×980 dynes. Thus the force of gravity—hence the weight of an object subject to that force—is proportional to the object's mass: doubling the mass doubles the weight. For this reason we often use the terms *mass* and *weight* interchangeably and (erroneously) speak of a "weight" of so many grams. This practice is permissible in everyday usage, but in scientific work we have to be careful, especially in calculations.

Although the gram and the kilogram are properly units of mass, in the English system the pound is a unit of weight. One pound equals 4.45×10^5 dynes. A body weighing 4.45×10^5 dynes has a mass as follows:

$$ m = \frac{F}{a} = \frac{4.45 \times 10^5 \text{ g-cm/sec}^2}{980 \text{ cm/sec}^2} = 454 \text{ g} $$

A weight of 1 lb, then, is equivalent to a mass of 454 g. Often we simply say that 1 lb equals 454 g.

Problem 6-3. A man's "weight" is 70 kg. He starts from rest and reaches a speed of 4 km/hr in a half second.
(a) What is his weight in pounds?
(b) What is his speed in miles per hour?
(c) What is his acceleration in km/hr-sec?
(d) In km/sec^2?
(e) In cm/sec^2?
(f) In ft/sec^2?
(g) How much force (in dynes) is required to produce this acceleration?
(h) Convert this to pounds of force.

Problem 6-4. A larger force unit sometimes used is the *newton*, which is the force required to accelerate a mass of 1 kg (1000 g) at the rate of 1 m/sec^2 (100 cm/sec^2).
(a) Calculate the number of dynes in 1 newton.
(b) How many newtons are equal to 1 lb?

VELOCITY VERSUS SPEED

So far we have merely alluded to the fact that a force acting on a body can change its direction of motion as well as its speed if the force is exerted in a direction other than the line of motion. For example, a stone thrown horizontally would continue at unreduced speed in that direction were it not for the force of gravity pulling it downward. Because of gravity, the stone's path will curve downward, and it will travel at increasing speed until it hits the ground.

If a force is exerted at right angles to a body's motion, it will produce a change of direction, but it cannot change the body's speed (so long as it remains at right angles to the motion). Thus the force of gravity causes a satellite to follow a curved path around the earth, but at a constant speed if the orbit is a circular one, because in this case, unlike that of the falling stone, the force remains at right angles to the path of the satellite. If we could suddenly remove the gravitional force, the satellite would fly off into space in a straight line tangent to the earth, at the same speed as before.

Similarly, if we whirl around our head an object attached to a string, the string acts like the force of gravity in the previous example. The string provides the centripetal force that pulls the object toward the center and causes it to follow a curved path. If the string breaks, the object will fly off at a tangent, in the direction of the motion it had at the moment the string broke. Both the orbiting satellite and the whirling object illustrate how a force can produce a change in direction without changing the body's speed.

We see, then, that a force acting on a body can change its speed, its direction, or both. The concept of *velocity* embodies both speed and direction. A change in the magnitude of either or both, therefore, is a change in velocity. Now we redefine *acceleration* to mean any change in velocity. Thus a circling satellite—even though it is traveling at constant speed—is undergoing a change in velocity, in this case meaning a change in direction; the satellite is constantly being accelerated by the force of gravity.

VECTORS AND VECTOR ADDITION[6]

If we attach two ropes to an object and pull in two directions (not 180° apart), it is clear that the resultant motion will not be in the direction of either rope but somewhere between. The resultant force, F_R, is a summation of the individual forces, F_1 and F_2.

We can illustrate this addition graphically by using arrows parallel to the directions of the forces they represent and whose lengths indicate the magnitudes of the forces. We call these arrows *vectors*. We can use vectors to represent anything that has both magnitude and direction: force, velocity, acceleration, momentum, electric field strength, etc. (Note that the following are *not* vector quantities: mass, speed, length, time, energy, and temperature.) We can add the two force vectors (F_1 and F_2) by

---

[6] This section can be omitted without loss of continuity.

completing a parallelogram and drawing in the diagonal to represent the resultant force, F_R (Fig. 6-2).

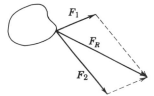

Fig. 6-2. Addition of force vectors by the parallelogram method.

An alternative way of adding vectors graphically is by drawing a triangle—equivalent to pretending that the two forces occur in sequence. The result is of course the same as by the parallelogram method (Fig. 6-3).

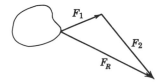

Fig. 6-3. Addition of force vectors by the triangle method.

Now let us analyze in more detail the path of a stone thrown horizontally. As the stone leaves the hand of the thrower it has a certain horizontal velocity, v_h, which we can represent by a horizontal vector. The length of the vector indicates the speed. After the stone leaves the hand there is no other force acting on it *in the horizontal direction*. Hence its *horizontal* speed remains unchanged at, say, 50 ft/sec.

However, the force of gravity gives the stone a downward acceleration (of 980 cm/sec², or 32 ft/sec²). At the end of the first second, the stone still has a horizontal velocity of 50 ft/sec and simultaneously a downward velocity of 32 ft/sec. The result is a path partly deflected downward. We can show this path by the addition of the two vectors v_h and v_g, whose lengths are in the ratio of 50:32; v_R represents the resultant velocity after one second (Fig. 6-4).

After two seconds v_h remains unchanged, but the force of gravity has continued to accelerate the stone downward; the downward velocity, v_g, has been increased by another increment of 32 ft/sec to 64 ft/sec. (Fig. 6-5). We see that now the path has curved more in the downward direction. Through repeated application of this process we can arrive at the path of such a projectile (Fig. 6-6).

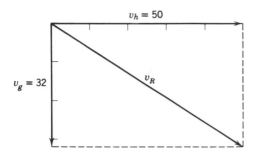

Fig. 6-4. Velocity of a freely falling body after one second.

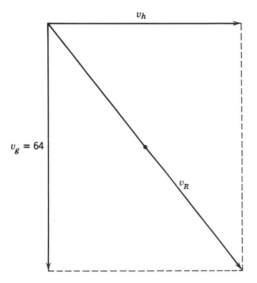

Fig. 6-5. Velocity of a freely falling body after two seconds.

Since the downward velocity component is independent of the horizontal component, if any, we conclude that a bullet merely dropped from our hand to the ground will hit the ground at the same time as another bullet shot simultaneously from the same height in a horizontal direction (if we can ignore air friction). This conclusion can be verified experimentally (Fig. 6-7).

An analogous situation is obtained when an airplane releases a bomb (Fig. 6-8). Plane and bomb have the same horizontal velocity component

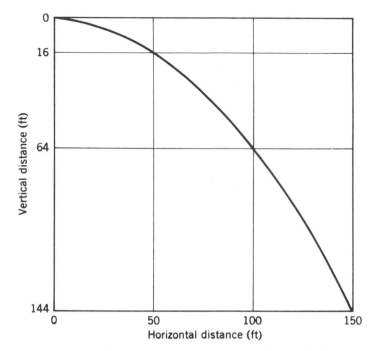

Fig. 6-6. Path during first three seconds of a falling body having a horizontal velocity component of 50 ft/sec.

at the time of release. The bomb curves downward, but if the plane's velocity is constant, the bomb always remains vertically under the plane. At the time of the impact, then, the plane, if it has maintained its velocity, is directly over the bomb. For low-flying planes releasing high-explosive bombs this creates a rather uncomfortable situation.

Problem 6-5. Can you suggest at least two different solutions to the foregoing "uncomfortable situation" of the low-flying bomber?

Problem 6-6. An earth satellite, as indicated previously, is held in orbit by the force of gravity. It therefore is being accelerated toward the earth, like the stone and like the bullet. Why doesn't it hit the earth at some point? (If it is high enough, where there is no atmospheric friction to slow it down, a satellite will orbit indefinitely. Our most famous satellite, the moon, has been doing just that for quite a while.) *Hint:* Remember that the earth is not a flat surface but a sphere.

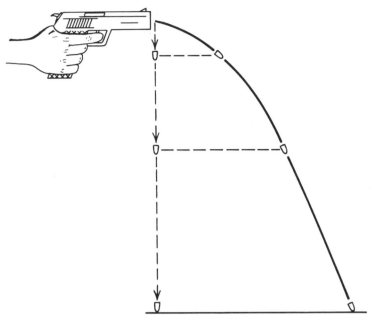

Fig. 6-7. A bullet shot horizontally reaches the ground at the same time as a bullet merely dropped simultaneously.

NEWTON'S THIRD LAW AND THE CONSERVATION OF MOMENTUM

We have implied that mass, or inertia, is the quality of a body that resists a change of motion or of velocity. More force is required to start a loaded automobile moving than an empty one, and more braking force is required to slow it down. Quantitatively, if we double the mass of the car, twice as much force will be required to stop it in the same distance or to change its direction (Newton's Second Law, $F = ma$). We say that a heavier car possesses more inertia.

In the case of a moving car, we also sometimes say that a heavy car possesses more *momentum* than a light one. However, in addition to mass, there is another factor affecting the momentum of the moving car—namely, its velocity. We know that it is harder to stop a fast-moving car than a slow one. The fast car possesses more momentum. Accordingly, we define the *momentum, p,* of a body as the product of its mass and its velocity:

$$p = mv$$

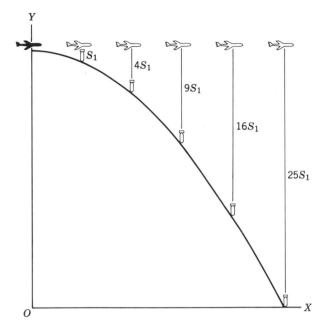

Fig. 6-8. A bomb released by a plane flying at constant speed remains vertically under the plane, until impact.

For example, a man weighing 75 kg and strolling at the rate of 2 km/hr has a momentum of $(75 \text{ kg}) (2 \text{ km/hr}) = 150$ kg-km/hr.

Problem 6-7. (a) What is the weight of the man in pounds?
(b) Convert his momentum into kg-m/sec.

Problem 6-8. (a) What is the momentum of a 3-oz bullet traveling at the rate of 2,000 ft/sec?
(b) Express this momentum in kg-m/sec and compare it to that of the man in Problem 6-7b. (The two momenta should be of the same order of magnitude.)

Since $p = mv$, and v is a vector quantity, momentum also is a vector quantity—that is, it possesses both magnitude and direction, with the p vector pointing in the same direction as the v vector.

With our definition of momentum we can now restate Newton's Second Law: The *rate of change* of a body's momentum is equal to the force acting on it. We can easily show that this is equivalent to our earlier formulation,

$$F = ma = m\frac{\Delta v}{\Delta t} = \frac{\Delta(mv)}{\Delta t} = \frac{\Delta p}{\Delta t}$$

where the symbol Δ indicates "change of . . ." (i.e., final state minus initial state) and Δt is the time interval involved.[7]

Problem 6-9. (a) Will a given force acting for a given time on a heavy body produce a greater or lesser change of momentum than when it acts for an equal time on a light body? (Look at $F = \Delta p / \Delta t$.)
(b) Which body will have the greater velocity?

If there is no force acting on a body, there can be no change in its momentum. Thus a system isolated from outside force will retain its momentum; whatever it has remains constant. This *Law of Conservation of Momentum* applies to a single body as well as to the total momentum of a collection of bodies.

In an isolated system, momentum is conserved.

We could also have arrived at this conclusion by considering Newton's *Third Law of Motion*, which states:

To every action there is always an opposed and equal reaction.

Or, when one body exerts a force on another body, the second body always exerts a force on the first body equal in magnitude and opposite in direction. This law can be demonstrated and verified experimentally. Qualitatively, we see it demonstrated in the recoil of a rifle or the backward motion of a rowboat as we try to step out of it (Fig. 6-9). Thus for any two interacting bodies there are two equal and opposite forces—F_1 acting on the one body, and F_2 on the other. These forces will produce equal accelerations—that is, changes of velocity in opposite directions; the relative magnitudes of the changes of velocity will depend on the relative masses of the two bodies:

$$F_1 = m_1a_1 \quad \text{and} \quad F_2 = m_2a_2 \quad \text{or}$$

$$F_1 = m_1\Delta v_1 = \Delta(m_1v_1) = \Delta p_1 \quad \text{and}$$

$$F_2 = \Delta p_2$$

[7] Instinct should tell us, and it is apparent from the equation above, that the longer a force acts on a body, the greater will be the change in its momentum. A force, F, acting for a time Δt produces a change of momentum, $\Delta p = F \Delta t$. The product $F \Delta t$ is also called *impulse*, a term used especially in the field of rocket propulsion.

Fig. 6-9. To every action there is an opposed and equal reaction!

Since F_1 and F_2 are equal but opposite ($F_1 = -F_2$), then $\triangle p_2 = -\triangle p_2$. From this we see that $\triangle p_1$ and $\triangle p_2$ simply cancel each other out, and there has been no change in the total momentum when two bodies interact—therefore, conservation of momentum (Fig. 6-10).

Fig. 6-10. This won't work.

Example 6-1. A chemical explosion propels a bullet weighing 0.5 oz from a rifle muzzle with a velocity of 2000 ft/sec. The rifle weighs 11 lb. (a) Calculate the momentum of both bullet and rifle. (b) Calculate the recoil velocity of the rifle.

Solution.

(a) $p_1 (= -p_2) = m_1v_1 = (0.5 \text{ oz})\left(\dfrac{1 \text{ lb}}{16 \text{ oz}}\right)(2000 \text{ ft/sec}) = \dfrac{1000}{16}$

$= 62 \text{ ft-lb/sec}$

(b) If $p = mv$, then $v = p/m = \dfrac{62 \text{ ft-lb/sec}}{11 \text{ lb}} \approx 6 \text{ ft/sec.}$

Problem 6-10. In Problems 6-7 and 6-8, you were asked to calculate, respectively, the momentum of a 75-kg man moving at 2 km/hr and of a 3-oz bullet traveling at 2000 ft/sec. (*Answers:* 42 kg-m/sec and approximately 50 kg-m/sec.) If the bullet slams into the man's chest and remains buried there, what happens to his forward speed?

A collection of bodies can be analyzed by considering separately the interactions of each pair of bodies in the collection. In the previous discussion we concluded that in any two-body collision the changes of momentum will cancel each other. Thus if many bodies within a system (say, a box) interact with each other but experience no external forces, the changes of momentum cancel out in pairs, and the total momentum of the system remains unchanged. Again we have deduced the conservation of momentum within an isolated system.

An interesting illustration of the conservation of momentum is furnished by an explosion in a closed container. If the walls of the container are not ruptured, the explosion will not move the container. The momentum vectors of all the big and little pieces propelled by the explosion simply add up to zero.

How, then, do we account for rocket propulsion, in which the effects of an explosion *do* propel the rocket? In this situation the explosion is

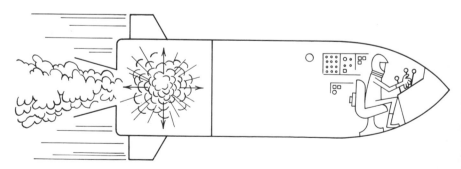

Fig. 6-11. Conservation of momentum: the backward momentum of the exhaust gases is equaled by the forward momentum gained by the rocket.

not contained; the gases are allowed to escape through a nozzle (Fig. 6-11). The backward momentum of the gases is equalled by the forward momentum of the rocket, analogous to the bullet and rifle (Example 6-1).

As a matter of fact, no internal explosion is required to obtain propulsion; witness the propulsive effect of an escaping jet of air from a child's balloon (Fig. 6-12). We can increase the propulsive impulse in a rocket ($Ft = m\Delta v$) by increasing the total mass of the exhaust gases and/or their velocity through the nozzle.

A more mechanical way of explaining rocket propulsion is through pressure considerations. A gas contained inside a box presses on all four walls equally; thus there is no net or resultant force (Fig. 6-13). If a section of one wall is removed, the total pressure on the remainder of that wall is reduced and no longer neutralizes the pressure on the opposite wall. The box will then move in response to the excess pressure.

Fig. 6-12. "Jet propulsion."

(a)   (b)

Fig. 6-13. Conservation of momentum: explosions in (*a*) a closed container and (*b*) an open container.

WORK

Work is defined as the product of a force, *F*, and the distance, *d*, that it moves:

$$W = Fd$$

For example, if in pushing a load we exert a force of 50 pounds through a distance of 8 feet, we have done $50 \times 8 = 400$ foot-pounds of work

Fig. 6-14.   Work equals force times distance.

(Fig. 6-14). Someone weighing 150 pounds who climbs 2 stories (approximately 20 feet) does 3000 foot-pounds of work. Other common units of work are the dyne-cm (called the *erg*) and the newton-meter (called the *joule*). All units of work involve force and distance.

Note that the force must act through a distance. If you merely push on a wall, no matter how hard, you are not doing any work in the above sense of work (Fig. 6-15). Likewise, a body moving through space at constant speed without any opposing force, such as friction, is not doing any work, no matter how fast it moves.

Fig. 6-15.   No work is being done.

Example 6-2.   Calculate the work done when a force of 24 dynes pushes against and moves a mass of 3 g for 4 sec.

Solution.

Since $F = ma$, the acceleration produced is:

$$a = \frac{F}{m} = \frac{24 \text{ g-cm/sec}^2}{3 \text{ g}} = 8 \text{ cm/sec}^2$$

If the body starts from rest, after 4 sec it will have attained a final velocity of

$$v = at = (8 \text{ cm/sec}^2)(4 \text{ sec}) = 32 \text{ cm/sec}$$

However, since it starts from rest and builds up velocity only gradually, the average velocity, $\bar{v}$, is only half of the final velocity

$$\bar{v} = \tfrac{1}{2}v \;(= \tfrac{1}{2}at) = 16 \text{ cm/sec}$$

This means that the body travels a total distance of

$$d = \bar{v}t \;(= \tfrac{1}{2}at \cdot t = \tfrac{1}{2}at^2)$$

$$= (16 \text{ cm/sec})(4 \text{ sec}) = 64 \text{ cm}$$

Thus the work done is

$$W = Fd = (24 \text{ dynes})(64 \text{ cm}) = 1536 \text{ ergs}$$

Problem 6-11. From previously introduced conversion factors calculate:
(a) how many ergs equal 1 joule (remember that a dyne is a g-cm/sec^2 and a newton is a kg-m/sec^2);
(b) the conversion factors between joules and foot-pounds (*see* p. 85 for the conversion of pounds to dynes).

KINETIC ENERGY

A moving body possesses energy of motion, or *kinetic energy* (from *kinētikos*, the Greek word for motion). As with momentum ($p = mv$), a heavy, fast-moving body has more kinetic energy than a light, slow-moving body. This implies that kinetic energy is some product of m and v. If we say that a body obtains energy of motion as result of the work done on it, and that energy is the ability to do work, then it becomes reasonable to express energy and work in the same units.

Example 6-3. If we assume that all the work done in Example 6-2 (p. 96) is transformed into energy of motion of the body—that is, kinetic energy—what will be the body's kinetic energy in terms of m and v?

Solution.

$$\text{Kinetic energy} = \text{work done}$$

$$KE = W = Fd$$

For F we substitute $F = ma$; and for d we substitute $d = \tfrac{1}{2}at^2$ (*see* Example 6-2).

Hence $$Fd = (ma)\,(\tfrac{1}{2}at^2) = \tfrac{1}{2}ma^2t^2$$

We also showed that the final velocity is

$$v = at$$

Thus $$a^2t^2 = v^2$$

and $$\tfrac{1}{2}ma^2t^2 = \tfrac{1}{2}mv^2$$

Therefore $$KE = \tfrac{1}{2}mv^2$$

This equation, then, is the definition of *kinetic energy:*

$$KE = \tfrac{1}{2}mv^2$$

Kinetic energy may be converted back to work; for instance, moving water can drive a turbine. The moving body's kinetic energy is then lowered by an amount equal to the work done; in other words, it is slowed down.

POTENTIAL ENERGY

Energy is the ability to do work, and sometimes a body has the ability to do work not only by virtue of its motion but also by virtue of its position. A compressed spring certainly has a potential for doing work; so does a body raised to a height. This energy resulting from position is called *potential energy.*

Various forms of energy are interconvertible. An excellent example is the pendulum. At the height of its swing it is momentarily standing still and possesses only potential energy by virtue of its raised position. As the pendulum swings down at increasing speed (accelerates), this potential energy is converted to kinetic energy. At the bottom of the swing the pendulum possesses only kinetic energy. As the pendulum starts to swing upward it slows down; kinetic energy is reconverted to potential energy (Fig. 6-16).

A parallel example is furnished by a bouncing ball (Fig. 6-17). We can easily show that the amount of work done in raising the ball to a given height—i.e., its potential energy at that height—is equal to its kinetic energy at the moment of impact after being dropped. The work done is force times distance. The force needed to raise a body is simply its weight, which equals its mass, m, times the gravitational acceleration, g, 980 cm/sec^2;

$$F = \text{weight} = mg$$

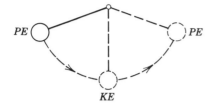

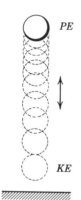

Fig. 6-16. Conversion of potential to kinetic energy and vice versa.

Fig. 6-17. Conversion of energy: a bouncing ball.

The distance is the height, h. Thus the potential energy is

$$PE = mgh$$

For example, a 20-g ball at a height of 40 cm possesses a potential energy of

$$PE = mgh = (20 \text{ g}) \, (980 \text{ cm/sec}^2) \, (40 \text{ cm}) = 7.8 \times 10^5 \text{ ergs}$$

After a time, t, when the ball has fallen back down from height h—that is, has traveled a distance, $d = h$—it will have acquired a velocity, $v = gt$, and kinetic energy.

$$KE = \tfrac{1}{2}mv^2 = \tfrac{1}{2}m(gt)^2 = \tfrac{1}{2}mg^2t^2 = mg(\tfrac{1}{2}gt^2)$$

In Example 6-2 (see p. 96), we showed that the distance, d, traveled by a constantly accelerating body is

$$d = \tfrac{1}{2}at^2$$

Thus the quantity $\tfrac{1}{2}gt^2$ in the foregoing equation represents the distance traveled by the falling ball—i.e., the height from which it fell. Therefore,

$$KE = \tfrac{1}{2}mv^2 = mgh = PE$$

Q.E.D.

Through arithmetic, we have shown that at the moment of impact the kinetic energy of the bouncing ball is equal to the potential energy it had at the top of its path. As with the pendulum, the process is then reversed and kinetic energy is reconverted to potential energy. Energy is not being lost; it is merely converted from one form to another.

Problem 6-12. The student may realize that for an instant after the impact, the bouncing ball is stopped dead, and it has neither kinetic energy nor the potential energy of elevated position (Fig. 6-18). What has happened to the energy of the ball?

THE MECHANICAL EQUIVALENT OF HEAT

The foregoing discussion hints at one of the most fundamental and far-reaching laws of nature—the Law of Conservation of Energy. However, the formulation of this law did not take place until the middle of the nineteenth century. The reason for the delay was the failure to recognize that heat is not a substance; rather, it is energy. For instance, the motion of any mechanical system, such as a pendulum or a bouncing ball, decreases steadily; eventually the system comes to a standstill. What happened to its energy? For years no one knew. Today we realize that, perhaps through friction, the mechanical energy is transformed into an equivalent amount of heat energy.

Probably the first qualitative experiments regarding the nature of heat were performed in 1798 by the Massachusetts-born Count Rumford, while he was working as a military engineer for the Bavarian government. Rumford had noticed the large amount of heat produced during the boring of a cannon and, after some experimentation, concluded that the heat resulted from the motion of the boring process. Today, we would call it friction.[8]

---

[8] Born as Benjamin Thompson in Woburn, Massachusetts, Count Rumford (1753–1814) had a varied career. During the American Revolution his sympathies lay with the losing side, and he left for England when the British evacuated Boston in 1776. Later he entered the employ of the Bavarian government for a time, during which he introduced the potato to Bavaria, served as minister of war and minister of police, saved Munich from invasion by the French, established workhouses for the poor, and achieved many other notable accomplishments. His discovery of the nature of heat is a good example of the often-cited scientific "fall-out" from military endeavors. When Rumford returned to London, he founded the Royal Institution, which later counted Michael Faraday and the chemist Humphry Davy among its more famous researchers. Rumford spent the later years of his life in Paris, where he married the widow of the great chemist Lavoisier. (Lavoisier had also ridden a wrong horse in a revolution—in his case the French Revolution—and lost his head as a result.)

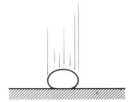

Fig. 6-18.   A bouncing ball at moment of impact.

Later, around 1845, the Englishman James P. Joule (1818–1889, a former pupil of Dalton) in a series of careful experiments determined the amount of heat energy produced by a given amount of mechanical work. He showed that a weight-driven paddle revolving in a tank of water raised the temperature of one pound of water 1°F for each 773 ft-lb of work done (Fig. 6-19). In other words, 773 ft-lb were shown to equal 1 British thermal unit (Btu). The best figure today for the mechanical equivalent of the Btu is approximately 778 ft-lb.[9]

It has since been shown through many other experiments that a definite quantity of heat is produced whenever a given amount of mechanical energy is converted to heat; and the heat energy can be reconverted to mechanical energy (e.g., through a steam engine).

Problem 6-13.   Calculate the number of calories equal to
(a) 1 Btu;
(b) 1 ft-lb;
(c) 1 joule (1 newton-meter).
(Check your answers against literature values.)

Problem 6-14.   (a) How many calories are required to raise the temperature of 1 g of water from 0°C to 100°C?
(b) How many calories are needed to heat 1 pound of water from 0° to 100°C?
(c) How many Btu are required to heat 1 pound of water from 32°F to 212°F?
(d) Convert your last answer to calories, using the conversion factor you obtained in Problem 6-13a. Does your result check with your answer to part (c) above?

[9] The *British thermal unit* (Btu) is defined as the amount of heat required to raise the temperature of 1 lb of water 1°F. The *calorie* is the amount of heat required to raise the temperature of 1 g of water 1°C (*see* Appendix 2 F).

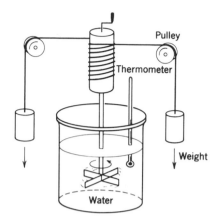

Fig. 6-19. Joule's paddle-wheel experiment.

CONSERVATION OF ENERGY

The establishment of heat as a form of energy and the quantitative determination of the mechanical equivalent of heat made it possible to formulate the *Law of Conservation of Energy:*

> *Energy can be neither created nor destroyed, but it can be converted from one form to another.*

We have discussed kinetic energy, potential energy, and heat (or internal energy). Other common forms of energy are electrical energy, radiant energy (light, X rays, radio waves), atomic energy, and chemical energy.

Problem 6-16. Try to complete the table below by naming at least one common device that converts one energy form into another. (Examples: a steam engine, which converts heat energy into mechanical energy; a gas flame, which converts chemical energy into heat energy.)

Into \ From	Kinetic	Potential	Heat	Electrical	Radiant	Atomic	Chemical
Kinetic	X						
Potential		X					
Heat			X				
Electrical				X			
Radiant					X		
Atomic						X	
Chemical							X

The two conservation laws, Conservation of Energy and Conservation of Momentum, enable us to understand many processes in nature, such as the behavior of bodies in collisions—automobiles, molecules, or nuclear particles. The simplest collisions are the so-called *elastic collisions,* involving no conversion of kinetic energy into other forms of energy; the total kinetic energy of the bodies before the collision is equal to the total kinetic energy of the bodies after the collision. Automobile collisions are usually *inelastic,* involving permanent deformation of fenders and sometimes generation of considerable heat (emotional as well as physical). On the other hand, collisions between billiard balls, between molecules, and between nuclear particles are frequently elastic, and their results are more easily predicted.

For example, in the elastic collision of two bodies of known mass (m and M), if we know their initial velocities (v_i and V_i), we can calculate their final velocities (v_f and V_f). From the Law of Conservation of Energy we get:

$$KE_{initial} = KE_{final}$$

$$\tfrac{1}{2}mv_i^2 + \tfrac{1}{2}MV_i^2 = \tfrac{1}{2}mv_f^2 + \tfrac{1}{2}MV_f^2$$

In order to calculate the two final velocities (two unknowns), we need a second equation, given to us through the Law of Conservation of Momentum:

$$p_{initial} = p_{final}$$

$$mv_i + MV_i = mv_f + MV_f$$

We now have two equations in two unknowns and can solve these by conventional methods.[10]

[10] A dramatic application of these conservation laws was James Chadwick's (English, 1891–   ) discovery in 1932 of a new subnuclear particle. In Chadwick's experiments, stationary protons (hydrogen nuclei) were hit by unknown invisible particles. After the collisions, the tracks of the protons could be seen (in a Wilson cloud chamber—see Chapter 10) and their velocities could be determined. On the basis of many such experiments and calculations involving the conservation laws, Chadwick concluded that these invisible and electrically neutral particles, which he called *neutrons,* had a mass almost equal to that of the protons.

In 1956 the existence of an even more interesting subnuclear particle was established by a group at the Los Alamos Scientific Laboratory: the *neutrino,* a particle without charge and without mass (!), which had been predicted more than twenty years earlier by Wolfgang Pauli (Austrian, 1900–1958) and Enrico Fermi (Italian, 1901–1954) on the basis of considerations involving conservation of momentum and energy.

HISTORICAL PERSPECTIVE OF NEWTON'S TIME

Newton was part of a broad intellectual revival in the latter half of the seventeenth century. The turmoil and devastation of the Thirty Years' War (1618–1648) had passed. The Italian Renaissance in the fifteenth century had liberated a remarkable outburst of creative energy, and the Protestant Reformation, initiated by Martin Luther's 95 theses in 1517, had shaken the power of the universal church and weakened the hold of tradition. More men began to think independently. Social, political, and economic changes were in the offing. Remember, this was the era of Milton, Molière, Racine, Swift, Spinoza, Locke, Wren, and Louis XIV of France. And science in the seventeenth century was not isolated from the general climate of inquiry and development.

Some of the earliest and most brilliant victories of the new scientific method were won in the field of astronomy. A new theory of the movement of the heavenly bodies by the Polish astronomer Nicolaus Copernicus (1473–1543) was published in the year of his death. Copernicus refuted the idea that the earth was the center of the universe and instead proposed that the earth, together with the other planets, revolved around the sun, thus considerably simplifying the picture. These movements of the planets were most carefully measured by the Swedish astronomer Tycho Brahe (1546–1601) and by Galileo. On the basis of their meticulous observations, Johannes Kepler (1571–1630), the German mathematician, physicist, astronomer, and astrologer,[11] worked out the geometry of planetary motions. He introduced the idea of elliptical orbits because of a minute discrepancy between his computations for circular orbits and Tycho's observations.

It was left to Newton (born the year Galileo died) to reduce Kepler's geometric descriptions of the planetary motions to a single, more comprehensive physical law, which explained the motions of the planets under the force of gravity as well as the motion of any other body when acted upon by external forces.

These developments in astronomy illustrate two habits that scientists had cultivated and to which they owed their amazing success. These were (1) the habit of basing all conclusions upon observation and experiment instead of appealing to authority or tradition, and (2) the habit of making quantitative observations whenever possible and interpreting these results through the newly developed language of mathematics.

---

[11] Kepler made necessity his excuse for a compromise with superstition: "Nature, which has conferred upon every animal the means of subsistence, has given astrology as an adjunct and ally of astronomy."

Milestones in the growth of mathematics in the seventeenth century include the introduction of decimals; the conception of logarithms in 1614 by John Napier (Scotch, 1550–1617);[12] the development of analytical geometry by Rene Descartes (French, 1596–1650); the translation of geometry into the language of algebra and vice versa; and the formulation of calculus about 1680, independently by Newton and Gottfried Wilhelm Leibniz (German, 1646–1716). (Newton needed calculus to handle problems in his new system of mechanics.)

The experimental method is illustrated by Copernicus, who rejected tradition, the opinions of learned men, the teachings of the church, and even the evidence of his own senses—all of which assured him that the earth is immovable and that the sun and planets move around the earth. But closer observation had shown that the planets move in the sky, and this could be reasonably explained only by a heliocentric system.

Other practitioners of the experimental method and contemporaries of Newton were the English chemist Robert Boyle, whom we have met before (1627–1691, definition of *element*); the Dutch physicist and mathematician Christian Huygens (1629–1695, diffraction of light); the English physicist Robert Hooke (1635–1703, wave theory of light); the English physician William Harvey (1578–1657, discoverer of blood circulation); the Dutch naturalist Anton van Leeuwenhoek (1632–1723, first observed bacteria under microscope); the German physicist Otto von Guericke (1602–1686, frictional electricity, air pump, manometer); the French mathematician and philosopher Blaise Pascal (1623–1662, barometer); and many others.

These scientists did not work in isolation from each other but exchanged information through correspondence and through journals published by newly founded societies, including an academy in Rome (1602); the Royal Society (founded 1662 in London); and the French Academy of Sciences (chartered by Louis XIV in 1666).

By the last decades of the seventeenth century, science and the scientific method had achieved a spectacular ascendency over other forms of thought. Amateur upon amateur fitted himself with the vocabulary and equipment of science. Many a member of nobility stocked his own observatory or laboratory. Even Charles II of England had his own laboratory for private scientific pursuits. Often these men, as patrons, supported men of science with money, influence, and governmental positions. Although conservatives feared that increased knowledge would lead to discontent and revolution, the supporters of science believed that the new

[12] It has been said that logarithms doubled the productive lives of astronomers by halving their labors through short cuts in cumbersome calculations.

knowledge and methods would strengthen and secure the state against such an eventuality. As we know today, science continued to grow, and humanity did benefit.

SUMMARY

This chapter has dealt with various aspects of motion and energy.

Speed is the distance traveled per unit of time. *Velocity* tells us not only the speed of a body but also the direction in which it is moving.

Acceleration is the rate of change of velocity. Whenever a body changes its speed, it is accelerating (positively or negatively). A body is also accelerating when it changes its direction of travel, even if its speed is constant.

The centimeter-gram-second (cgs) units of velocity and acceleration are cm/sec and cm/sec^2, respectively.

Newton's three *Laws of Motion* are as follows.

1. If no force acts on a body, it will continue in its state of rest or of constant velocity.

2. A force acting on a body produces an acceleration that is proportional to the force and inversely proportional to the mass of the body:

$$F = ma$$

3. In the interaction of two bodies, the forces exerted on each other by the two bodies are equal in magnitude and opposite in direction.

The First Law introduces the concept of *inertia*, the tendency of a body to persist in its state of rest or of uniform motion. The Second Law introduces *mass* as the quantitative measure of inertia.

The *weight* of a body is not an inherent quality but the force with which the earth attracts that body.

Force can be measured by the observed acceleration it produces in a body of known mass ($F = ma$). Hence the cgs unit of force is g-cm/sec^2, called the *dyne*. The *newton* is kg-m/sec^2, which is an mks (meter-kilogram-second) unit, an alternative metric-system unit. 4.45 newtons, or 4.45×10^5 dynes, equal 1 lb.

Any quantity that has both magnitude and direction can be represented by a *vector*. Vector quantities can be added graphically or with the help of trigonometric formulas. Vector quantities include, among others, velocity, acceleration, force, and momentum.

Momentum is the product of the mass of a body and its velocity. The Law of the Conservation of Momentum can be tested experimentally and is implicit in Newton's Third Law of Motion.

Work is defined as the product of a force and the distance through which it acts. *Energy* is the ability to do work. We speak of *kinetic, potential, heat* (internal), *electrical, atomic,* and *chemical* energies, and many others. Within an isolated system, energy can be converted from one form to another, but it cannot be created or destroyed; energy therefore is conserved. The cgs unit of energy is the *erg;* the mks unit of energy is the *joule;* a heat unit of energy is the *calorie.*

ANSWERS TO SELECTED PROBLEMS

6-1. 1 mph $\approx$ 1.5 ft/sec $\approx$ 45 cm/sec.

6-2. 1 ft/sec² $\approx$ 30 cm/sec².

6-3. (a) $(70 \text{ kg})\left(\dfrac{2.2 \text{ lb}}{\text{kg}}\right) = 154$ pounds.

(b) $\left(\dfrac{4 \text{ km}}{\text{hr}}\right)\left(\dfrac{0.62 \text{ mi}}{\text{km}}\right) = 2.5$ mi/hr.

(c) 4 km/hr/½ sec = 8 km/hr-sec.

(d) $\left(\dfrac{8 \text{ km}}{\text{hr-sec}}\right)\left(\dfrac{\text{hr}}{3600 \text{ sec}}\right) = 2.2 \times 10^{-3}$ km/sec².

(e) 220 cm/sec².

(f) 7.2 ft/sec².

(g) $F = ma = (7 \times 10^4 \text{ g}) (220 \text{ cm/sec}^2) = 1540 \times 10^4$ g-cm/sec²
     $= 1.54 \times 10^7$ dynes.

(h) $(15.4 \times 10^6 \text{ dynes})\left(\dfrac{\text{pounds}}{4.45 \times 10^5 \text{ dynes}}\right) = 35$ pounds.

6-4. (a) $F = ma$; 1 newton $= (1000 \text{ g}) \times (100 \text{ cm/sec}^2) = 1 \times 10^5$
     g-cm/sec² $= 10^5$ dynes.

(b) 4.45 newtons = 1 pound.

6-7. (b) $\left(\dfrac{150 \text{ kg-km}}{\text{hr}}\right)\left(\dfrac{10^3 \text{ m}}{\text{km}}\right)\left(\dfrac{\text{hr}}{3.6 \times 10^3 \text{ sec}}\right) = 42$ kg-m/sec.

6-8. (b) $\left(\dfrac{6000 \text{ oz-ft}}{\text{sec}}\right)\left(\dfrac{30 \text{ g}}{\text{oz}}\right)\left(\dfrac{\text{m}}{3.3 \text{ ft}}\right)\left(\dfrac{\text{kg}}{1000 \text{ g}}\right) \approx 50$ kg-m/sec.

6-11. (a) 1 joule = 1 newton-meter = (kg-m/sec²) (m) = 1000 g $\times$ 10²
     cm $\times$ 10² cm/sec² = 10⁷ (g-cm/sec²) (cm) = 10⁷ dyne-cm
     = 10⁷ ergs.

(b) 1 joule = $(10^7 \text{ dyne-cm})\left(\dfrac{\text{pounds}}{4.45 \times 10^5 \text{ dynes}}\right)\left(\dfrac{\text{ft}}{30 \text{ cm}}\right) = 0.74$
     ft-lb.
     1 ft-lb = 1.36 joules   (1/0.74 = 1.36).

6-13. (a) 1 Btu = energy taken up by 1 lb of H_2O for each °F.
1 cal = energy taken up by 1 g of H_2O for each °C.
1 lb = 454 g (tending to make the Btu larger):
°F = 5/9°C (tending to make the Btu smaller).
Thus 1 Btu = 454 × 5/9 cal = 252 calories.

(b) 1 ft-lb = 1/778 Btu = 1/778 × 252 cal = 0.324 calorie.

(c) 1 joule = 0.74 ft-lb = 0.74 × 0.324 cal = 0.239 calorie.

Chapter 7

EARLY DEVELOPMENTS IN ELECTRICITY

The years from 1650 to 1750 were a period of intense interest in attaining a better understanding of nature in general. New theories were proposed about matter (Boyle's definition of element), about heat (the phlogiston theory), and about motion, gravitation, and light (by Newton and others). Electricity was no exception.

Why, in a book on chemistry, should there be a chapter on electricity? What has electricity got to do with chemical reactions? A great deal. For one thing, chemical reactions can be used to produce electric current, as in the electric battery and in the modern fuel cell, which directly and efficiently converts the energy of a combustion process into electrical energy. Conversely, electric currents can be used to carry out chemical reactions. For example, electric current can be used to decompose water into hydrogen and oxygen, and molten sodium chloride into sodium and chlorine. Many metals, such as aluminum, are produced principally by an electrolysis process. Electroplating is another example of a chemical reaction effected by electricity. It is probably correct to say that all chemical reactions are basically electrical in nature.

In fact—and this is the aspect of electricity relevant here—matter itself is fundamentally electrical in nature. The discovery of electrolysis in 1800 (about the time that Dalton formulated his atomic theory) was the first concrete evidence of the electrical nature of matter. In later chapters we deal with further researches, most of them around the turn of the twentieth century, that led to a clearer picture of the arrangements of positive and negative charges in atoms.

STATIC ELECTRICITY

The ancient Greeks knew that rubbing amber with wool or fur caused the amber to attract lightweight objects. Today we say that rubbing amber causes it to acquire an electric charge (from the Greek word *elektron,* meaning amber). Not much was done about investigating this phenomenon until about 1600, when an English physician, William Gilbert (1540–1603), performed a series of experiments to find out more about the nature of electricity. Let us follow through some experiments analogous to those that Gilbert and his contemporaries performed.

If we suspend a pith ball, a piece of aluminum foil, or some other light object by a thread and then hold it near a piece of amber that has been rubbed with fur, we note that amber attracts the ball or foil. But if the amber is allowed to touch the ball or foil, it will thenceforth repel the ball or foil (Fig. 7-1).

A rubbed glass rod will also attract a pith ball (Fig. 7-2). But material that is attracted by the amber will be repelled by the glass, and vice versa. If we suspend the amber and bring the glass rod near it (or vice versa), we observe an attraction between the two (Fig. 7-3). On the other hand, two rubbed glass rods will repel each other, and two pieces of rubbed amber will also repel each other.

We conclude that rubbing some objects gives them what we call an

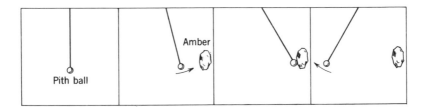

Fig. 7-1.

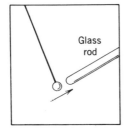

Fig. 7-2.

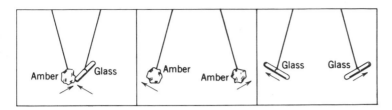

Fig. 7-3.

"electric charge," and that there seem to be two kinds of charge—that on glass and that on amber. Strangely, we can produce both kinds by rubbing different objects. We can show by further experimentation with many, many other objects that only these two kinds of charge exist. The behavior of the two kinds of charge is summarized in the following law:

Like charges of electricity repel each other; unlike charges attract each other.

Apparently the charges we put on the amber or on the glass can be transferred to other objects, such as a pith ball or a piece of aluminum foil (evidenced by the repulsion after the rod touched the object). However, the charge on the amber or glass rod is exhaustible; that is, we can charge only so many other objects with it before the rod itself must be recharged. In other words, the electricity is somehow "drained away," much as a fluid is.

Further experimentation reveals that objects possessing opposite kinds of electricity, if allowed to touch, will neutralize each other. This suggests that the two kinds of electricity, which we seem to have separated by rubbing, are able to flow back together again when the objects touch. It is not even necessary for the two bodies actually to touch. The flow of electricity can take place through a third body, such as a wire, which touches the other two bodies. On investigating further, we find that some objects will allow electricity to flow while others will not. We classify the former substances as *conductors* and the latter as *nonconductors,* or insulators, of electricity.

Experiments with static electricity and the observations we have described quite naturally led to the two-fluid theory of electricity, proposed in 1733 by the French chemist C. Du Fay (1698–1739). According to this theory there are two types of electricity—resinous (from the Latin word *resin,* for amberlike plant secretions) and vitreous (from the Latin word *vitrum,* meaning glass).

In 1747 a one-fluid theory was proposed by Benjamin Franklin (1706–1790, the first American physicist and the first American to gain worldwide fame, earned through his work on electricity). According to Franklin there is only one kind of electric fluid; charged bodies have either an excess or a deficit of this fluid. (This "model" of electricity led to his famous kite experiment, in which he withdrew electricity from a cloud.)

Franklin's idea has withstood the test of time and is the basis of our present theory of electricity. Careful measurements have shown that when a glass rod is rubbed with silk, a "vitreous" charge appears on the rod and a "resinous" charge of equal magnitude appears on the silk. This suggests that the rubbing process does not create charges but merely transfers charges from one object to another; that is, charge has been conserved in this process. Many parallel observations led to the Law of Conservation of Charge. No exceptions have ever been found, on either the macroscopic or the atomic scale.[1]

Reviewing our experiments in the light of the law that unlikes attract each other, we wonder why the amber was able to attract a *neutral* object, one that hadn't been rubbed or touched. (A neutral object is one that has equal amounts of positive and negative charges.) Keeping in mind that electricity seems to be mobile, we can come up with a reasonable explanation. Suppose that bringing a negative rod up to a neutral body attracts the positive electricity in the neutral body and repels the negative electricity, as shown in Fig. 7-5. The attractive force to the nearby positive charge should then be stronger than the repulsive force to the distant negative charge, and thus the ball moves toward the rod. Charge separation in a neutral body induced by a nearby charged body is called *induction*.

If this description is correct, we ought to be able to use induction to bring about a permanent separation of charge, simply by breaking the neutral object apart in the presence of a nearby charge. A more convenient way of inducing permanent charge separation is to bring two neutral objects into the vicinity of a charged rod, as shown in Fig. 7-6, and then simply separate them while they are still in the presence of that

[1] At that time, Franklin rather arbitrarily decided that it was the glass that had the excess electric fluid, and therefore he assigned a positive sign (+) to vitreous electricity and a negative sign (−) to resinous electricity. This turned out to be an unfortunate choice since today we know that the positive charge on the glass results from a *dearth* of negative electrons. Franklin's arbitrary assignment of positive to the glass has made the conventional designation of "current flow" opposite that of the electron flow (Fig. 7-4). Nevertheless, we still follow Franklin's practice of using a negative sign for the amberlike electricity and a positive sign for the glasslike electricity.

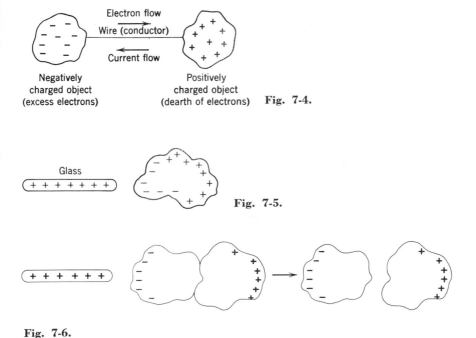

Fig. 7-4.

Fig. 7-5.

Fig. 7-6.

charged rod. The result should be one positively and one negatively charged object. A simple experiment would verify our prediction.

A consequence of this phenomenon of induction is that, since like charges repel each other, any excess charges on a hollow body will move as far away from each other as possible. Thus they will be on the outer surface, leaving the inside of the body perfectly neutral (Fig. 7-7). For this reason, passengers in an airplane passing through an electric storm are perfectly safe from the effects of lightning (provided, of course, that the airplane itself is not damaged by the storm).

Fig. 7-7.

COULOMB'S LAW OF ELECTROSTATIC FORCES

We have noted that these attractive and repulsive effects, like the force of gravity, are experienced through intervening space without direct contact; and we also noted that the electrical forces decrease with increasing distance, also like gravity. We can measure quantitatively the relation of the forces to the distance.

In 1785 the French physicist Charles A. Coulomb (1736–1806), using his newly developed torsion balance (Fig. 7-8), found that electrostatic forces decrease with the square of the distance between the bodies involved ($F \propto 1/r^2$) and are proportional to the charges (q_1 and q_2) on the two bodies. Hence

$$F \propto \frac{q_1 q_2}{r^2}$$

We call this the *Inverse-Square Law of Electrostatic Forces*. Note that if the two charges are of like sign (both positive or both negative) the force is repulsive (indicated by a resultant positive sign). If the two charges have opposite signs ($+$, $-$) the force is attractive (negative sign).

The Inverse-Square Law of Electrostatic Forces had actually been derived previously by the English scientist Henry Cavendish (1731–1810), the discoverer of hydrogen. But, like so much of Cavendish's work, this law was not published during his lifetime; hence the credit of discovery went to others, for a discovery is of little use if it is kept secret.

At this point the student might wonder about the nature of these forces that are felt through intervening space. At first glance they seem to be a property of the air or medium between the two objects. For example, the strength of the force is different for different media—air or various types of gases. But in fact these forces exist even across a vacuum, again like the force of gravity. Even today the exact nature of these forces is still speculative and far from understood.

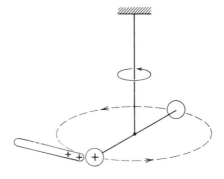

Fig. 7-8.

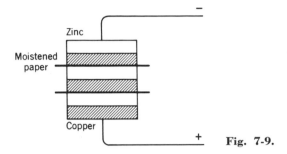

Fig. 7-9.

ELECTRIC CURRENT

We have seen that charges can flow from one body to another, either directly or through an intermediate conductor. This flow of electric charge is called a *current*. But currents from statically charged bodies are of short duration, and for a long time there was no reliable source of continuous current.[2]

In 1800 an Italian physicist, Alessandro Volta (1745–1827), invented the electric battery, following up experiments on "animal electricity" by his countryman Luigi Galvani (1737–1798), a professor of anatomy. Volta's battery consisted simply of a pile of alternating zinc and copper plates separated by blotters soaked in a salt solution (Fig. 7-9). This pile was found to generate electric current; that is, chemical energy was being converted into electrical energy.[3]

The Law of Conservation of Charge was shown to be still valid: Tests revealed that one side of the pile was positively charged, while the other side was negatively charged.

UNITS OF ELECTRICITY

Two independent sets of definitions of electrical units are in common usage today—one based on the forces between static charges, the other based on the forces between current-carrying wires.

The first set of definitions is based on Coulomb's Inverse-Square Law,

$$F \propto \frac{q_1 q_2}{r^2}$$

[2] A frictional electric machine was invented in 1633 by Otto von Guericke (German physicist, 1602–1686). This machine, however, proved to be little more than a toy.

[3] You can demonstrate a one-cell pile by placing a piece of aluminum foil in your mouth. It will have no taste until it touches a metal filling in a tooth. At that time current will flow, which you can detect by a sour taste and perhaps a slightly painful feeling in your tooth.

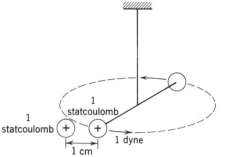

Fig. 7-10.

A *statcoulomb* is defined as the amount of charge that, when placed at a 1-cm distance from a charge of the same sign and magnitude, will repel the latter charge with a force of 1 dyne (Fig. 7-10). The statcoulomb is the cgs (centimeter-gram-second) electrostatic unit of charge.

The second set of definitions, based on the forces between wires carrying current, defines an *ampere* as an unvarying current that, if present in each of two parallel conductors of infinite length and 1 meter apart in empty space, causes each conductor to experience a force of exactly 2×10^{-7} newton (or 0.02 dyne) per meter of length (Fig. 7-11). From this definition of a unit of *current* we can define a new unit of electric *charge:* The *coulomb* represents the amount of electricity or charge that flows in 1 second through a wire carrying 1 ampere. An ampere is one coulomb per second.

Note the distinction between coulomb and ampere: The coulomb represents a *quantity* of charge (analogous to, say, a cubic foot of water); the ampere represents a *flow* of charge or current (analogous to, say, 150 cubic feet of water per minute).

$$\text{current} = \frac{\text{charge}}{\text{time}}$$

1 ampere = 1 coulomb/second

total charge flowed = current × time of flow

total coulombs = amperes × seconds

Both the statcoulomb and the coulomb are units of charge, based on two independent definitions. The relation between the two, found experimentally, is:

1 coulomb = 2.998×10^9 statcoulombs

In this text we will use mostly the coulomb rather than the statcoulomb.

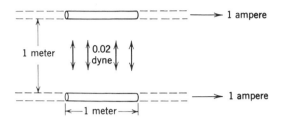

Fig. 7-11.

ELECTROLYSIS

Volta showed that there is a relation between a chemical reaction and electricity. If a chemical reaction can produce an electric current, can an electric current produce a chemical reaction?

Within weeks of Volta's invention, the answer was found through the exploitation of an accidental discovery by two English chemists, William Nicholson (1753–1815) and Anthony Carlisle (1768–1840). While experimenting with electric current from the voltaic pile, they added a drop of water in order to get better contact between two wires. They immediately noticed the formation of gas at the juncture. Chemical tests revealed some of the gas to be hydrogen, which had been discovered by Cavendish only 34 years earlier. More precise measurements disclosed that water decomposed to yield two volumes of hydrogen and one volume of oxygen. Such an electrical decomposition process is called *electrolysis* (Fig. 7-12).[4]

The foregoing incident once more illustrates the role of accidental discoveries in science. In the hands of untrained, unobservant, or less astute persons, the formation of gas might have gone unnoticed. As Louis Pasteur has pointed out, chance favors the mind that is prepared. Nicholson and Carlisle not only noticed the evolution of the gas but also recognized its significance and followed up their discovery with further experimentation. Most important, they communicated their results to others.

We must also realize that scientific discoveries are not made in a technological and cultural vacuum; they are the children of their time. For one thing, they depend on related previous work—in this case, the invention of the battery by Volta. Second, there must be the proper "climate"

[4] Write a balanced equation for the electrolysis of water. Note, however, that this work was done before the final formulation of Dalton's atomic theory and before Avogadro's hypothesis.

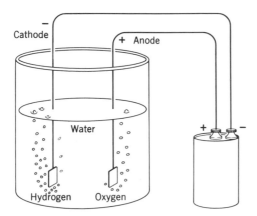

Fig. 7-12.

for recognition of the discovery, follow-up, and eventual acceptance of any findings or theories evolved. This climate was lacking in the case of Avogadro's hypothesis, and as a result that important idea lay dormant for nearly fifty years. But the time was ripe for Nicholson and Carlisle's discovery. Almost immediately, practical use was made of electrolysis by the English chemist Humphry Davy (1778–1829). In 1807 Davy isolated two new elements, sodium and potassium, through electrolysis of their molten hydroxides. Today many elements—aluminum, sodium, magnesium, chlorine, and others—are manufactured primarily by electrolytic processes.

The realization that substances can be decomposed by electric currents led Berzelius in 1811 to formulate his electrochemical theory of the chemical bond (mentioned briefly in Chapter 3). According to Berzelius, the atoms of the various elements are characteristically positively or negatively charged. Atoms of unlike charges are attracted to each other and thus form molecules. Today we realize that Berzelius' theory is essentially correct for some compounds.[5]

Much later, in 1887, a Swedish chemist, Svante Arrhenius (1859–1927) published a comprehensive theory to explain electrolysis among other things. According to his theory compounds exist in the molten state or in solution as positive and negative *ions*. During electrolysis these ions are attracted to the oppositely charged electrodes, where they accept the re-

[5] Berzelius' theory could not explain the formation of diatomic molecules of *elements*, such as H_2 and O_2, in which the atoms presumably have similar charges. Because of this difficulty, most scientists rejected the idea of polyatomic elements as proposed by Avogadro, also in 1811 (*see* Chapter 3).

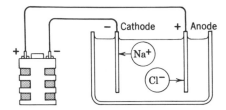

Fig. 7-13.

quired charges to become neutral atoms. For example, sodium chloride, NaCl, exists in the molten state as Na^+ and Cl^- ions. The sodium ions are attracted to the cathode ($-$) and the chlorine ions to the anode ($+$) (see Fig. 7-13 for labels), where they are neutralized and appear as sodium metal and chlorine gas (Cl_2).

FARADAY'S QUANTITATIVE EXPERIMENTS WITH ELECTROLYSIS

A great leap forward in our understanding of the structure of matter and the nature of electricity came through the quantitative experiments with electrolysis conducted by the great English chemist and physicist Michael Faraday (1791–1867). Around 1833, in experiments with the electrolysis of solutions and molten salts, Faraday found that a fixed amount of charge must flow to obtain a given amount of any particular substance by electrolysis. Furthermore, he found that to obtain one mole (that is, 1 gram-molecular weight, or Avogadro's number of particles) of *any* substance requires either 96,500 coulombs, or a small-integer multiple thereof. For convenience, we call this quantity of charge a *faraday* (f).

$$1 \text{ faraday} = 96,500 \text{ coulombs}$$

For example, it takes 1 faraday (96,500 coulombs) of electricity to obtain 1 gram of hydrogen (i.e., 1 mole of hydrogen atoms), or 35.5 g of chlorine, or 23 g of sodium, or 108 g of silver (AW 35.5, 23, and 108, respectively).

$$HCl \; + 1f \rightarrow 1/2 \, H_2 \; (1 \text{ g}) + 1/2 \, Cl_2 \; (35 \text{ g})$$

$$NaCl + 1f \rightarrow Na \; (23 \text{ g}) \quad + 1/2 \, Cl_2 \; (35 \text{ g})$$

However, it requires 2 faradays (193,000 coulombs) to obtain a gram-atomic weight of magnesium (24.3 g), or calcium (40 g), or oxygen (16 g). And it requires 3 faradays to obtain a gram-atomic weight of aluminum (27 g).

$$MgO + 2f \rightarrow Mg \ (24 \ g) + 1/2 \ O_2 \ (16 \ g)$$

$$H_2O + 2f \rightarrow H_2 \ (2 \ g) \ + 1/2 \ O_2 \ (16 \ g)$$

$$AlCl_3 + 3f \rightarrow Al \ (27 \ g) \ + 3/2 \ Cl_2 \ (106 \ g)$$

The quantity of charge required to obtain a particular element is generally the same regardless of the source of the element; for example, 1 mole of hydrogen atoms requires 1 faraday whether the hydrogen is obtained from acid solution, from brine, or from any other source; 35.5 g of chlorine requires 1 faraday whether the chlorine is obtained from molten sodium chloride, from potassium chloride, or from brine. In each case, it takes 96,500 coulombs to deposit 1 gram-atomic weight. And it is of no consequence whether we use a weak current for a long time (say, 3 amperes for 9 hours) or a strong current for a correspondingly shorter time (e.g., 27 amperes for 1 hour):

$$3 \ \frac{coulomb}{sec} \times 9 \ hr \times 3600 \ \frac{sec}{hr} \approx 97,000 \ coulombs$$

$$27 \ \frac{coulomb}{sec} \times 3600 \ sec \approx 97,000 \ coulombs$$

Nor is the required amount affected by other conditions of the experiment, such as the concentration of solutions, temperature, voltage, size of electrodes, etc.

These observations support the idea that atoms in solution and in molten salts all have the same charge or multiples thereof. The atoms of elements requiring the minimum amount of electricity—namely, 96,500 coulombs—have the smallest charge say, 1; those requiring twice 96,500 coulombs have double charges, etc. The charged atoms are called *ions*. The deposition of the element at either the positive or the negative electrode tells us that the charge on the ion was negative or positive, respectively (since unlike charges attract each other). We indicate the amount of charge by the number of superscript pluses or minuses following the symbol of the element. For example, hydrogen, sodium, silver, magnesium, calcium, and aluminum are found as positively charged ions, with either one, two, or three charges: H^+, Na^+, Ag^+, Mg^{++}, Ca^{++}, Al^{+++}. Chlorine, bromine, and sulfur are found as negatively charged ions: Cl^-, Br^-, S^{--}. We say that the ionic charge of sulfur, for example, is minus two.

This idea opens up an interesting possibility. If we know the number of atoms in a gram-atomic weight (Avogadro's number), we can calculate the amount of charge on a single atom (i.e., ion). Thirty years after

Faraday's original experiments with electrolysis, Loschmidt determined Avogadro's number, N, to be 6×10^{23} (see Chapter 4). Now, if 96,500 coulombs is required to discharge N atoms, the charge on a single atom is 96,500 divided by 6×10^{23}. The result is 1.6×10^{-19} coulomb; this is the charge on a singly charged ion, such as H^+ or Na^+.

IONIC CHARGES AND CHEMICAL FORMULAS

In accordance with the electrochemical bonding theory of Berzelius (p. 118), charges on atoms would account for the combining powers of atoms (see Chapter 5) in at least some compounds, and would therefore also explain the observed formulas. For example, if sodium exists as a positively charged ion (Na^+), a chlorine ion with a negative charge (Cl^-) can unite with it and yield the neutral compound sodium chloride, NaCl. It would take *two* chlorine ions to neutralize a doubly charged calcium ion (Ca^{++}), and we find in fact that calcium chloride has the formula $CaCl_2$. Similarly, aluminum chloride is $AlCl_3$.

SUMMARY

In this chapter we traced the early developments that indicated that (neutral) atoms contain positive and negative charges. In other words, matter appeared to be partly electrical in its makeup.

The first systematic observations were of necessity concerned with static electricity and led to the following conclusions.

1. There seem to be only two types of charge in nature, which we arbitrarily label "positive" and "negative." A body that has equal amounts of positive and negative charges is electrostatically neutral.

2. The production of a positive charge always is accompanied by the production of an equal amount of negative charge, and vice versa. Apparently in "producing" charge we are merely separating positive and negative charges from each other; that is, we are not creating charges. Thus follows the Law of Conservation of Charge—electric charge can be neither created nor destroyed.

3. Like charges repel each other, unlike charges attract each other.

4. Quantitatively, these forces of attraction or repulsion are directly proportional to the magnitudes of the charges and inversely proportional to the square of the distance between them.

5. Electric charges apparently flow through certain materials, which we call *conductors* of electricity. Materials unable to conduct electricity are called *insulators*.

With Volta's invention of the battery in 1800, scientists started to investigate the properties of electric currents. Almost immediately they discovered that some solutions and molten substances can conduct electric currents, but decompose in the process. This electrical decomposition process, called *electrolysis,* hinted at the electrical nature of chemical bonds.

Berzelius came up with the theory (1811) that molecules are formed by electrostatic attractions between positively and negatively charged atoms. (Charged atoms are called *ions.*)

Subsequent measurements by Faraday (1833) indicated that all ions in solutions and in molten salts have identical amounts of charge (either + or −) or else small multiples thereof. This discovery coupled with the bonding theory of Berzelius made it possible to explain the combining powers of atoms in certain compounds and therefore also the molecular formulas of these compounds.

From Faraday's measurements and Loschmidt's determination of Avogadro's number (1865), scientists could calculate the minimum charge on a single atom (such as H^+ or Na^+)—1.6×10^{-19} coulomb.

PROBLEMS

Problem 7-1. (a) How many coulombs are required to liberate by electrolysis 1 g of hydrogen from water? (Does it matter that the evolved hydrogen gas is diatomic—H_2?)

(b) How many coulombs are required to liberate 1 gram-molecule (that is, 1 mole, or 2 g) of hydrogen gas?

(c) How many faradays are required in (a) and (b)?

(d) What is the charge of the hydrogen ion?

Problem 7-2. (a) How many faradays are required to liberate 1 gram-atom (16 g) of oxygen from water?

(b) What presumably is the ionic charge of oxygen in water?

(c) How many faradays are required to liberate 1 gram-molecule (1 mole, or 32 g) of oxygen?

Problem 7-3. (a) How many faradays are required to liberate 1 gram-atom of iron from ferrous chloride, $FeCl_2$? (The ionic charge of chlorine is minus one; i.e., it exists in solution as Cl^-.)

(b) How many faradays are required to liberate 1 gram-atom of iron from ferric chloride, $FeCl_3$?

(c) What is the combining power of iron in each case? (There are many elements that have variable combining powers—*see* Chapter 5).

Problem 7-4. (a) How many faradays are required to deposit 1 gram-atom of hydrogen by the electrolysis of hydrochloric acid, HCl?

(b) How many grams of hydrogen are deposited by 1 faraday?

(c) What is the total charge, in faradays, of 1 gram of hydrogen ions?

(d) What is the ratio of the mass to the charge of H^+ ions, in grams per faraday?

Problem 7-5. (a) What would be the ratio of the mass to the charge of sodium (Na^+) ions, in grams per faraday?

(b) Of silver (Ag^+) ions?

(c) Of chlorine (Cl^-) ions?

Chapter 8

THE ELECTRON

In the foregoing chapter we traced the developments leading to Faraday's experiments (1833) with electrolysis, which showed that ions in solution and in molten salts all have the same electrical charge or a simple multiple thereof (later calculated as 1.6×10^{-19} coulomb). This finding indicated that electricity comes in discrete quantities, or "packages," when found on atoms.

Scientists began to wonder whether this packaging of electricity is a general phenomenon. Is electricity grainy ("atomistic") like matter, rather than continuous like a fluid? If so, is the smallest "lump" of electricity the charge on an atom (1.6×10^{-19} coulomb), or does this quantity of charge represent several of the smallest possible charges?

In this chapter we discuss more conclusive evidence of the "atomicity" of electricity and the determination of the smallest electrical charge in nature.

INTERACTION OF ELECTRIC CURRENTS AND MAGNETS

In 1820, even before Faraday's electrolysis experiments, the Danish scientist Hans Oersted (1777–1851) found that a magnetic needle is deflected by an electric current in a nearby wire (Fig. 8-1). Three-quarters of a century later, this interaction between magnets and moving charges was utilized, as we shall see, in an experiment to "weigh" electric charges."[1]

---

[1] Only charges *in motion* have magnetic properties. There is no interaction between magnets and static electricity: Only charges affect charges, and only magnets affect magnets. The interaction between electric currents and magnets has been put to use in electric motors and generators.

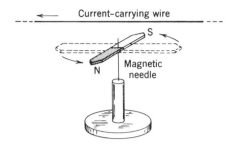

Current-carrying wire

S

N

Magnetic
needle

Fig. 8-1. A magnetic needle is deflected by a nearby electric current.

CATHODE RAYS

Scientists had observed that two wires carrying electric current produce a spark when separated, and that this spark jumps farther in rarified air. Faraday, in 1838, decided to carry this observation one step further—to see how the spark would behave in the absence of air. He sealed two wires into opposite ends of a glass tube and pumped out as much air as he could. As Faraday reduced the air pressure inside the tube, it began to conduct current and a greenish glow appeared inside.[2]

After Faraday, further advance along this line had to await the development of a better vacuum pump. This was accomplished in 1855 by the German glassblower Heinrich Geissler (1814–1879). When the physicist Julius Plücker (1801–1868) used Geissler's evacuated glass vessels for experimentation (1858), he obtained a more extended glow than Faraday had, and he also observed that the glass itself seemed to glow, or "fluoresce," under the action of electrical currents. By 1875 the English physicist William Crookes (1832–1919) had devised a still better evacuated tube (the Crookes tube) for studying these phenomena (*see* Fig. 8-2).

Let us summarize some additional observations that had been obtained by about that time and show what deductions can be drawn from them.

1. *Observation.* The positive electrode (called the *anode* need not be at the opposite end of the glass tube from the negative electrode (called the *cathode*); it can be any place else, even at the side of the glass tube, as illustrated in Fig. 8-3. In such an arrangement the glass wall opposite the cathode emits a glow, and a solid object placed in front of the cathode casts a sharp shadow on the opposite glass wall.

---

[2] Faraday's glowing tube was the grandfather of the modern neon tube. Today we vary the colors of the glow in such tubes by using various residual gases—e.g., greenish from argon, red from neon, blue from mercury vapor, yellow from sodium vapor.

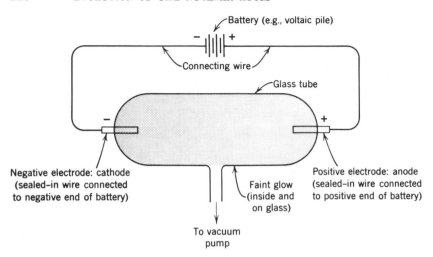

Fig. 8-2. A typical cathode-ray tube.

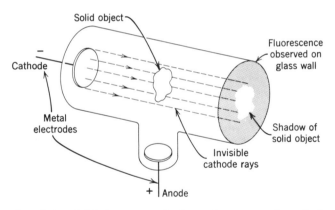

Fig. 8-3. Solid objects in cathode-ray beams cast sharp shadows.

Deduction. The (invisible) electric current is emitted from only one of the electrodes, the cathode, and travels in straight lines (sharp shadow). Consequently, these currents were called cathode rays. *(In modern television sets, the image on the screen is produced by cathode rays.)*

2. **Observation.** The rays are deflected by a magnetic field, as shown in Fig. 8-4.

Deduction. This observation dovetails with the observed interaction of electric currents and magnets first reported by Oersted in 1820.

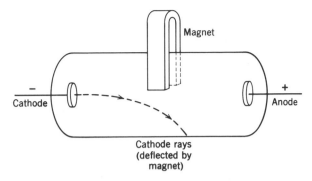

Fig. 8-4.   Magnets deflect cathode rays.

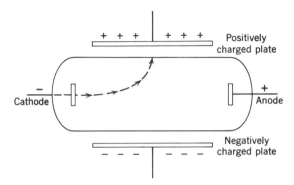

Fig. 8-5.   A positively charged plate attracts cathode rays.
(A negatively charged plate repels cathode rays.)

3. **Observation.** The rays are attracted by positively charged electric plates and repelled by negatively charged plates (Fig. 8-5).

Deduction. *The rays carry a negative charge.*[3]

4. **Observation.** A slightly concave, dish-shaped cathode can be used to produce a converging beam of cathode rays; that is, the rays can be focused in this way (Fig. 8-6). (In this respect cathode rays differ from light emitted by a glowing body, for such light cannot be focused in this way.)

[3] The flow of electricity in the cathode tube, from the negative to the positive electrode, once again contradicts Benjamin Franklin's conception (*see* footnote, p. 112). Nevertheless, as we indicated in Chapter 7, we still use his convention that electric current flows from positive to negative even though we now know that the effect of current is produced by a flow of electrons from negative to positive.

Fig. 8-6. Cathode-ray beams can be focused.

Deduction. The rays are emitted perpendicular to the surface of the cathode (and continue in straight lines unless deflected by magnetic or electric fields). Many scientists suspected that cathode rays might be streams of negatively charged particles, although a significant number of scientists believed that cathode rays were some form of ultraviolet light or other electromagnetic radiation. (The argument for the charged-particle model of cathode rays was clinched by J. J. Thomson in 1897, as we shall see in the next section.) In 1891 an Irish physicist, G. J. Stoney (1826–1911), proposed the name electrons for these presumed particles (from the Greek word for amber, the original source of "negative" electricity—see Chapter 7).

THE MASS-TO-CHARGE RATIO OF THE ELECTRON

In 1897 the Englishman J. J. Thomson (1856–1940), a stout defender of the charged-particle model of cathode rays, came up with a unique method of "weighing" these particles. Naturally, he could not determine their mass directly, even on the most delicate of chemical balances. Instead he devised a clever indirect method. In a series of measurements the effects of known magnetic and electric forces on a cathode-ray beam were observed. Calculation of the electron mass in essence, then, involved a measured acceleration (namely, a deflection) and application of Newton's Second Law of Motion—$F = ma$, or $m = F/a$ (*see* Chapter 6).

Thomson's actual method was the following. He first applied a magnetic field of known strength to a cathode ray and measured the amount of beam deflection (*see* Fig. 8-4). Then he brought the beam back onto a straight path by additionally applying an electric field of known strength (*see* Fig. 8-5). From the measured deflection and the known magnetic and electric field strengths, the electron mass, m, could then be calculated—if its charge were known. Since the charge, e, of the electron was not determined until 1910 (*see* next section), all Thomson could calculate was the ratio of mass to charge, m/e.

By measuring experimentally the mass-to-charge ratio for the electron, Thomson definitively established the charged-particle model. Consequently, he is frequently called the discoverer of the electron.

Significantly, Thomson obtained the same m/e ratio no matter what electrode material and what residual gas he used in the cathode-ray tube, and no matter what the strength of the beam was. This meant that all cathode-ray particles were alike. Thomson correctly interpreted this observation as indicating that these particles, or electrons, are found in all matter.

The presence of negative electrons in matter implies the presence also of positive particles, since atoms and molecules normally are neutral. The discovery of positive particles is described in the section on canal rays below.

The modern value for the mass-to-charge ratio of the electron is

$$5.7 \times 10^{-9} \text{ gram per coulomb,[4] or}$$

$$5.5 \times 10^{-4} \text{ gram per faraday}$$

One faraday (96,500 coulombs) of negative electricity thus has a mass of only about one two-thousandths of a gram (0.0005 gram per faraday). To Thomson, this value was surprisingly small. He knew that in electrolysis experiments one faraday of electricity can discharge 1 g of hydrogen atoms, or 23 g of sodium atoms, or 108 g of silver atoms (*see* Chapter 7, p. 119). These are mass-to-charge ratios of 1, 23, and 108 grams per faraday, respectively (*see* Problem 7-5). The smallness of the m/e ratio of the electrons means that electrons are thousands of times lighter than atoms, or that their charges are thousands of times greater than those of atoms, or a combination of the two possibilities. Thomson himself leaned toward the first explanation, the smallness of the electron mass. A dozen years later his educated guess was proved correct when Millikan determined the charge of the electron.

THE CHARGE OF THE ELECTRON

Once Thomson had measured the electron's mass-to-charge ratio, an independent determination of its charge would then permit calculation

[4] Expressing this ratio as its reciprocal, we obtain a charge-to-mass ratio of 1.8×10^8 coulombs per gram—quite a charge! Converting coulombs to statcoulombs and remembering the definition of the statcoulomb (*see* Chapter 7, p. 116), we can calculate the hypothetical repulsive force between two grams of electrons placed one centimeter apart: 1.8×10^8 coulombs $\times 3 \times 10^9$ statcoul/coul $= 5.4 \times 10^{17}$ statcoulombs; $F = (5.4 \times 10^{17})^2 \approx 3 \times 10^{35}$ dynes, or 7×10^{29} pounds!

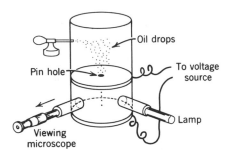

Fig. 8-7. Millikan's oil-drop apparatus (schematic).

of its mass. The American physicist Robert A. Millikan (1868–1953) made this possible in 1910 through his celebrated oil-drop experiment.

Millikan knew that oil drops sprayed from a small atomizer usually acquire a small charge,[5] just as does amber or glass rubbed with a cloth. If these oil drops are sprayed into the space between oppositely charged plates, negatively charged drops will be attracted to the positive plate and repelled by the negative one, while positively charged oil drops will travel in the opposite direction (Fig. 8-7).

The electric force, F_e, on a drop is a product of the electric field strength, E, and the charge on the drop, q:

$$F_e = Eq$$

The electric field strength can be calculated from the applied voltage, V, and the measured distance, d, between the two plates:

$$E = V/d$$

There is, of course, also a constant downward force on each drop exerted by the attraction of gravity—namely, the drop's weight, W. The weight of a drop can be determined from its speed of fall when no electric field is applied. In a vacuum such a drop would accelerate steadily under the action of gravity, increasing its velocity indefinitely (*see* Chapter 6, p. 84). But in air, the retarding force of friction also increases steadily with the drop's increasing speed. Soon the frictional force balances the force of gravity, and a steady rate of descent is reached. (This is fortunate; otherwise falling raindrops coming from a great height would have disastrous effects!) The opposing frictional force of the air depends on the size of the drop. Knowing this relationship, we can clock

---

[5] They can also be charged by placing some radioactive material in their vicinity. This material will cause the air molecules to become charged ("ionized"), and the oil drops pick up these charges by collision with the air molecules (*see* Chapter 9, especially Problem 9-1).

a droplet's rate of fall when its steady state is reached and then calculate the size of the drop. From that, in turn, we can calculate its weight, because we know the density of the particular oil we are using (weight = volume × density).

In one version of the Millikan experiment the downward force of gravity on the oil drop is balanced by an upward electrical force between the two charged plates. Thus weight equals upward force,

$$W = F_e = Eq$$

The charge on the drop then equals the ratio of W to E, both of which have been determined.

$$q = W/E$$

Proceeding in this manner in a large number of experiments, Millikan measured the charges on many oil drops. He then used the following reasoning to deduce the charge of the electron. If a negative charge on an oil drop results from the acquisition of excess electrons, the charge on any particular oil drop should be some multiple of a single electron's charge. In a large number of experiments, the smallest observed charge would probably be that of one electron, and the charges on the other drops would be multiples of that charge. Moreover, since positive charges on drops would be due to the loss of one or more electrons, they also would be multiples of the smallest charge. (Note the similarity to Cannizzaro's method of determining atomic weights, described in Chapter 5.)

The smallest charge, positive or negative, obtained by Millikan in this way was 1.6×10^{-19} coulomb. That no smaller charge has ever been obtained gives us confidence in assuming that this is the smallest charge that can exist in nature. And the fact that all other observed charges are whole-number multiples of this charge confirms our belief that electricity, both positive and negative, comes in packages. The smallest package of negative electricity, then, is the electron, with a charge of 1.6×10^{-19} coulomb (4.8×10^{-10} statcoulomb).

This numerical value has been corroborated by other independent experiments. One of these we discussed previously in connection with Faraday's quantitative experiments with electrolysis (Chapter 7, p. 121). Faraday found in 1833 that the minimum amount of electricity required to neutralize and deposit one mole of any element (i.e., Avogadro's number of atoms, N) is 96,500 coulombs, or one faraday. After N had been determined by Loschmidt in 1865, it became possible to calculate the minimum charge of one atom: 96,500 coulombs divided by 6×10^{23}. The result is 1.6×10^{-19} coulomb. This supports the idea that ions are formed by an atom losing or gaining electrons.

We can now return to the value of m/e (5.7×10^{-9} gram per coulomb) and use it to calculate the mass of the electron:

$$m = (m/e)\ (e) = (5.7 \times 10^{-9}\ \text{gram/coulomb})\ (1.6 \times 10^{-19}\ \text{coulomb})$$
$$= 9 \times 10^{-28}\ \text{gram}$$

This is 1/1840 of the mass of the hydrogen atom, or 1/1840 atomic mass units.

CANAL RAYS: POSITIVE PARTICLES

Experimentation with cathode rays led to an obvious question. If we can produce negatively charged particles in a cathode-ray tube, can we not produce *positively* charged rays in a similar manner? In 1886 the German physicist Eugen Goldstein (1850–1930) discovered that faintly luminous rays seem to come through holes in a perforated cathode into the space behind the cathode (away from the anode, *see* Fig. 8-8). He called these rays *Kanalstrahlen,* or "channel-rays," because they apparently are emitted by the holes or channels in the cathode. They came to be known in English as *canal rays.*

These rays were shown to be positive by magnetic and electric deflection. In 1897 both J. J. Thomson and W. Wien (German, 1864–1928) independently determined mass-to-charge ratios of canal rays and found the ratios to be several thousand times larger than the m/e of electrons. In other words, positively charged particles are apparently much heavier than electrons. Further, while the m/e of cathode rays (electrons) is independent of the cathode material and of the residual gas in the tube, the m/e of canal rays depends on the type of gas in the tube.

The explanation of this phenomenon is as follows. Electrons en route from the cathode to the anode collide with molecules of the residual gas in the tube, knock off electrons, and thus cause them to be positively charged ("ionized"). These gas ions are then attracted to the negatively charged cathode. Some of them pass through the holes in the cathode and are seen as a stream behind it. That is, the positive canal rays consist of gas atoms (or molecules) that have lost an electron through colli-

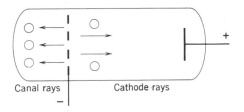

Canal rays   Cathode rays

Fig. 8-8. Canal rays in a cathode-ray tube.

sion with the cathode ray and thus have become positive ions; the negative cathode rays comprise electrons only.

CONCLUSION

The experiments described in this chapter suggest that atoms (and molecules), which are normally neutral, contain positive and negative particles and that the positive particles are much heavier than the negative electrons.

The significance of Thomson's and Millikan's work cannot be overemphasized. We indicated in the last chapter that Faraday's electrolysis experiments had yielded mass-to-charge ratios of ions; and when Faraday's results were coupled with Loschmidt's subsequent determination of Avogadro's number, the charge on a single ion could be calculated (*see* especially Problems 7-4 and 7-5). Granted, Faraday's findings hinted at the "graininess" of electricity, but they still left some doubt. Beyond that, they did not answer the question whether the charge on singly charged ions represents the *smallest* charge in nature or several such charges. Nor did they tell us whether packages of electricity are energy or matter.

Thomson showed that the negative electricity of cathode rays consists of particles that have a definite mass; that these particles apparently are present in all matter; and that they are all alike—there is only one kind of electron in nature. Millikan showed that the charge of this electron actually seems to be the smallest charge we can find in nature.

Problem 8-1. We noted that J. J. Thomson also measured the deflections of positive beams (canal rays) and calculated their m/e ratios. With oxygen in the cathode-ray tube, several beams are obtained.

(a) The m/e of one of the beams is 16 g per faraday. What presumably are the particles in this beam?

(b) Other beams have m/e values of 16/2 and 16/3 g per faraday. What do you think are the particles in these beams?

(c) Another beam has an m/e of 32 g per faraday. What are the particles in this beam?

(d) Would you expect to find other beams? Explain.

(e) When helium (AW 4) is in the tube, positive particles having an m/e of 2 g per faraday are observed. What are these particles?

(f) If chlorine gas were in the tube, what m/e beams would you expect to find? (*See* Chapter 4, p. 48.)

(g) In view of your answer to (f), can you suggest some practical applications of this experimental method?

Chapter 9

RADIOACTIVITY AND X RAYS

In addition to electrons and canal rays, the cathode-ray tube gave birth to X rays, which in turn led to the discovery of radioactivity. All four discoveries had a profound effect on our understanding of atomic structure.

X RAYS

In 1895 (two years before Thomson's "discovery" of the electron) the German physicist Wilhelm Konrad Röntgen (1845–1923), like so many of his contemporaries, was experimenting with that fascinating new gadget, the cathode-ray tube. Röntgen was interested in the fluorescence (glow) caused by the cathode rays when they hit the wall of the tube. While working in his darkened laboratory, he noticed a glow coming from some chemicals at a considerable distance from the cathode-ray tube and not in the direct path of the cathode-ray beam. The chemicals glowed only when the cathode-ray tube was in operation. Apparently this fluorescence was caused not directly by the cathode rays themselves but by some sort of invisible secondary rays, which Röntgen called X (for unknown) rays. These X rays were able to pass not only through the glass of the cathode-ray tube and several meters of air but also through many other materials such as paper, cardboard, thin metal foil, flesh, walls, and quite thick blocks of wood. On the other hand, bones, thin sheets of lead, and thicker sheets of most other metals were found to be relatively impervious to these rays. Using shadows cast by objects, Röntgen traced the X rays back to their source, the target area in the cathode-ray tube, where the cathode-ray beam strikes the glass wall or the anode (*see* Fig. 9-1).

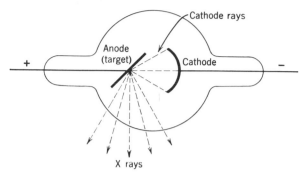

Fig. 9-1. A simple X-ray tube.

Röntgen immediately realized the significance of his discovery and experimented further. He observed that his X rays excited fluorescence in a number of chemicals and could fog (expose) photographic plates in a dark room, right through protective wrapping.[1] Like cathode rays, X rays travel in straight lines, casting sharp shadows, for example, of bones. But, unlike cathode rays and canal rays, X rays cannot be deflected by electric or magnetic fields. This observation indicates that X rays do not consist of charged particles. It was then suggested that X rays might be some sort of ultraviolet (light) waves but with properties quite different from those of ordinary ultraviolet waves. In 1912 this theory was proved correct by the German physicist Max von Laue (1879–1960), who showed by means of crystal diffraction that X rays are indeed "light" waves of very short wavelength (*see* Chapter 13).

X rays derive their energy from the kinetic energy of the cathode-ray electrons which are suddenly stopped when they hit the target in the cathode-ray tube. The higher the original speed and therefore the kinetic energy of the electrons, the higher the energy and the penetrating power of the resultant X rays.[2]

[1] Well before 1895, scientists working with cathode rays knew that photographic plates kept near their equipment would frequently be fogged. But, as Pasteur would have pointed out, their minds were not prepared. In fact, a number of other investigators had the discovery of X rays within their grasp but somehow failed to perceive the significance of their various observations. For example, in 1880 Goldstein, the discoverer-to-be of canal rays, reported that fluorescent screens within cathode-ray tubes were excited even when they were shielded from the cathode rays but exposed to the glow from the tube's glass walls. And J. J. Thomson in 1894 reported fluorescence of materials *outside* cathode-ray tubes.

[2] Because this type of X ray results from the stopping, or braking, of the electrons, it is often called *Bremsstrahlung*—from the German word *bremsen*, meaning "braking."

Even though they did not understand the exact nature of the newly discovered X rays, people did not hesitate to put them to practical use—often to their personal detriment since, as we know today, X rays can cause burns and even induce cancer. Cathode-ray equipment was common in university laboratories; thus X rays were readily accessible to physicians, once they had learned of their existence. Within weeks of Röntgen's first publication of his discovery, the broken arm of a patient in Hanover, New Hampshire, was set with the help of X rays. Soon X rays were used regularly for a variety of medical purposes, especially in diagnosis. Today they are commonly applied in industry in the detection of flaws or cracks in metal castings.

Beyond such practical applications, X rays have turned out to be a powerful research tool in the hands of scientists. Von Laue's discovery of X-ray diffraction in 1912 has given us a means of measuring the distances between atoms in crystals and determining their geometric arrangements. Also, it was work with X rays that led to the discovery of atomic numbers in 1913 (see Chapter 11), thus advancing our understanding of atomic structure. And the most immediate, albeit indirect, result of the understanding of atomic structure was the discovery of radioactivity, within months of Röntgen's discovery.

THE DISCOVERY OF RADIOACTIVITY

Early in 1896, upon hearing of Röntgen's discovery, the French physicist Henri Antoine Becquerel (1852–1908) decided to investigate further. Becquerel had long-standing interest in the phenomena of fluorescence and phosphorescence;[3] his late father had been an eminent contributor to the field. What intrigued him was the fluorescent glow from the X ray-emitting target in the cathode-ray tube. He wondered whether fluorescent glows in general contain X rays.

Accordingly, Becquerel tested various fluorescent materials, among them uranium, to determine whether they emitted X rays. He proceeded as follows. He wrapped a photographic plate in several sheets of thick black paper to protect it from exposure to direct sunlight. He then placed some of the uranium compound on top of the package and exposed the whole thing to the sun for several hours, to stimulate fluorescence. When he developed the photographic plate he saw the silhouette

[3] A phosphorescent substance, in the strict sense of the word, differs from a fluorescent one in that it continues to glow for a while after removal of the exciting radiation such as ordinary sunlight. A phosphorescent substance may continue to glow in the dark for several seconds, or even hours, after having been exposed to light.

of the fluorescent substance on the photographic plate; the plate had been exposed right through the protective paper. Becquerel tentatively concluded that the uranium compound had emitted X rays along with the visible fluorescence.

But Becquerel was a careful man. He wanted to exclude the possibility that the photographic image had been caused, not by X rays, but by chemical vapors emanating from the fluorescent substance when it had been heated by sunlight. He therefore performed a "control experiment." To block possible chemical vapors, he interposed a thin sheet of glass between the fluorescent material and another photographic package. When he again obtained positive results, he logically concluded that the fluorescent substance was emitting very penetrating radiations—in other words, X rays.

Notice the interplay between theory and experiment. Becquerel had hypothesized that fluorescence, caused by exposure to visible light, contained X rays. This hypothesis suggested an experiment, and the experiment had the predicted results. But was his theory *really* proved? Becquerel had failed to run another control experiment.

Fortunately, an accident of nature provided the missing experiment. An overcast sky caused Becquerel to postpone further experimentation, and he temporarily stored his wrapped photographic plates and uranium compound in a dark drawer. We pointed out that Becquerel was a careful man. When he was ready to resume his experiments several days later, he first checked to see whether his plates were still intact. He therefore developed one of them. Lo and behold! The silhouettes of the uranium crystals appeared even more clearly than on the plates exposed to light. Apparently the penetrating rays were emitted even in the dark, without fluorescence.

Further experimentation confirmed the suspicion that this penetrating radiation is not connected with fluorescence or phosphorescence. Whereas phosphorescence decreases rapidly after the exciting light source is removed, this penetrating radiation continued unabated even when the material had not been exposed to light. What is more, Becquerel found this radiation activity in some uranium compounds that did not fluoresce at all; on the other hand, some fluorescent materials, such as compounds of calcium and zinc, did *not* exhibit this radiation. The presence of uranium was the only factor common to the various experiments that yielded this radiation. Moreover, the intensity of the radiation was found to be independent of the chemical combination involved; it depended only on the percentage of uranium in the compound. Therefore, this radiation activity must be a property of the element uranium itself.

THE DISCOVERY OF RADIUM

The next obvious question was whether radiation activity is exhibited by elements other than uranium. Systematic examination of a large number of minerals and elements was begun by Marie Sklodowska Curie (1867–1934), later joined by her husband, Pierre Curie (1859–1906), a colleague of Becquerel in Paris. The Curies soon showed that the previously known element thorium and its compounds are also radioactive. Thus they demonstrated that radioactivity is not a unique property of the element uranium.[4]

In working with the uranium ore pitchblende, they found that the crude ore was much more radioactive than its small content of uranium could account for. Therefore the ore must contain some other substance even more radioactive than uranium. After a painstaking task of chemical separation, they managed in 1898 to isolate this new radioactive element, which they called polonium, in honor of Mme. Curie's native Poland. A few months later, even more heroic efforts enabled them to isolate another, much more intensely radioactive element, which they named radium.[5]

The isolation of radium is interesting for its relation to the periodic table. Pitchblende uranium ore contains barium, which the Curies chemically removed as barium chloride ($BaCl_2$). To their surprise, most of the radioactivity went with the barium compound instead of remaining with the uranium. Since barium itself is not radioactive, the Curies concluded that the radioactivity resided in another element present as an impurity. This element must be chemically similar to barium—i.e., in the same chemical family, or vertical column, of the periodic table. In fact, there *was* an appropriate gap in the periodic table, directly under barium—occupied today by the element radium.

Originally, in 1898, the Curies separated only a trace amount of radium in impure form and confirmed the presence of this new element by means of its spectrum.[6] Only in 1902 did they manage to isolate enough radium

[4] G. C. Schmidt (1865–1949) in Germany made the same discovery independently at about the same time. It was the Curies who coined the word *radioactivity*.

[5] Marie Curie received two Nobel prizes for her work: the first in physics in 1903 for the discovery of radioactive elements, jointly with her husband and Becquerel; the second in chemistry in 1911 for the isolation of polonium and radium. Marie Curie was the only person to have received two Nobel prizes until Linus Pauling was awarded the Nobel peace prize in 1962 after receiving the prize in chemistry in 1954. Only three other women ever received Nobel prizes in physical sciences, and one of them is a daughter of the Curies. Irene Joliot-Curie, together with her husband, Frederic Joliot, received the 1935 Nobel prize in chemistry for work in the artificial production of radioactive elements.

[6] The principle of this technique is discussed in Chapter 14.

(0.1 g of $RaCl_2$) to determine its atomic weight and some of its properties. To obtain this small amount of radium, the Curies had started with several tons of crude ore. It was not until 1910 that Mme. Curie succeeded in isolating pure radium metal—by electrolysis of molten radium chloride.[7]

THE NATURE OF RADIOACTIVE RADIATIONS

The early work of the Curies brought about renewed interest in radioactivity. Ernest Rutherford (1871–1937), a young New Zealander working at Cambridge University with J. J. Thomson, started to investigate the ionizing properties of Becquerel's penetrating rays and continued this work after moving to Montreal. In 1899 he found that there were actually two kinds of rays produced by radioactivity. He called them *alpha rays* and *beta rays*.

Alpha rays are stopped easily by a sheet of paper or by thin (0.001 inch thick) metal foils.[8] Beta rays can penetrate foils up to 0.1 inch thick. A third, still more penetrating, constituent of radioactive radiation was found the following year (1900) by the French scientist Paul Villard (1860–1934). These so-called *gamma rays* are more penetrating than even X rays from cathode-ray tubes and can easily pass through lead several centimeters thick.

Alpha and beta rays are deflected by magnetic and electric fields. The directions of their paths indicate that alpha rays carry positive charges, while beta rays are negative. Gamma rays, like Röntgen's X rays, are not affected by electric or magnetic fields. Consequently, charged plates can be used to separate the three types of radiation (*see* Fig. 9-2).

Beta rays were soon shown to have the same m/e as cathode rays. This means that beta rays are simply very energetic, high-speed electrons.

Gamma rays were recognized as highly penetrating X rays. They are "light" waves of even shorter wavelengths than ordinary X rays.

The true nature of alpha rays was first suggested by Rutherford in 1906, and he succeeded in proving his hypothesis in 1909. Rutherford

[7] Radium is a rare and very precious metal. In 1921 President Warren G. Harding presented Mme. Curie with a gram of radium, which at that time was worth approximately $100,000. Five hundred tons of ore from Colorado had been processed to obtain that gram of radium. By 1940 a total of only about a thousand grams of radium (2 lb) had been produced. Today's major suppliers are Canada and Katanga, in the Congo.

[8] The penetrating power of alpha rays depends somewhat upon the radioactive element that is their source. For example, alpha rays emanating from radium have a range in air of about 34 mm; rays from polonium have a range of about 39 mm.

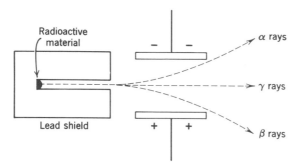

Fig. 9-2.   Separation of radioactive radiation by means of an electric field.

demonstrated that neutralized alpha particles have a spectrum identical to that of helium gas, first obtained in 1895 (*see* Chapter 14). He further showed that alpha particles have the same mass-to-charge ratio (2 g per faraday) as canal-ray particles present in helium-filled cathode-ray tubes. In other words, alpha particles are doubly charged helium ions; that is, helium atoms that have lost two electrons, He^{++}.

In 1911 Rutherford and his coworkers determined Avogadro's number through an experiment involving alpha particles. What they did, roughly, was to seal some radioactive material that was a strong source of alpha particles into a glass tube with very thin walls. This tube in turn was sealed into a thick-walled tube, and the space between the two was pumped out (*see* Fig. 9-3). The alpha particles were able to pass through the thin wall of the inner tube but were stopped by the thick outer wall. Here they picked up two electrons, probably from the glass, and became ordinary helium gas:

$$He^{++} + 2e^- \rightarrow He$$

Rutherford could measure the amount of helium collected in a given period of time. He could also count the individual alpha particles emitted by a radioactive source, with the help of a scintillation counter: When an alpha particle strikes a screen coated with a fluorescent substance such as zinc sulfide, the fluorescent material gives off a tiny flash of light, which can be seen through a low-power microscope. From these measurements he could calculate Avogadro's number—that is, the number of helium atoms in 4 g of helium (*see* Problem 9-2).

✧   ✧   ✧

The foregoing developments—all children and grandchildren of the cathode-ray tube—set the stage for next great step forward—the use of

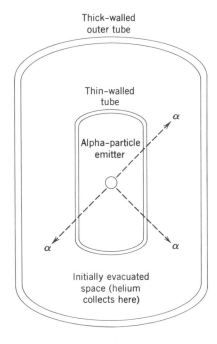

Fig. 9-3. Rutherford's apparatus for determining Avogadro's number.

these radioactive particle streams to learn more about the structure of the atom.

PROBLEMS

Problem 9-1. Millikan's oil-drop experiment (Chapter 8) clearly shows that a particular drop in the microscope's field of view, when exposed to radioactive radiation or X rays, can pick up or lose charges only in definite packages. To answer the following questions, assume that a negative charge on a drop represents an excess of electrons, whereas a positive charge is due to a dearth of electrons.

(a) Propose a *very simple* mechanism whereby a drop of whatever charge, when exposed to radioactive radiation, can decrease its negative charge or increase its positive charge (i.e., lose electrons). This simple mechanism will *not* account for the frequently observed reverse phenomenon—a decrease of positive charge or an increase of negative charge (i.e., a *gain* of electrons).

(b) We noted in Chapter 8 (Footnote 5) that drops generally gain or lose charge through the capture of a positive or negative charge from ionized atoms or molecules of the surrounding gas. X rays or gamma rays ionize the gas or air, and the drop then picks up one or more of the ions. How, by this mechanism, can a drop that is already positively charged become *more* positive and how can a negatively charged drop collect *more* negative charge, in view of the fact that like charges repel each other? In suggesting an answer, take into account the fact that gas molecules are in constant motion.

Problem 9-2. In an experiment using Rutherford's apparatus (*see* p. 141), a total of 6.0×10^{17} alpha-particle emissions are counted over a six-month period. These alpha particles penetrate into the initially evacuated space between the two tubes. The total mass of these alpha particles (measured as helium gas) is 3.85×10^{-6} g (i.e., 3.85 micrograms). (a) What fraction of a mole of helium has been collected? (b) From these data calculate Avogadro's number.

Answer.

(a) One mole of He has a mass of 4.00 g.

Hence 3.85×10^{-6} g represents $3.85 \times 10^{-6}/4.00$, or 0.96×10^{-6} mole.

(b) If 0.96×10^{-6} mole contains 6.0×10^{17} atoms, one mole would contain

$$\frac{1}{0.96 \times 10^{-6}} \times 6.0 \times 10^{17}, \text{ or } 6.2 \times 10^{23} \text{ atoms. This is Avogadro's}$$

number according to these data.

Avogadro's number could also have been calculated in one step as follows:

$$N = 6.0 \times 10^{17} \times \frac{4.00}{3.85 \times 10^{-6}} = 6.2 \times 10^{23}$$

(The literature value of Avogadro's number is 6.02×10^{23}.)

Chapter 10

EARLY MODELS OF THE ATOM

By 1900 it was clear that electrons are a fundamental constituent of matter, because they could be produced with any electrode material in cathode-ray tubes and they had the same mass-to-charge ratio no matter how they were obtained. (Remember that the electron charge itself was determined by Millikan only in 1910.)

When negative electrons are removed from atoms, positively charged ions remain, because atoms as a whole are neutral. We have seen that this was the explanation of Goldstein's canal rays in cathode-ray tubes, and it was actually shown in 1897 that the positive particles of canal rays have the masses of ions. Even before that, there were indications that atoms contain both positive and negative electricity: During electrolysis, different substances are collected at the positive and negative electrodes. Further evidence that matter contains both positive and negative particles came in 1899 with Rutherford's discovery of positive alpha and negative beta particles. But how are these positive and negative particles arranged within the atom?

THE RAISIN-PUDDING MODEL

In 1904 both J. J. Thomson and independently another Thomson—William Thomson, Lord Kelvin (1824-1907)—published papers picturing the atom as a sphere of positive electricity in which enough electrons were imbedded to make the atom neutral—much like raisins in a pudding (Fig. 10-1).

Although this model could explain some of the known properties of atoms, new experimental evidence soon forced its abandonment (*see* p. 147 and Chapter 15). Meanwhile scientists were asking how big such atoms might be.

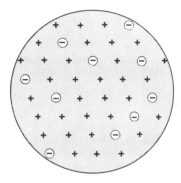

Fig. 10-1. The raisin-pudding model of an atom.

THE SIZE OF ATOMS

We can calculate the diameters of atoms in solids and liquids by assuming that their atoms actually touch each other, indicated by the very low compressibilities of these substances. For example, mercury, a liquid metal, has a density of 13.6 g per cubic centimeter; thus 1 g of mercury has a volume of 1/13.6 cc. Avogadro's number of mercury atoms (that is, 1 mole, or 200.5 g) will then have a volume of 200.5 times as much:

$$\text{volume of 1 mole of Hg} = \frac{200.5 \text{ g}}{13.6 \text{ g/cc}} = 14.7 \text{ cc}$$

The volume of *one* mercury atom, then, is 14.7 cc divided by Avogadro's number:

$$14.7 \text{ cc}/6.02 \times 10^{23} = 2.44 \times 10^{-23} \text{ cc}$$

$$= 24.4 \times 10^{-24} \text{ cc}$$

If, for simplicity's sake, we assume that these atoms are tiny cubes, the length (or diameter) of an atom would equal the cube root of 24.4×10^{-24}, or 2.9×10^{-8} cm.[1]

We must use different methods to determine atomic diameters in gases, because here the atoms do not touch each other, as indicated by the very high compressibility of gases. One such method involves "collision cross-sections." The details of this technique are beyond the scope of this book. We can, however, suggest a simple analogy. If we shoot a rifle

---

[1] This is also expressed as 2.9 angstroms. The angstrom, abbreviated A, was named in honor of the Swedish spectroscopist A. J. Ångström (1814–1874). It is used in spectroscopy and also is a convenient unit for atomic dimensions. 1 A = 10^{-8} cm. Note a similar calculation was performed for Problem 4-8.

blindfolded into a forest, the average path length that the bullet will travel until it hits a tree ("mean free path") depends on the average distance between the trees and the average width of the tree trunks. The more closely grouped the trees are and/or the larger their trunks, the shorter will be the mean free path. If we can somehow measure the mean free path *and* the average distance between tree trunks, we can then calculate the average width of the tree trunks. For gaseous atoms or molecules, we know the average distances between them because we know the number per given volume: Avogadro's number of molecules per mole—i.e., per 22.4 liters;[2] and we can experimentally measure the mean free path by various methods—for example, in the cloud chamber described in the next section. Atomic diameters calculated in this manner are compatible with those derived from solids and liquids.

Still other methods lead to similar results. For example, the later-developed method of X-ray diffraction (Chapter 13) yields very accurate values for solids. It turns out that atomic diameters measured by these methods range from about 1×10^{-8} to 5×10^{-8} cm (1 to 5 A).

THE SIZE OF ALPHA PARTICLES

With his invention of the "cloud chamber" in 1911, the Scottish physicist C. T. R. Wilson (1869–1959) made it possible to see the tracks left by subatomic particles, such as alpha and beta particles. The principle of the Wilson cloud chamber is relatively simple. You have seen the condensation of the moisture in your breath on a cold day. Imagine the surprise of some scientists in Alaska when, on a frigid day, their breath did not condense into fog. To condense readily, water vapor needs some sort of particles, or *nuclei of condensation,* and in the clean Alaska atmosphere these apparently were absent.

These condensation phenomena form the basis of the Wilson cloud chamber. It is well known that compression of gases usually warms them and that expansion cools them. For example, air compressed at the base of a tire pump is warmed, and carbon dioxide is cooled as it emerges from the orifice of a fire extinguisher and comes out as Dry Ice. In the cloud chamber, water vapor in an enclosed volume is cooled by sudden expansion to a point below its condensation temperature, called the *dew point.* The water vapor would now condense, but needs some condensation nuclei upon which to form droplets. If a charged particle enters the chamber, it will collide with air and water molecules and leave a series of ions along its trajectory. These ions, acting as condensation nuclei,

[2] Recall the discussion of the *molar volume* in Chapter 4.

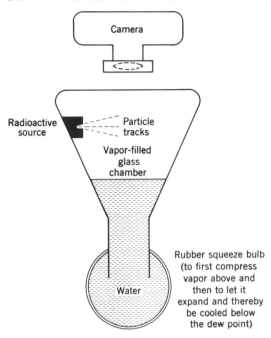

Fig. 10-2.   A simple cloud chamber.

now cause a series of tiny water droplets to appear. Thus they make the track of the charged particle visible (*see* Fig. 10-2).

Alpha rays emitted by radioactive materials, you may recall, are simply doubly charged positive helium ions. In the cloud chamber their paths show up as straight, fairly long trajectories, indicating relatively few collisions with air and water-vapor molecules. From the paucity of collisions scientists concluded that alpha particles are *very* small compared to atoms and most other ions.

RUTHERFORD'S EXPERIMENT

Although the experiments described in this section were begun two years before Wilson's development of the cloud chamber, we discuss them out of chronological sequence because Rutherford subsequently used Wilson's results to interpret his experimental findings.

Rutherford had heard of some experiments by the German physicist Philipp Lenard (1862–1947), who found in 1903 that beta rays (high-velocity electrons from radioactive sources) passed through thin metal foils with surprising ease. A few beta particles, however, were deflected, some of them quite sharply.

Rutherford decided to repeat these experiments, using alpha rays instead. He chose alpha rays for several reasons. First, alpha particles penetrate thin foils, but they are stopped by slightly thicker foils, which however do permit the passage of beta particles. Second, alpha particles are easy to observe because they produce tiny sparks on a zinc sulfide screen and thus can be counted.[3] Therefore, Rutherford would easily be able to distinguish between those particles that passed through, those that were slowed, those that were absorbed entirely, and those that were deflected by various angles. Finally, because of their high kinetic energy and momentum (owing to their relatively high mass), alpha particles are not affected much by collisions with mere electrons.

Rutherford and his students Hans Geiger (of Geiger counter fame, German, 1882–1947) and Ernest Marsden (English, 1889–   ) began experiments about 1909. They directed streams of alpha particles with velocities of approximately 10,000 miles per second at very thin sheets of gold foil (see Fig. 10-3). Even though the foils had thicknesses of several hundred atoms, most of the particles simply passed right through with almost undiminished speed. This would have been strange enough; but, even more surprising, some alpha particles bounced back almost in the direction from which they had come. As Rutherford put it, this result was about as unanticipated "as if you had fired a 15-inch shell at a piece of tissue paper, and it came back and hit you."

These observations are totally incompatible with the raisin-pudding model of atoms, which in solids would be closely packed, like marbles in a box. In the first place, those alpha particles passing through the foil would have been greatly slowed, if not stopped outright, by their passage through several hundred pudding interiors. In the second place, alpha particles would not bounce back from such a raisin-pudding mass; at best, they would be only slightly deflected: Since the atom as a whole is neutral, there would be no interaction between the atom and the alpha particles until they had actually penetrated the positive spheres; then there would be interactions with the light electrons and the positive sphere. But the spongy positive sphere would have only little effect on high-speed alpha particles; likewise, since electrons are so much lighter than alpha particles (1/7,000), they wouldn't be able to cause much deflection either (considering conservation of momentum). Accordingly, the raisin-pudding model had to be abandoned.

[3] See p. 140.

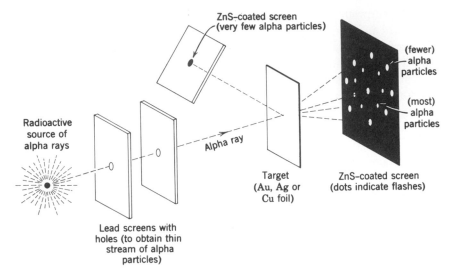

Fig. 10-3. Rutherford's alpha-ray-scattering experiment.

RUTHERFORD'S NUCLEAR MODEL OF THE ATOM

Then how did Rutherford explain the fact that some of the alpha particles rebounded? They must have hit something hard and massive. Apparently the mass of the atom is not spread throughout the atom but is concentrated in very small centers, or *nuclei*.

From the percentage of alpha particles deflected and their size (determined by cloud-chamber observations), Rutherford calculated the diameter of the metal's nucleus. He came up with 10^{-12} centimeters (10^{-4} A)—that is, 1/10,000 of the diameter of the atom itself. In other words, most of the mass of the atom is concentrated in a tiny nucleus, which has only one-trillionth the volume of the atom.

Additional considerations led Rutherford to the conclusion that the nucleus is not neutral but has a charge. If this were not so, alpha particles passing very close by the nucleus would not be deflected at all. If, however, the nucleus does have a charge, even near misses would be deflected slightly—either attracted by a negative charge or repelled by a positive charge. Experimental measurements involving near misses confirmed that the nucleus indeed has a charge, but they gave no direct indication whether the charge is positive or negative.

Rutherford decided that the nuclear charge must be positive on the basis of the following circumstantial evidence. First, the ease with which electrons are removed from atoms—by means of cathode rays, the photo-electric effect (see Chapter 17), and just ordinary ionization processes; this easy removal indicates that the negative electrons occupy the outer parts of the atoms. Since the atom as a whole is neutral, the nucleus must accordingly be positive. Second, negative radiations from atoms (cathode rays, beta rays) are lightweight electrons, whereas positive emanations, notably alpha rays, are heavy and could have come only from the massive nucleus; therefore, again, the nucleus must be positive.

Indeed, we can conclude that alpha particles themselves are nuclei. To begin with, Rutherford has shown that alpha particles are simply helium atoms that have lost two electrons—He^{++} (see Chapter 9, p. 140). Experimental evidence reveals that helium loses two electrons rather easily, but never three or more electrons; He^{+++} does not exist. Apparently helium that has lost two electrons has been stripped down to the nucleus. This conclusion is supported by cloud-chamber observations, which indicate that alpha particles are very small compared to atoms (see p. 146).

Rutherford could compute an approximate value for the charge on the nuclei in his foils. He based his calculations on Newton's Laws of Motion (Chapter 6), Coulomb's Inverse-Square Law of Electrostatic Forces (Chapter 7), the percentage of scattered alpha particles in his experiments, and their angles of deflection. For gold, Rutherford obtained a charge about 80 times that of the hydrogen nucleus. When silver and copper foils were substituted in the experiments, the results indicated charges of 50 and 30 units respectively.[4]

These charges are roughly half the atomic weights of gold, silver and copper. Similarly the alpha particle (charge +2, mass 4). At first glance, this consideration suggested that alpha particles might be the basic building blocks of atoms, an idea that was supported by the observation that radioactive elements emit alpha particles. Although it was an appealing concept, several facts dispelled such a simple notion (see Problem 10-2). Two years later, a more precise method for determining nuclear charges (pp. 155-159) revealed that these charges usually are not equal to exactly one-half of the atomic-weight values.

At any rate, Rutherford's alpha-ray scattering experiments indicated that the atom consists of a very dense positive nucleus with enough electrons outside the nucleus to make the atom as a whole neutral. But what would keep the outer negative electrons in the atom from being attracted by the positive nucleus and simply falling in, thus causing the collapse of the atom? Well, what keeps the moon from being attracted by the

---

[4] In 1967 a similar, remotely controlled alpha-ray scattering experiment was performed on the moon by Surveyor 5, to analyze lunar soil samples.

earth's gravity and falling in, or the earth from falling into the sun? The answer is that these heavenly bodies are not stationary but orbiting, and the force of gravity is offset by the centrifugal force. Analogously, Rutherford proposed that, in atoms, the outer electrons simply orbit around the nucleus.

SUMMARY

The Rutherford model of the atom, which had emerged by about 1911, had the following features: (1) a mass concentrated almost entirely in a tiny, dense, positively charged nucleus; (2) enough negative orbiting electrons outside the nucleus to make the atom as a whole neutral; (3) a nuclear charge (and therefore a number of orbiting electrons) that increases with increasing atomic weight; and (4) an atomic diameter about 1×10^{-8} to 5×10^{-8} cm (1 to 5 A) and a nuclear diameter on the order of 10^{-12} cm (10^{-4} A); thus the atom is mostly empty space.

This model of the atom explained both of Rutherford's principal observations: (1) the fairly easy passage of most alpha particles through thin metal foils; (2) the deflection of some alpha particles through very sharp angles. At the same time, the model brought with it a new difficulty, which we shall discuss in Part Three. Meanwhile, scientists wanted to know the precise number of positive charges and therefore planetary electrons in each atom. (Rutherford's experiments provided only approximate values.) This queston, among others, is taken up in the next chapter.

PROBLEMS

Problem 10-1. (a) What is the mass in grams of 1 mole (1 gram-atom) of hydrogen *atoms*?

(b) What is the mass of a single hydrogen atom?

(c) The diameter of the hydrogen atom is approximately 1 angstrom, that is, 10^{-8} cm. Assuming that the hydrogen atom is a little cube, what is its volume in cubic centimeters?

(d) What is the density of the hydrogen atom in grams per cubic centimeter?

(e) What is the density of liquid water in grams per cubic centimeter? (Recall the definition of the gram—Appendix II).

(f) How does the density of the hydrogen atom compare quantitatively with the density of water?

(g) If the radius of the nucleus is 1/10,000 the radius of the hydrogen atom, what is the relationship of the volume of the nucleus to the volume of the atom?

(h) If the mass of the nucleus is essentially the same as the mass of the atom, what is the density of the nucleus in grams per cubic centimeter?

(i) What would be the weight of 1 cubic centimeter of such nuclear material in grams, in kilograms, and in tons? (A metric ton equals 1000 kg.)

(j) The densest material known on earth is osmium, at 23 grams per cubic centimeter. One of our nearby stars, the companion of Sirius, was found to have a density of 60,000 grams per cubic centimeter. How do you explain this value in the light of the foregoing questions?

Problem 10-2. Many radioactive elements emit alpha particles (with a mass of 4 atomic mass units and a charge of +2). Their mass and high kinetic energy indicate that these particles came out of atomic nuclei. Could these alpha particles be the elemental building blocks of all nuclei? (*See* p. 149.) Cite at least two pieces of evidence *against* that theory. *Hints:* What other radioactive emanations are there? (All radioactive emanations originate in the nucleus.) Also, look at a table of atomic weights.

ANSWERS TO SELECTED PROBLEMS

10-1(d). $D = \text{mass/volume} \approx \dfrac{1}{6 \times 10^{23}} \Big/ 10^{-24} \approx 1.7 \text{ g/cc.}$

10-1(h). $D \approx 2 \times 10^{12} \text{ g/cc.}$

10-1(i). Two million tons!

Chapter 11

INVESTIGATING THE NUCLEUS

THE DISCOVERY OF ISOTOPES

In earlier chapters we mentioned the existence of isotopes—that is, *two or more forms of an element that have almost identical chemical properties but slightly different atomic weights.* Through examination of radioactive-decay products, isotopes were first discovered in 1913 independently by the English chemist F. Soddy (1877–1956) and the American T. W. Richards (1868–1928). They found that through the emission of radioactive particles uranium is eventually converted to lead with an atomic mass of 206, whereas thorium is converted to lead-208. This would explain the fractional atomic weight of lead as found in nature, 207.2, as due to a mixture of lead-206 and lead-208.

Problem 11-1. (a) If naturally occurring lead were a mixture of only two isotopes, lead-206 and lead-208, what would be the percentages of these isotopes in natural lead?
(b) Given the fact that natural lead (207.2 amu) actually is a mixture of *three* isotopes (206, 207, and 208), can you calculate the percentage of each?

A second discovery of isotopes, not related to radioactivity, came through J. J. Thomson's work with canal rays in 1913. You will recall that canal rays are positive rays that emerge through channels in perforated cathodes in a direction opposite to that of the cathode rays. J. J. Thomson and others determined the mass-to-charge ratios of these positive particles, as they had done with electrons, and concluded that canal rays have masses considerably greater than electrons do—masses very similar to those of

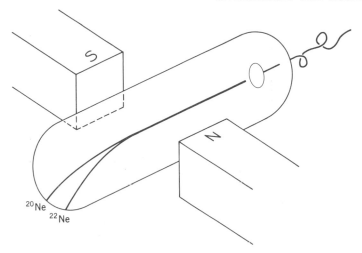

Fig. 11-1. Thomson's discovery of neon isotopes.

atoms themselves. In other words, canal rays are positively charged gas ions, created by the high-speed electrons of cathode rays colliding with molecules of the residual gas in the canal-ray tube. These positive ions are attracted to the negative cathode. Those that collide with the cathode will pick up an electron there and become neutral atoms or molecules, while those that miss the electrode and pass through the holes will emerge as a stream of canal rays.

While working with neon, a monatomic gas, Thomson discovered that the canal-ray beam, when bent by a magnetic field, was split into *two* beams, with masses corresponding to 20 and 22 amu (Fig. 11-1). From this fact he concluded that natural neon, atomic weight 20.2, is actually composed of two isotopes of atomic weights 20 and 22.

Problem 11-2. Calculate the percentages of neon-20 and neon-22 in ordinary neon, AW 20.2.

This modified canal-ray tube is the grandfather of the modern mass spectrometer. F. W. Aston (American, 1877–1945) improved the device, and by 1919 it had become a precision instrument. In that year he showed that the fractional atomic weight of chlorine, 35.45 amu, is due to the contributions of Cl-35 and Cl-37.

The modern mass spectrometer—which can vary in size from a roomful of equipment to a tiny gadget carried aloft by a satellite—has proved to be a most versatile instrument. Not only does it tell us about the existence of isotopes, but it can determine their masses to five decimal places. It also has become a powerful tool in analytical chemistry. By using an electron beam to fragment chemical compounds into their various constituent units, we can learn much about molecular structure. Additionally, by correlating the energy of the electron beam with the amount of disruption it produces in the chemical bond, we can obtain an idea of bond strengths.[1]

Perhaps the most striking isotope discovery was that of an isotope of hydrogen with twice the mass of an ordinary hydrogen atom. For this work in 1932 Harold C. Urey (American, 1893–   ) was awarded the Nobel Prize. This form of hydrogen became so important that it was given a special name, *deuterium;* it has many practical and theoretical applications. It should be added that the discovery of deuterium was made not through mass spectrometry but through optical spectroscopy (*see* Chapter 18, p. 251).

Today, isotopes are used as tools for chemical analysis, for following chemical reactions, for tracing the paths of various substances through the human body, and for determining the ages of archeological discoveries and even of the earth itself.

The discovery of isotopes immediately contributed a minus as well as a plus to our understanding of atomic structure. On the one hand, the existence of isotopes explained fractional atomic weights (*see* Chapter 4, p. 48). On the other hand, the existence of elements with several atomic weights further confused the significance of atomic weight. You will recall from Chapter 5 that, with only a few unexplained exceptions, atomic weights had been the basis of the arrangement of the elements in Mendeleev's periodic table (1869). If an element can have several atomic weights, then atomic weight apparently is *not* the basic key to the properties of elements, as expressed by the periodic table.[2]

We have pointed out that several irregularities in the periodic table had already indicated to Mendeleev that atomic weight is not the major factor in determining the properties of elements. For example, in order to classify argon, atomic weight 39.9, and potassium, atomic weight 39.1, into their proper chemical families, he had to reverse their positions. Furthermore, different elements may have the same atomic weight; for example, the common isotopes of calcium and argon both have mass 40.

What aspect of the atom, then, is the key factor in determining its chemical properties? Rutherford's work on alpha-ray scattering gave a

---

[1] The mass spectrometer is discussed in greater detail in Chapter 27.

[2] The word *isotope* was coined by Soddy in 1913 and comes from the Greek, meaning "the same place" (occupying the same place in the periodic table).

hint. He found that in most cases the positive charge on nuclei increases with increasing atomic weight. Could this positive charge be the important criterion?

ATOMIC NUMBERS

Originally, the atomic number of an element was merely a serial number indicating its sequential position in the periodic table. In 1913 these numbers ran from 1 for the lightest element, hydrogen, to 92 for the heaviest then-known element, uranium.

In 1913–1914 the 26-year-old English physicist H. G. J. Moseley (1887–1915) made a discovery while systematically examining the X rays

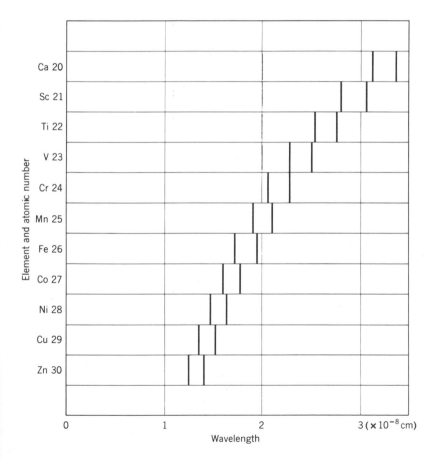

Fig. 11-2. X-ray spectra of various elements.

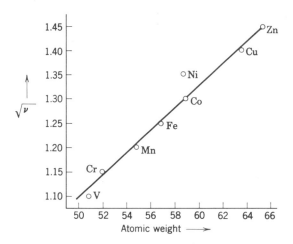

Fig. 11-3. Plot of the square root of observed X-ray frequencies versus atomic weights. Note that it is impossible to pass a straight line through all points.

emitted by different metals used as targets at the anode in modified cathode-ray tubes.[3] Max von Laue's discovery of crystal diffraction in 1912 proved that X rays are light waves of very short wavelengths. Crystal diffraction also provided a means of measuring the wavelengths or frequencies of X rays.[4]

Moseley found that X rays of definite frequencies were emitted by the target materials he was studying, and that different targets produced X rays of different frequencies (Fig. 11-2). In general, frequencies increased with increasing atomic weights (*see* Table 11-1).[5] In the hope of discovering a systematic relationship, Moseley tried various graphic approaches. When he plotted the square root of the observed X-ray frequencies against atomic weights, most of his points lay along a straight line (*see* Fig. 11-3). There were, however, a significant number of experimental points falling off the straight line. He tried several other plots with no better results. However, when he used atomic numbers—that is, the sequence numbers in the periodic table—instead of atomic weights, as the basis for his abscissa, *all* points lay along a straight line (*see* Fig. 11-4). He

[3] Moseley had been working with Rutherford. He was killed in action in World War I at the age of only twenty-eight.

[4] See Chapters 9 and 13.

[5] Actually, there are several X-ray frequencies emitted by each element. We are here concerned with the strongest emission in each case.

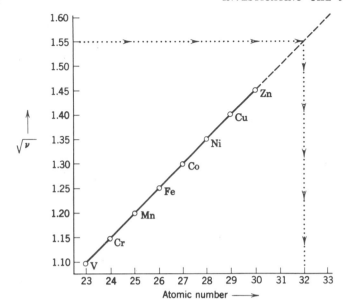

Fig. 11-4. Plot of the square root of observed X-ray frequencies versus atomic numbers. All points lie along a straight line. The dashed line is an extrapolation of the straight-line plot. The dotted lines indicate how the measured X-ray frequency of germanium ($\nu = 2.39 \times 10^{18}$ sec^{-1}, $\sqrt{\nu} = 1.55 \times 10^{9}$) is used to determine its atomic number—32 (*see* explanation in text).

therefore concluded that it is the atomic number rather than the atomic weight that is of prime importance in determining X-ray properties.[6]

From his graph Moseley could obtain an algebraic relationship between X-ray frequency and atomic number. He found that the following formula fitted his observed plot.

$$\sqrt{\nu} = 4.98 \times 10^{7}(Z - 1)$$

where ν is the observed X-ray frequency of a particular element, in cycles per second, and Z is the atomic number of the element.[7] Another way of putting the formula is:

$$\nu = 2.48 \times 10^{15} (Z - 1)^{2}$$

[6] In searching for a relation between the X-ray frequencies and *some* property of the atom, Moseley obtained a hint from Niels Bohr's (1865–1962) contemporary theory (1913) about the atom. In Bohr's theory the positive charge, Z, on the nucleus played a fundamental role in his attempt to explain the emission spectra of atoms. The development of Bohr's theory is the subject of Part Three.

[7] Note the analogy to the simple formula from your analytical geometry course in high school: $y = mx + b$, where m is the slope of the straight line and b is its intercept on the y axis.

Table 11-1. Wavelengths and Frequencies of X-ray Spectra of Several Elements[a]

(1) Element	(2) Observed frequency, ν (cycles/sec)	(3) Atomic weight (amu)	(4) Atomic number, Z	(5) $\sqrt{\nu}$
Vanadium, V	1.20×10^{18}	50.9	23	1.10×10^9
Chromium, Cr	1.31	52.0	24	1.15
Manganese, Mn	1.43	54.9	25	1.20
Iron, Fe	1.55	55.8	26	1.25
Cobalt, Co	1.68	58.9	27	1.30
Nickel, Ni	1.81	58.7	28	1.35
Copper, Cu	1.95	63.5	29	1.40
Zinc, Zn	2.09	64.4	30	1.45

[a] Note that the numbers in column 2 increase steadily going down the table, but the atomic-weight values of column 3 do not increase steadily (there is a reversal between cobalt and nickel). Apparently there is no simple direct relationship between X-ray frequency and atomic weight (also *see* Fig. 11-3). On the other hand, even a cursory examination indicates a direct relationship between the atomic numbers of column 4 and the square roots of the observed frequencies in column 5 (also *see* Fig. 11-4).

For example, for iron, atomic number 26, the formula predicts a frequency of

$$\nu = 2.48 \times 10^{15}(26 - 1)^2 = 2.48 \times 10^{15} \times 625$$

$$= 1.55 \times 10^{18}$$

This calculated frequency agrees with the observed frequency in Table 11-1.

Moseley's discovery for the first time provided a method of *directly* determining the atomic number of an element as follows: Obtain the element's X-ray spectrum; from the appropriate measured frequency, calculate the atomic number or obtain it graphically. For example, let us say that examination of the X-ray spectrum of the element germanium yields a frequency of 2.39×10^{18} cycles per second. We can calculate its atomic number by solving the first of our two formulas for Z and then substituting the measured frequency:

$$Z = 1 + \sqrt{\nu}/(4.98 \times 10^7) = 1 + 1.55 \times 10^9/(4.98 \times 10^7)$$

$$= 1 + 31 = 32$$

We can also obtain this result graphically, by (a) extending the diagonal solid line of Fig. 11-4 (the extension is indicated by a dashed line); (b) finding $\sqrt{\nu} = 1.55$ on the ordinate (vertical) axis; (c) drawing a horizontal (dotted) line from 1.55 until it meets the dashed line; (d) from there dropping a vertical (dotted) line down to the abscissa (horizontal axis). At that point, the atomic number reads 32.

Problem 11-3. Given 2.55×10^{18} cycles per second for the measured X-ray frequency of an element, determine its atomic number both graphically and by formula. What is the name of the element?

It should be evident that Moseley's discovery can be used as the basis for a new operational definition of atomic number. When the elements are arranged in the periodic table in the sequence of these new atomic numbers rather than atomic weights, the earlier-mentioned discrepancies disappear.

Since the atomic number derived strictly from X-ray measurements corresponded closely to the approximate nuclear charge that could be calculated from Rutherford's alpha-ray-scattering experiments, Moseley correctly surmised that his atomic number, Z, is equal to the positive charge on the atomic nucleus.

NUCLEAR STRUCTURE[8]

With existence of the nucleus clearly established, scientists were wondering what its structure might be. In 1815 William Prout had noted that most atomic weights appeared to be integral multiples of the atomic weight of hydrogen and he suggested the hydrogen atom as the basic building block of atoms (Chapter 4, p. 48). Prout's theory had to be abandoned in 1860 when J. S. Stas proved that the atomic weights of several elements were not integrals at all (e.g., boron 10.8 and chlorine 35.5).

But Prout's theory was revived in modified form after the discovery of isotopes in 1913. Now fractional atomic weights were explained as the result of mixtures of isotopes having integral atomic weights (e.g., boron-10 and boron-11). So, once more, a whole-number relationship was noted, only this time it applied to charges as well as masses: Nuclear masses and charges both appeared to be integral multiples of the mass and charge of the hydrogen nucleus—which is called a *proton*. Accordingly, Rutherford suggested the proton as a basic building block of all nuclei.

[8] This section may be omitted without loss of continuity.

Rutherford also postulated the existence of a neutral particle with a mass similar to that of the proton. This neutral particle would account for the nuclear mass number being greater than the charge number in all atoms except ordinary hydrogen. It would also explain the existence of *isotopes—two or more atoms of an element that have the same atomic number (or nuclear charge) but different atomic weights* (that is, different numbers of neutral particles in the nucleus). According to Rutherford's theory, then, the ordinary hydrogen nucleus would simply be a proton; that of heavy hydrogen, or deuterium (*see* p.   ), a proton plus a neutral particle. The helium nucleus, or alpha particle (charge 2, mass 4), would consist of 2 protons and 2 neutral particles. Ordinary lithium (charge 3, mass 7) would consist of 3 protons and 4 neutral particles; and so on.

Problem 11-4. Refer to a modern table of atomic weights and atomic numbers (*see* endpaper) and describe, according to Rutherford's theory, the composition of the following atoms in terms of protons, neutral particles, and orbiting electrons. (Remember that the atom as a whole is neutral.)

beryllium	chlorine-35
boron-10	chlorine-37
boron-11	lead-206
carbon	lead-207
nitrogen	lead-208
oxygen	silver
fluorine	gold
neon	radium
	uranium

In essence, Rutherford's ideas of nuclear structure form the basis of modern theory. The first direct evidence for the existence of protons in nuclei of atoms other than hydrogen was obtained by Rutherford himself in 1919. Using a cloud chamber, he observed that protons were ejected when certain elements—nitrogen, aluminum, fluorine, etc.—were bombarded with high-speed alpha particles. Neutral particles with zero charge and approximately the same mass as the proton were discovered in 1932, and named *neutrons*, by the English physicist Chadwick, a former assistant to Rutherford. Chadwick's search, you may recall (Chapter 6, p. 103) was guided principally by the Law of Conservation of Momentum.

Problem 11-5. Rutherford originally suggested that his neutral particle might be a proton-electron pair in close association. This might account for the emission of beta rays (high-speed electrons) from some radioactive nuclei. However, also discovered in 1932 by C. D. Anderson (American, 1905–   ) was the *positron*, a particle with a mass equal to that of the electron but with positive charge—in other words, a positive electron. Instead of considering the nucleus as made up of protons and proton-electron pairs (neutrons), would it not be just as reasonable to consider the *neutron* as the basic building block and the nucleus as made up of neutrons and neutron-positron pairs? On the basis of the following table of particle masses, argue the pros and cons of

(a) considering the neutron as a proton-electron pair, and

(b) considering the proton as a neutron-positron pair.

Particle	Charge (coulombs)	Mass (g)	Mass (amu)
Proton	$+1.60210 \times 10^{-19}$	1.67252×10^{-24}	1.007277
Neutron	0	1.67482×10^{-24}	1.008665
Electron	-1.60210×10^{-19}	9.1091×10^{-28}	5.4860×10^{-4}
Positron	$+1.60210 \times 10^{-19}$	9.1091×10^{-28}	5.4860×10^{-4}
H atom	0	1.67343×10^{-24}	1.007825

Actually neither idea has been found tenable, for reasons beyond the scope of this text. Today the neutron is accepted as a particle in its own right.

Problem 11-6.   (a) When a radioactive radium atom emits alpha particles (helium nuclei), how many protons and neutrons will remain in the nucleus?

(b) What, then, is the atomic number of that atom?

(c) Into which element is radium transformed by the emission of an alpha particle?

(d) With the help of a periodic table, predict the combining power of this new element. Is it a metal or a nonmetal? A solid, a liquid, or a gas?

(e) Comment briefly but critically on the definition of *element* given in Chapter 1, p. 5.

SUMMARY OF PART TWO

As envisioned about 1913 the atom consisted of a tiny but massive nucleus surrounded by distant orbiting electrons, much like a miniature solar system. The electrons possessed negative electric charge, whereas the nucleus possessed positive charge and apparently consisted of positive as well as neutral particles.

In tracing the evolution of the discovery of this nuclear atom, which is largely electrical in nature, we began with early researches in static electricity, among them Coulomb's Inverse-Square Law of Electrostatic Forces (1785). After Volta's invention of the electric battery in 1800, scientists began to investigate the properties of electric currents.

The subsequent discovery, also in 1800, that electric currents can decompose matter—in a process called electrolysis—hinted strongly at the electrical nature of chemical bonds. Faraday's measurements in 1833 indicated that at least some substances consist of positively and negatively charged atoms, called ions, and that the charges on these ions come in definite-sized packages. Thus Faraday's experiments in electrolysis suggested both the electrical nature of atoms as well as the atomistic nature ("graininess") of electricity.

The atomicity of electricity was confirmed through Thomson's identification in 1897 of the electron as a particle with a definite mass-to-charge ratio and through Millikan's determination in 1910 of its charge. Millikan's value of the electron's negative charge is equal in magnitude to the charge of singly-charged ions (positive or negative), which can be calculated from Faraday's electrolysis measurements and Avogadro's number (first determined by Loschmidt in 1865). Apparently this quantity of charge is the smallest charge that is found in nature.

There is much additional evidence that electricity is contained within the atoms, such as: (1) emission of electrons in the form of cathode rays by all kinds of electrode material (Plücker, 1858; Crookes, 1875; et al.); (2) the discovery of positive ions (canal rays) in cathode-ray tubes (Goldstein, 1886); (3) emissions of positive and negative particles from radioactive substances (Rutherford, 1899).

Work with cathode rays led to the accidental discovery of X rays (Röntgen, 1895), which in turn triggered the discovery of radioactivity (Becquerel, 1896); and it was the employment of alpha rays from radioactive sources that led Rutherford to realize in 1911 that atoms consist of tiny nuclei and lots of empty space.

The presence and the number of positive and neutral particles in the nucleus were revealed through the discovery of isotopes and of atomic numbers, both in 1913.

Isotopes were found in radioactive-decay products by Soddy and Richards and in canal-ray beams by Thomson. Isotopes not only explained fractional atomic weights but also helped eliminate atomic weight as the principal criterion for differentiating elements (since an element could have several atomic weights, and different elements could have identical atomic weights).

Atomic numbers were determined through Moseley's clever detective work involving X rays from cathode-ray tubes. These atomic numbers then replaced atomic weights as the index criterion for sorting out the elements, eliminating some discrepancies in the periodic table. Even more important, Moseley's atomic numbers were recognized as the positive charges of atomic nuclei.

In a nutshell, the nuclear atom as it was envisioned by 1913 had a diameter of about 1 to 5 angstroms and a tiny nucleus (diameter on the order of 10^{-4} A) consisting of Z positive particles (protons) and a similar number of neutral particles (neutrons) of approximately the same mass.[9] This nucleus contained most of the mass of the atom. The electroneutrality of the atom as a whole was preserved by Z negative, lightweight particles (electrons)—1/1840 of the proton's mass—moving in unspecified orbits around the nucleus.

The elucidation of the atom's electron structure—which hopefully would explain the properties of the elements and their periodicity—is taken up in Part Three.

ANSWERS TO SELECTED PROBLEMS

11-1(a). Let the percentages equal a and b, respectively.
   Then $(206a + 208b)/100 = 207.2$
   Since b = 100-a, the equation becomes:
   $206a + 208 (100-a) = 207.2 \times 100$
   $206a - 208a = 20{,}720 - 20{,}800$
   $2a = 80$; a= 40%, b = 60%.
   Or, to simplify we could reduce all weights by 200. Then, $6a + 8 (100-a) = 7.2 \times 100$; $-2a = 720-800$; $2a = 80$; a = 40%.

11-2. Ne-20, 90%; Ne-22, 10%.

11-3. Element 33, As.

11-6. $^{226}_{88}\text{Ra} - ^{4}_{2}\alpha \rightarrow ^{222}_{86}\text{Rn}$

---

[9] The actual discovery and naming of the positive and neutral particles occurred between 1919 and 1932.

ADDITIONAL READING FOR PART TWO

Anderson, David L. *The Discovery of the Electron: The Development of the Atomic Concept of Electricity,* Princeton, N.J.: D. Van Nostrand Company, 1964 (paperback). Includes a good brief history of the discovery of X rays, some of it in Röntgen's own words.

Astin, Allen V. "Standards of Measurement." *Scientific American,* June 1968.

Halliday, David, and Robert Resnick. *Fundamentals of Physics,* New York: John Wiley and Sons, 1970. A college physics textbook in two volumes.

Lemon, Harvey B. *From Galileo to the Nuclear Age,* Chicago: University of Chicago Press, 1946 (paperback).

McDougal, W. *Fundamentals of Electricity.* American Tech. Soc., 1954.

Overman, Ralph T. *Basic Concepts of Nuclear Chemistry.* New York: Reinhold Publishing Corp., 1963 (paperback).

Peierls, R. E. "The Atomic Nucleus." *Scientific American,* January 1959; Freeman Reprint No. 235.

Roller, Duane, and Duane H. D. Roller. *The Development of the Concept of Electrical Charge* (Case 8 of the Harvard Case Histories in Experimental Science). Cambridge: Harvard University Press, 1954.

Also see some of the references cited at the end of Part One, particularly: Young, *The Mystery of Matter,* which contains a brief history of the discovery of radioactivity written by Alfred Romer and a short account by Marie Curie of the discovery of radium.

PART THREE

Elucidation of the Electron Structure of the Atom

The foregoing chapters trace the developments that led to the nuclear model of the atom in 1913—an atom consisting of a dense nucleus of positive charge Z, surrounded by Z negative electrons in unspecified orbits. Since chemical reactions involve the making and breaking of bonds between atoms, it is presumably the parts of the atoms that touch each other—that is, the outer parts—that play the important roles and therefore govern chemical properties—bond strength, relative stability, etc. Consequently, chemists were vitally interested in learning more about the arrangement and properties of the orbiting electrons.

An ingenious and very useful model of these electron orbits was proposed by Niels Bohr in 1913, the same year that Moseley published his work on atomic numbers. The Bohr model of the atom, which has been so helpful to chemists, was devised not to meet the needs of chemists but primarily to explain the optical spectra of elements. Even though the principal events in the developmental chain took place concurrently with those of the period described in Part Two, they can be detailed as a separate development.

Inasmuch as most of the evidence supporting Bohr's atom model came from a study of atomic spectra, Part Three begins with an attempt to understand the nature of light and radiations in the electromagnetic spectrum. We shall examine the basic implications and applications of such phenomena as reflection, refraction, diffraction, and interference. We shall find out what a spectrum is and how we can use the spectroscope, not only for chemical analysis, but also for investigating the internal structure of the atom.

Throughout Part Three it is fascinating to observe the continual interplay between observation and theory, culminating in Bohr's clever atom model.

Chapter 12

THE NATURE OF LIGHT

NEWTON'S EXPERIMENTS WITH PRISMS

In 1666 Isaac Newton (1642–1727), then only twenty-four years old, began a systematic study of light. The first telescope had been built by Galileo about half a century earlier. To meet the demands of the increasingly exact science of astronomy, Newton tried to improve optical lenses to obtain better images, but was rewarded by only slight improvement. He conjectured that the trouble did not lie in the lenses but in light itself, and thus he began his exploration into the nature of light.

It was well known in the seventeenth century that sunlight passing through glass or water frequently yielded the various colors of the rainbow; it was assumed that the colors were somehow produced by the glass or the water. Newton, however, through a carefully planned series of experiments, showed that the colors are ingredients of sunlight itself.[1]

First, he admitted a beam of sunlight into a darkened room through a small round hole in a window shade and passed the light beam through a glass prism set up before a screen; as expected, this arrangement yielded the well-known colors of the rainbow on the screen (Fig. 12-1). Then, removing the screen, he passed the colors through a second, inverted prism; the colors thereupon recombined into white light (Fig. 12-2). Similar results were obtained with white light from other sources, such as candles.

To convince himself that colors are inherent in white light itself and not somehow manufactured by prisms, Newton performed two control

[1] The author is proud to report that the first recorded observation of the solar spectrum was made by one of his ancestors, also recognized as the world's first fermentation chemist. (*See* Genesis 9:13-21.)

167

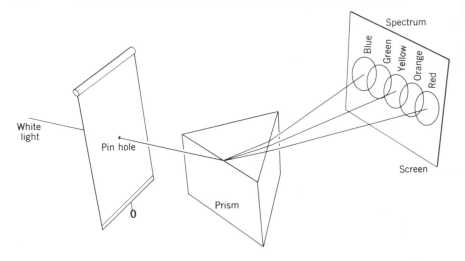

Fig. 12-1. Refraction and dispersion of white light by a prism.

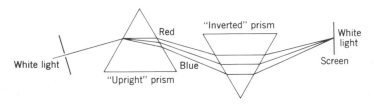

Fig. 12-2. Recombination of colors by a second prism to form white light.

experiments. In the first control experiment he used the inverted-prism arrangement pictured in Fig. 12-2, but interposed a screen with a small hole between the two prisms, so that only one of the colors—say, red—could pass through the inverted prism (Fig. 12-3). This setup did *not* yield white light from the second prism, unlike the setup of Fig. 12-2. Evidently the second prism could not manufacture white light from light of a single color.

In a second control experiment Newton reinverted the second prism into the "upright" position (Fig. 12-4). The color of the light passing through this prism remained unaltered. Evidently prisms do not produce rainbow colors from a one-colored ("monochromatic") light ray.

On the basis of these experiments Newton concluded that white light is a mixture of many colors and that prisms merely break up, or *disperse,* this mixture into a band of colors, now known as a *spectrum.*

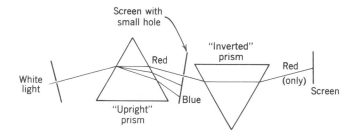

Fig. 12-3. A second prism cannot form white light from red light alone.

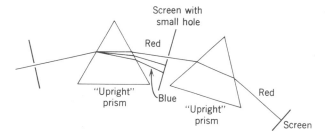

Fig. 12-4. A prism cannot produce a spectrum from light of one color. That is, a prism cannot disperse monochromatic light.

Prisms disperse white light into a spectrum by bending light. Blue light is bent more than red and thus the two colors appear at different locations on the screen (Fig. 12-1). When red light is passed through the prism (Fig. 12-3), the screen will have a red image of the original round hole. With blue light, a blue image will appear at a slightly shifted position. But when white light is passed through, the screen will have overlapping, varicolored images of the round hole, as in Figure 12-1.[2]

THE CORPUSCULAR THEORY

Newton was essentially an experimental scientist, and because his observations have withstood the test of time, they are now regarded as scientific "laws,"—e.g., his famous laws of motion. But he was skating on thin ice when he offered a theoretical explanation for his observations with prisms. Newton thought of light as a stream of tiny particles, or

[2] This discussion will help us understand the discovery made by Wollaston and Fraunhofer, who replaced the round entrance hole with a thin slit (Chapter 14).

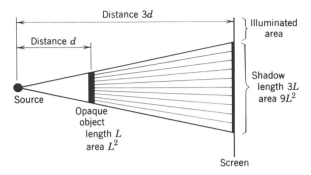

Fig. 12-5. Rectilinear propagation of light and the Inverse-Square Law of Light Intensity.

corpuscles, sent out by the emitting body and traversing space at a very high but finite speed.[3] Such a picture allows a number of logical predictions which are in accord with observation.

First of all, the corpuscular theory logically predicts rectilinear propagation of light; i.e., the little particles would travel out from the source in straight lines. We know this is true from everyday experience. For example, objects illuminated by a point source of light cast sharp shadows (Fig. 12-5).

The Inverse-Square Law of Light Intensity follows directly from Figure 12-5. We know that light intensity decreases with increasing distance from the source, such as a candle. (This principle does not apply to collimated or focused-light beams.) A light beam, or stream of particles, sent out spherically by a source covers a certain area at a given distance from the source, for example, a square surface of length L and area L^2. Then at triple the distance, the beam will have fanned out to illuminate a surface of length $3L$, and therefore area $(3L)^2$, or $9L^2$. In other words, at triple the distance, the same number of particles are spread over an area nine times as big with only one-ninth the intensity. The *Inverse-Square Law of Light Intensity* states that

> *light intensity varies inversely with the square of the distance from the light source.*

The well-known *Law of Reflection—*

> *angle of incidence equals angle of reflection—*

follows clearly when we think of light as particles, or perhaps as tiny tennis balls (Fig. 12-6).

[3] At about this time, in 1676. Olaf Roemer made the first determination of the velocity of light (*see* next section).

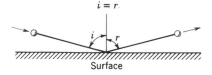

Fig. 12-6. The Law of Reflection: angle of incidence equals angle of reflection.

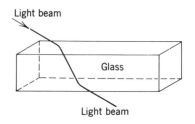

Fig. 12-7. Refraction of a light beam by a flat piece of glass. Note that the emerging beam is displaced from the entering beam but is parallel to it.

How did Newton explain *refraction,* the bending of light as it passes from one medium into another? As light particles traveling in air approach a denser medium, they would be attracted by it and would speed up. An obliquely approaching particle would thus be deflected toward the surface. In Figure 12-7 the deflection is downward, toward the perpendicular, like the path of a falling body. On emerging from the denser medium, the beam would once more be deflected toward the surface— this time away from the perpendicular, so that the emerging beam in Figure 12-7 follows a path parallel to the entering beam.

If this explanation were correct, light would travel faster in a denser medium: A denser medium's attraction would speed up approaching particles and slow them down again after emergence. Light thus would travel most slowly in a vacuum, faster in air, and still faster in glass or water. This conclusion is of course subject to experimental verification, but almost two centuries passed before experimental methods were developed to allow a test of Newton's prediction. In 1850 Foucault measured the speed of light in various media (*see* p. 173) and found Newton to be wrong. With Foucault's work the particle, or "corpuscular," theory became completely untenable.

THE SPEED OF LIGHT

In the seventeenth century, scientists wondered whether light travels with finite speed or propagates instantaneously. Galileo attempted to measure the speed of light but failed.

The question was settled in 1676 by the Danish astronomer Olaf

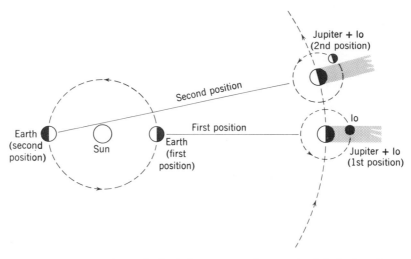

Fig. 12-8. Roemer's method of determining the speed of light by observing the eclipses of Jupiter's moons.

Roemer (1644–1710), who obtained a fairly accurate value by a simple method. Io, one of the orbiting moons of Jupiter (discovered by Galileo in 1610, on the first evening he used his newly constructed telescope) disappears behind the parent planet at regular intervals averaging 42.5 hours. This time interval is slightly greater during half of our earth year, and slightly shorter during the other half. Roemer found that the accumulated delay (relative to the yearly average) during the slower half year is 22 minutes (16.6 minutes by modern measurements). He ascribed this delay to the added distance the light had to travel as the earth moved in its orbit around the sun, from the point closest to Jupiter to the point farthest away. In other words, 22 minutes represented the time required for light to travel the additional distance across the earth's orbit (*see* Fig. 12-8).

From the then-accepted value of the earth's orbital diameter (174 million miles), Roemer calculated the speed of light, c, as follows:

$$c = \frac{\text{distance}}{\text{time}} = \frac{174 \times 10^6 \text{ miles}}{22 \times 60 \text{ seconds}} = 132{,}000 \text{ miles per second}[4]$$

[4] A value of the earth's orbital radius—i.e., the earth-sun distance—had been obtained only four years earlier in a rather interesting manner. In 1618 Johannes Kepler had devised a scale model of the solar system (Fig. 12-9). From this model the various distances in the solar system could be computed as soon as one distance had been measured. The required measurement was made in 1672 by the Italian astron-

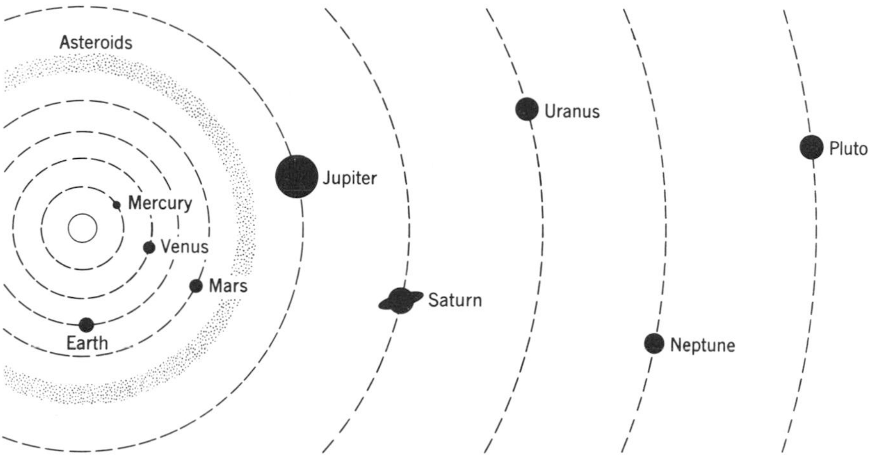

Fig. 12-9. The solar system.

Problem 12-1. From today's more accurate figures of 186 million miles for the earth's orbital diameter and 16.6 minutes (1000 seconds) for the time variation in the satellite's eclipses, calculate the correct speed of light. What was the percentage of error in Roemer's value?

Many other scientists have since measured the speed of light and obtained values in essential agreement with Roemer's. The first purely terrestial determination of the velocity of light was made in 1849 by the French physicist A. H. L. Fizeau (1819–1896). His method (*see* next paragraph) was improved in the following year by another Frenchman, J. B. L. Foucault (1819–1868), who found that light travels more slowly in water than in air. This finding was in definite conflict with the prediction of the corpuscular theory but was in accord with that of the wave theory (*see* pp. 184–186). Foucault's measurement was able to convert to the wave theory even the staunchest adherents of the old corpuscular theory.[5]

---

omer G. D. Cassini (1625–1712), who determined the Earth-Mars distance by the parallax method. This method involved measurement of the apparent displacement of Mars, relative to the stars, as seen from two locations on Earth (like the apparent displacement of objects viewed first through one eye, then through the other). The Earth-Mars distance could then be computed by a simple trigonometric formula. From the Earth-Mars distance the Earth-Sun distance was calculated as 87 million miles. Today's figure is 39 million miles, verified by several other methods of measurement.

[5] From the observation that light is bent toward the perpendicular on entering a denser medium (say, water) Newton had predicted that light corpuscles travel faster in that medium—and later was proved wrong. (Footnote cont. on next page.)

The Polish-born American physicist A. A. Michelson (1852–1931) further improved the Fizeau-Foucault method. From about 1880 Michelson devoted his life to ever more accurate determinations of the speed of light. Michelson's method was as follows. A beam of light is reflected from a rotating octagonal mirror (e.g., on Mt. Wilson, California) to a plane mirror several miles away (Mt. San Antonio, 22 miles distant) and from there is reflected back to the octagonal mirror and into a telescope or detecting instrument (Fig. 12-10). In order to be able to reflect the beam into the telescope, the octagonal mirror must be in the proper posi-

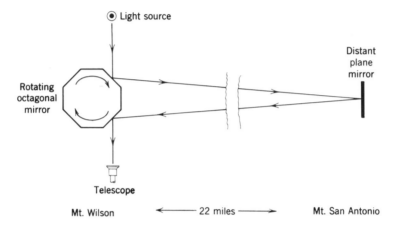

Fig. 12-10.   Michelson's method of measuring the speed of light.

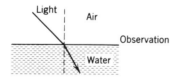

Let us do the converse. From the new observation that light travels more slowly in water, let us try to predict how it should be bent *if* it were corpuscular. How is a big particle, such as a ball, deflected when entering water? *Away* from the perpendicular!

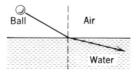

The observed deflection of the ball is opposite that observed for light beams. Conclusion: Light does *not* behave like balls or particles.

tion. That is, it must have rotated a certain fraction of a turn during the time the light traveled to and from the distant mirror. From the values of the mirror's rotational rate at synchronization and the exact path length of the light, it is easy to calculate the speed of light.

Today's best value of the speed of light, c, is:

$$2.9979250 \times 10^{10} \text{ cm/sec}$$

In round figures, this is

$$3 \times 10^{10} \text{ cm/sec, or } 186,000 \text{ miles/sec}$$

"The speed of light, c," always refers to the speed in vacuum or free space, unless it is specifically stated otherwise. This speed in vacuum is independent of the color of the light.[6]

In media other than a vacuum, the speed of light is less than c. The greater the "density" (from an optical point of view) of the medium, the less the velocity. Light travels fastest in a vacuum; in air, light is slowed down by less than 0.03%; but in diamond, one of the optically "densest" substances, the speed of light is reduced to almost half.

Two points should be emphasized: The speed of light in a given medium is constant; and light, having entered a medium, regains its original speed after emerging from the medium. These statements may be deduced from two observations on refraction: The amount of bending (refraction) due to the slow-down effect depends on the optical "density" of the medium, but not on its thickness; and the angles of bending at entry and exit are equal but opposite.

The precise speed of light in various media (but not in a vacuum) depends on the color of the light. In most substances colors near the red end of the spectrum travel faster than colors near the blue end. In water the speed of red light is about 2.5×10^{10} cm/sec; that of blue light, 2.0×10^{10} cm/sec. Thus in refraction blue light usually is bent more than red light, as Newton observed (Fig. 12-1).

DIFFRACTION

Rectilinear propagation, the Inverse-Square Law of Light Intensity, reflection, refraction, and dispersion can be explained somehow by Newton's corpuscular model of light. But they do not prove the correctness

[6] Further experimental evidence (chiefly the famous Michelson-Morley experiment of 1887) indicates that this speed also is independent of the motion of the source or the observer. This observation became one of the cornerstones of Einstein's Theory of Relativity (1905). According to that theory no signal can be transmitted by any means whatsoever, in free space or in a material medium, at a speed greater than the speed of light, c. The speed of light in a vacuum, c, is considered one of the fundamental constants of nature.

of the model. As a matter of fact, experiments suggesting the model was wrong had already been performed before Newton began his work on light. These experiments, posthumously published by an Italian professor of mathematics, Francesco Grimaldi (1618–1663), revealed that light does *not* always travel in exactly straight lines.

In one of Grimaldi's experiments a narrow beam of sunlight, admitted through a small hole in a window shade, was passed through another tiny hole in a screen. The illuminated area on the opposite wall was slightly larger than could be explained by rectilinear propagation (Fig. 12-11).

In another experiment the shadow cast by a small object was smaller than expected (Fig. 12-12), indicating that light had been bent around the edges of the object. This bending of light by the edges of obstacles is known as *diffraction* (not to be confused with refraction, which is the bending of light as it passes from one transparent medium to another). Grimaldi also noticed that the areas AC and BD in both experiments (Figs. 12-11 and 12-12) were fringed with colors. It appears, then, that dif-

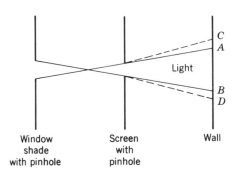

Fig. 12-11. What Grimaldi observed: light is bent around the edges of obstacles—the observed area of illumination (CD) is larger than the predicted area (AB).

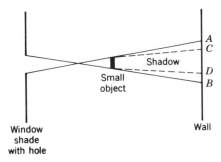

Fig. 12-12. Light is bent around the edges of obstacles: The observed shadow area (CD) is smaller than predicted (AB).

fraction, as well as refraction, can result in dispersion of light (production of spectra).[7]

From the foregoing discussion we see that the Law of Rectilinear Propagation is only an approximation, subject to restrictions and refinements discovered by further experimentation. This is so with many of our "laws" of nature, among them: the Law of Conservation of Mass (which neglects the mass equivalence of energy, $E = mc^2$); the Law of Constant Composition (which ignores the existence of isotopes); the equal-volume-equal-numbers principle (which is true only for low-pressure gases); the statement that atomic weights are whole numbers (exceptions were explained as mixtures of isotopes, but more refined mass-spectrometric work has shown that the atomic weights of even isotopically pure elements are not exact integers).

THE WAVE THEORY

Newton was content to ascribe Grimaldi's observation of the very slight bending of light as it passed the edge of an obstacle to some sort of deflection of the particles. Grimaldi extracted a different interpretation from his experimental results—a wave theory of light; but he was not taken seriously, perhaps because from his grave he could not compete with Newton's prestige. The Englishman Robert Hooke (1635–1703) independently discovered diffraction and also proposed a wave theory. And in 1678, the Dutch mathematician and physicist Christian Huygens (1629–1695) cited the extremely high speed of light (determined by Roemer in 1676) as an argument against Newton's corpuscular theory and presented a more comprehensive wave theory that satisfactorily explained the known properties of light.[8]

The real clue to the wave theory is the phenomenon of diffraction. A detailed explanation of diffraction would become too involved, but we can explain it qualitatively by analogy to water waves. We take diffraction of water waves for granted. For example, waves bend around a small obstacle in their path—its "shadow" does not extend indefinitely (Fig. 12-13). The diffraction of light waves by a small aperture is analogous to a water wave passing through a small gap in a sea wall; the wave in the gap forms a hump, or swell, which then spreads out once

[7] You can see a diffraction pattern if you look at a bright light through a very narrow slit—e.g., the space between two almost-touching fingers or a razor cut in an index card. You will see a series of alternating light and dark bands and, if you do it carefully, some tinges of color.

[8] Waves can propagate much faster than the medium in which they travel. For example, winds seldom exceed 50-100 miles per hour, but sound waves move through air at approximately 750 miles per hour.

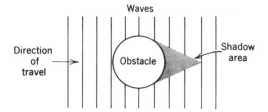

Fig. 12-13. Water waves bend around obstacles. Undisturbed (shadow) area does not extend indefinitely.

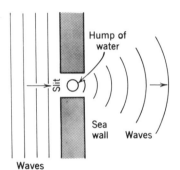

Fig. 12-14. Water waves spread out again after passing through a gap in a sea wall.

more after passing through the gap (Fig. 12-14). The smaller the hole relative to the *wavelength* (i.e., the distance interval between crests or between troughs), the *greater* the angle of diffraction, θ (*see* Figs. 12-15 and 12-16). In order to produce noticeable diffraction, the size of the hole must be of approximately the same order of magnitude as the wavelength.

Problem 12-2. (a) Would waves of greater wavelength be diffracted more or less or the same by a given opening?
(b) According to the wave theory, the wavelength of red light is approximately twice that of blue light. Which would be bent more by the diffraction process? How does this compare with the bending produced by refraction?
(c) Explain why Grimaldi observed colors in his experiments. Describe the sequence of these colors on the basis of your answer to (b).

INTERFERENCE

The next hundred years brought no significant experiments on light, until Young's discovery of interference in 1800. Thomas Young (1773–

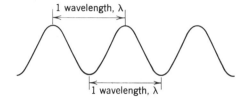

Fig. 12-15. A wavelength is the distance between two crests or two troughs.

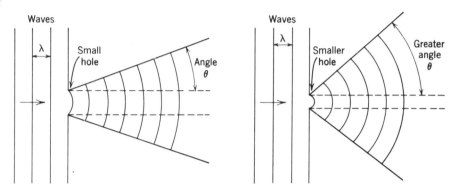

Fig. 12-16. The smaller the aperture relative to the wavelength, λ, the greater is the angle of diffraction, θ.

1829), an English physician and amateur physicist, was interested in the physiology of the eye. Believing the corpuscular theory to be inadequate, he performed an experiment that tipped the balance in favor of the wave theory for a century (until the discovery of the photoelectric effect by Hertz in 1887 and its explanation by Einstein in 1905—*see* Chapter 17).

Young passed a beam of light from a small hole through *two* pinholes. Where the light from the two pinholes overlapped on a screen, alternating bands of light and darkness appeared (Fig. 12-17).

In the area of overlap the two beams of light somehow interact to produce light and dark bands. This interaction of light beams, usually from a common source (the first pinhole), is called *interference*. In some regions the light beams seem to cancel each other (!), resulting in darkness; this is called *destructive interference*. In between, they reinforce each other in *constructive interference*.[9]

[9] You can see what Young saw if you squint through your eyelashes at a thin-line source of light parallel to your lashes, such as the space between two almost-touching fingers or a razor cut on an index card (*see* footnote 7). The slit corresponds to Young's first hole, and the spaces between your eyelashes represent multiple pinholes.

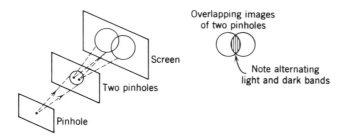

Fig. 12-17. Young's experiment with two pinholes, demonstrating interference of light waves.

Even though the corpuscular theory was able to explain the various phenomena previously described (including diffraction, after a fashion), it simply could not explain destructive interference. How can light plus light equal darkness? How could two light corpuscles cancel each other? Young reached the conclusion that light has wave character.

But the time still was not ripe for discarding the corpuscular theory. (Remember, at that same time Dalton was advancing his corpuscular theory of matter.) Interference of light was rediscovered in 1815 by the French physicist Augustin Fresnel (1788–1827), and in 1817-1819 he and Young separately developed a mathematical wave theory to explain diffraction and interference. Even so, some scientists refused to abandon Newton's corpuscular theory of light until Foucault measured the speed of light in water in 1850.

It is one of the paradoxes in the development of science that the great genius Newton, who did so much in the field of experimental optics, could through a wrong theory delay our understanding of his discoveries.[10] It should be added, however, that Newton did not *push* his theoretical views. If his theory really did retard progress in optics, the fault lay with those who attached too great a weight to his opinions, perhaps because they had retained something of the medieval respect for authority.

Today we consider the observation of diffraction and interference phenomena as sufficient evidence to establish the existence of waves.

[10] Dalton, the father of the atomic theory, also retarded the further development of the atomic theory for nearly a half century by misinterpretating the Law of Combining Gas Volumes, and during that time even caused scientists to doubt the reality of atoms themselves.

INTERFERENCE EXPLAINED

We can best explain interference phenomena by again referring to water waves. Water waves entering two channels will simply travel along. If these two channels then recombine into a single channel, two extreme situations may occur.

First, if the separate channels are of equal length, the crests of waves which enter the two channels simultaneously will arrive at the junction at the same time—i.e., *in phase*. They will reinforce each other and result in waves of greater intensity, or amplitude—constructive interference (Fig. 12-18).

Problem 12-3. What happens if the two channel lengths differ by one wavelength? By two wavelengths? By three?

Second, if one path is longer than the other by exactly half a wavelength (or 3/2, or 5/2, etc.), the trough of a wave from one path will meet a crest from the other; the waves are *out of phase* and will cancel or destroy each other—destructive interference. As a result, the water in the combined channel will be relatively quiet (Fig. 12-19).

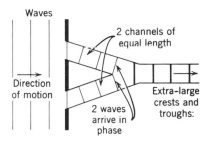

Fig. 12-18. Constructive interference of water waves.

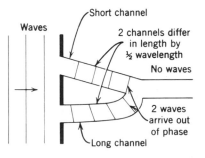

Fig. 12-19. Destructive interference of water waves.

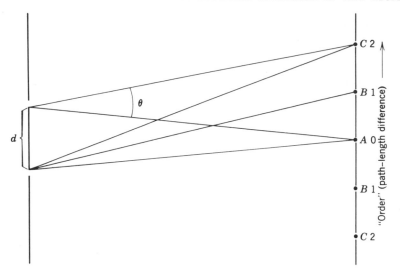

Fig. 12-20.   Explanation of interference pattern in Young's experiment.

In Young's two-hole experiment the bright portions on the overlapping images on the screen are caused by two light beams whose path lengths are either equal or differ by whole wavelengths. Midway between each pair of bright areas are dark areas due to light waves arriving out of phase, when the path lengths differ by 1/2, 3/2, 5/2, etc. wavelengths. In Figure 12-20, A, B, and C represent bright areas on the screen. At A the two light paths from the two are equal in length; at B they differ by one wavelength; at C, by two wavelengths; etc. In the area halfway between A and B, the path-length difference is 1/2 wavelength; between B and C, 1 1/2 (i.e., 3/2), etc. (A is called the zero-order image; B and C are called, respectively, first-order and second-order images.)

Through trigonometry it can be shown that at B the difference in path-length from the two holes or slits (namely, one wavelength, λ) equals d sin θ, where d is the distance between the pinholes or slits and θ is the angle of diffraction. For any image—say, the nth-order image—the formula becomes

$$n\lambda = d \sin \theta$$

Measuring the diffraction angle θ thus enables us to determine wavelengths. Young actually used this method to calculate the wavelengths of various colors of light. He found blue light to have the shortest wavelength, and red the longest:

blue—4×10^{-5} cm $= 4,000 \times 10^{-8}$ cm $= 4,000$ A

red—8×10^{-5} cm $= 8,000$ A[11]

THE DIFFRACTION GRATING

Our discussion of interference thus far has assumed light of only one wavelength; i. e., *monochromatic light.* If white light is used instead, the different wavelengths, λ, will reinforce each other at slightly different positions, wherever $n \lambda = d \sin \theta$, or $\sin \theta = n \lambda/d$. Thus each area ($B$, C, etc.) will have a spread of colors, in other words a spectrum.

We have, then, two ways of dispersing light and producing a spectrum: (1) via refraction by a prism; and (2) via diffraction and interference effects, by passing light through holes or slits.

Problem 12-4.   (a) In the first-order spectrum at B of Figure 12-20, which color is on the side toward A and which is toward C?
(b) What is the color of the light spot at A? (Does a spectrum appear at A?)

The spectrum from a double-hole or double-slit arrangement has some serious limitations. First, two small slits allow only very little light to pass through. Second, the available light is distributed over several orders of images. Thus any one spectrum is weak and difficult to observe.

The intensity of the spectra is improved when more than two slits are used, as was shown around 1820 by the German optician Joseph Fraunhofer (1787–1826). He made *transmission gratings* as fine as 200 "slits" per centimeter by winding thin silver wire around two fine parallel screws (Fig. 12-21). Light passing through these multiple "slits" between the wires yielded brighter and more sharply defined diffraction patterns.

Later, Fraunhofer devised a ruling machine containing a diamond stylus, with which he scratched up to 3000 lines per centimeter on a sheet of glass. Such a transmission grating allows light to pass through the clear glass between the scratches. The smaller the distance between adjacent scratches, the more spread-out the spectrum will be; for this reason, gratings with 20,000 lines per centimeter are in common use today. For best results the grating interval should be of approximately the same order of magnitude as the wavelengths of the light to be dispersed (*see* p. 178).

[11] (One angstrom equals 10^{-8} cm—*see* p. 144.) The wavelengths of other colors of the visible spectrum lie between these extremes.

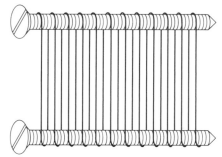

Fig. 12-21. Fraunhofer's first diffraction grating: silver wire wound around two screws.

Problem 12-5. (a) In Fraunhofer's wire grating of 200 wires per centimeter, what was the interval between wires in centimeters?

(b) What is the spacing of a 3000-lines-per-centimeter grating, in centimeters? In angstroms?

(c) What is the angstrom interval of a 20,000-per-centimeter grating? How does this compare to the wavelength of red light?

Diffraction effects and light dispersion may also be obtained with a *reflection grating.* In such a grating light from a single source is reflected by a group of narrow mirrors, instead of being transmitted through a group of narrow slits. Reflection gratings can be made by ruling lines on silvered-glass surfaces. The spaces between lines represent multiple mirrors, analogous to the multiple slits of diffraction gratings. Intensity loss by absorption is less in reflection gratings because the light does not have to pass through glass or other material.[12]

REFRACTION EXPLAINED BY WAVE THEORY

Two phenomena—the bending of light toward the perpendicular and its deceleration upon entering a denser medium—when considered together conflict with the corpuscular theory. How, then, are they explained by the wave theory?

We define a *wave front* as the tangent drawn to the foremost points of propagating waves. The wave front of a series of ocean waves is the first wave itself (Fig. 12-22). Fig. 12-23 shows wave fronts at two successive

[12] A long-playing record will serve as a crude reflection grating. Hold an LP record almost edgewise so as to reflect light from a bright source, and you will see several spectra. Peacock feathers and some insect wings have ribbed structures that also produce colors by acting as diffraction gratings.

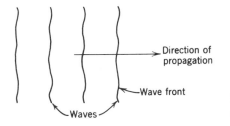

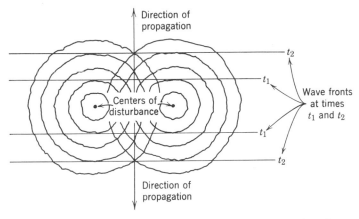

Fig. 12-22. Wave front of a series of ocean waves.

Fig. 12-23. Wave fronts at two successive times of a disturbance originating from two points.

times of a disturbance originating from *two* points. Note that wave fronts are straight lines that travel in a direction perpendicular to themselves.

Light, as we have learned, travels more slowly in a dense medium such as glass. Let us consider what happens when the wave front of a beam of light passes through a flat piece of glass where both surfaces are perpendicular to the beam (Fig. 12-24). The lines representing successive wave fronts at equal time intervals are compressed in the glass and return to their original spacing outside. In a way, this is like the compression and re-expansion of a column of automobiles on a highway as they pass through an obstruction, such as a tunnel or a bridge.

The beam is deflected when the surfaces of the glass are not perpendicular to the direction of propagation. In that case, one side of the beam reaches the surface first and is retarded before the other; the opposite takes place when the beam emerges (Figs. 12-25 and 12-26).

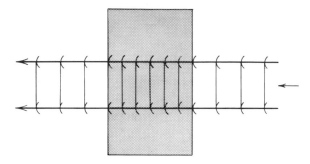

Fig. 12-24. A beam of light passing through a planar, perpendicularly oriented piece of glass. Successive wave fronts are compressed and reexpanded. Note that there is no refraction (bending) and consequently no dispersion.

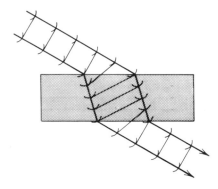

Fig. 12-25. Refraction of light by a planar piece of glass. Note that the entrance and exit beams are parallel (also *see* Fig. 12-7).

Different colors, or wavelengths, of radiation are retarded by different amounts when they enter another medium, and thus are bent by different amounts. This fact accounts for the dispersal of white light into its constituent colors by refraction in a prism.

Problem 12-6. Will a beam of white light be dispersed into a spectrum by a flat piece of glass whose surfaces are not perpendicular to the beam, as in Figure 12-25? Do you normally see spectral colors when you look at a street light through a window pane?

SUMMARY

In 1666 Newton began a series of experiments on the *refraction* of light through prisms. He concluded that white light is a mixture of many

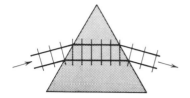

Fig. 12-26.   Refraction of light by a prism.

colors; prisms bend blue light more than red light and thus *disperse* white light into a band of colors called a *spectrum.*

To explain the known properties of light, Newton proposed a *corpuscular theory,* according to which a light beam consisted of a stream of tiny particles, or *corpuscles,* which travel at very high speeds.

The *speed of light* was first determined by Roemer in 1676 on the basis of astronomical observations. The modern value for the speed of light in a vacuum is approximately 3×10^{10} cm/sec, or 186,000 miles/sec.

In media other than a vacuum, light travels more slowly, as shown in 1850 by Foucault's measurement of the speed of light in water. Foucault's finding caused final abandonment of Newton's corpuscular theory.

While Newton was experimenting with prisms, Grimaldi and Hooke discovered the phenomenon of *diffraction,* the bending of light around obstacles. They explained their observations on the basis of *light waves* rather than corpuscles. Their *wave theory* could explain the observation that in diffraction (but not in refraction) red light is bent more than blue light—red light has longer *wavelength* than blue light.

Although a corpuscular theory might, with great difficulty, explain diffraction, it could not at all explain the phenomenon of *interference,* discovered by Young in 1800. Even though at that time some scientists were slow to accept the wave theory, today we consider the observation of diffraction and interference phenomena as sufficient proof of the existence of waves.

In 1820 Fraunhofer invented the *diffraction grating,* with which he could obtain high-quality spectra.

IMPORTANT TERMS

Refraction   The bending of light when it passes obliquely from one transparent medium to another.

Diffraction   The bending of light by edges of obstacles, such as a knife edge or the edges of a hole or a slit.

Interference   The mutual reinforcing (*constructive interference*) or weakening (*destructive interference*) of two or more light beams. Constructive interference is due to addition of simultaneously arriving crests (or troughs); destructive interference is due to partial or complete cancellation of a crest by a simultaneously arriving trough.

Monochromatic light Light having a single wavelength or frequency; that is, one color of the spectrum. The phenomena of refraction, diffraction, and interference can be observed with monochromatic as well as polychromatic (white) light.

Dispersion The separation of polychromatic light into its constituent colors. It can be accomplished through refraction (e.g., by prisms) or through diffraction and interference (e.g., by slits or gratings). Monochromatic light cannot be dispersed.

Spectrum The band of colors produced by the dispersion of polychromatic light, comprising violet (shortest wavelength), blue, green, yellow, orange, and red light (longest wavelength).

ADDITIONAL PROBLEMS

Problem 12-7. Knowing the results of Grimaldi's experiment (light does not always travel in straight lines), show how both the wave and the corpuscular theories can account for this behavior.

Problem 12-8. Thomas Young discovered that overlapping beams of light produce darkness in some places (destructive interference). Why did this disprove Newton's corpuscular theory?

Problem 12-9. Explain interference of waves.

Problem 12-10. White light is a mixture of many colors. Certain objects appear colored when white light passes through them. What has happened when a piece of glass appears blue when you look through it? Which colors of light were absorbed and which transmitted? If you look at a red object through this blue glass, what color will it appear to be?

Problem 12-11. Can you offer a hypothesis that explains *both* the redness of the setting sun and the blueness of the daytime sky? (Look up *Rayleigh scattering* and related topics in a reference book.)

Problem 12-12. An organic chemist, a physical chemist, and an analytical chemist decided to measure the speed of sound. Each climbed to a different mountain top. Once on top, the organic chemist fired a gun, and each of the others, from his mountaintop, fired his own gun when he heard the report. The organic chemist then recorded the time lapses until *he* heard the answering shots. That evening, over some apple cider diluted with ethanol, the physical chemist calculated the results.

(a) The analytical chemist's answering shot was heard by the organic chemist 11 seconds after his own shot. The two mountain tops were exactly one mile apart. What was the calculated speed of sound?

(b) The organic chemist and the physical chemist were exactly five miles apart, and the recorded time interval was 49 seconds. What was the calculated speed of sound?

(c) The discrepancy between the two experimental results completely baffled the organic chemist. (Remember, it had been a hard, hot day, and that evening he had done his best to quench his thirst with cider!) The physical chemist was about to devise some very profound theories, when the analytical chemist, a very practical man, yelled, "Eureka!" What was his simple (and correct) explanation?

(d) Galileo had tried a similar experiment to determine the speed of light. Why had he failed?

Problem 12-13. The speed of light in vacuum is constant, independent of wavelengths. If this were not so, and, say, blue light traveled faster than red light, what would you see during the course of a total eclipse of the sun (the moon slowly moving between the observer and the sun)? Describe the appearance of the sun at the beginning of the eclipse, during totality, and again as the sun emerges from behind the moon.

Problem 12-14. A *light-year* is the *distance* light travels in one year. Calculate the approximate number of kilometers and miles in a light year.

Problem 12-15. How many seconds are there in a light year? (What are the *dimensions* of the light year? Of the second? How many cubic centimeters and how many grams are there in a light year?)

Problem 12-16. (a) What is the circumference, C, of a circle of 1-inch radius, r? ($C = 2\pi r$)

(b) What is the length of a 60° segment, S, of this circumference?

(c) What is the straight-line distance between the ends, A and B, of the segment? (Even simpler: You have an isosceles triangle with an apex of 60 degrees; therefore all angles are 60 degrees, and it is really an equilateral triangle.)

Problem 12-17. A beacon on a light tower is rotating clockwise at the rate of one revolution per second. Two ships, at A and B, are both 1 mile from the tower, T, and form an angle ATB of 60 degrees.

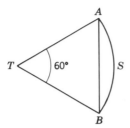

(a) What is the distance between the ships?

(b) How long does the light beam require to sweep from A to B? (It is revolving in that direction.)

(c) What is the speed of the light beam (mi/sec) in going from A to B?

(d) Two other ships, D and E, are 3 miles from the tower and behind ships A and B as seen from the tower. How far apart are D and E? (If necessary, make a small scale drawing.)

(e) What is the time required for the light to sweep from D to E?

(f) What is the speed of the light beam in going from D to E?

(g) Answer questions (d), (e), and (f), if the two observers D and E are 1000 miles (instead of 3) from the light tower.

(h) Answer these same questions for observers placed 100,000 miles from the beacon. That is a very small distance by astronomical standards; the earth-moon distance, for instance, is 240,000 miles, and the earth-sun distance is almost 400 times as great.)

(i) How can you reconcile your last answer with the generally accepted principle (from Einstein's Special Theory of Relativity): "No signal can be transmitted *by any means whatsoever*, in free space or in a material medium, at a speed faster than the speed of light (approximately 186,000 mi/sec)"? (*See* footnote 6, p. 175.)

ANSWERS TO SELECTED PROBLEMS

12-5.   (a) 5×10^{-3} cm.

(b) 33,000 A.

12-12.   (a) 960 feet/sec.

12-16.   (a) $C = 2\pi r = 2 \times 3.14 \times 1 = 6.28$ miles.

(b) $S = (60/360)\ 6.28 \approx 1.05$ mile.

(c) 1 mile exactly (radius is 1 mile).

12-17.   (c) 6 mi/sec.

(f) 18 mi/sec.

Chapter 13

THE ELECTROMAGNETIC SPECTRUM

In the last chapter we confined our discussion to visible light. Although our eyes are rather sensitive instruments, they can see only a small part of the spectrum. The visible region extends roughly from the violet and blue end at about 4000 angstroms to the red end at about 8000 angstroms. Maximum sensitivity lies somewhere in the middle, dropping off sharply at both ends.

Other detectors are not so limited as the human eye. The photographic method, for example, has considerable range in the shorter wavelengths beyond the violet, into the ultraviolet and X-ray regions; heat detectors can pick up longer wavelengths beyond the red end of the visible spectrum (infrared radiation); and other detectors can pick up still longer wavelengths (microwaves and radio waves). These other radiations in the *electromagnetic spectrum* differ from visible light principally in their wavlengths and frequencies. We shall discuss the distinctive features of each of these regions, but first let us look at the features common to all.

RELATIONS AMONG WAVELENGTH, FREQUENCY, WAVE NUMBER, AND ENERGY: UNITS OF MEASUREMENT

We know that light or electromagnetic radiation is actually energy propagated through space with wave characteristics. With waves in general we associate several quantities.

The *wavelength*, λ, is the distance interval between adjacent crests, or the equivalent interval between adjacent troughs (*see* Fig. 12-15).

The number of waves arriving at or passing a given point in a given interval of time is called the wave *frequency*, ν. For visible light, the frequency is of the order of 10^{14} waves per second. We call this *cycles per second* (abbreviated *cps,* sometimes written as sec⁻¹) or *hertz* (abbreviated *Hz*).

It should be evident that if so many waves pass a point in a given period of time, and each wave has a certain length, then their speed— i.e. the distance traveled per unit time—is the product of the frequency times the wavelength.

$$v = \nu\lambda$$

In a vacuum the speed of all electromagnetic radiation is constant:

$$c = \nu\lambda = 3.00 \times 10^{10} \text{ cm/sec}$$

In any other medium the speed is less than 3×10^{10} cm/sec, but always $v = \nu\lambda$.

What happens to frequency and wavelength when light enters a denser medium? As an analogy, imagine cars traveling along a highway at velocity v, well spaced out, and passing a given point with frequency ν (so many cars each minute). Along the way is a tunnel, which creates a slowdown inside but does not cause a slowdown in the spaced-out line of cars approaching the tunnel. Then cars enter the tunnel with frequency ν. They also leave the tunnel with frequency ν. If this situation were not so, if the latter frequency were less, fewer cars would leave the tunnel than entered it; more and more cars would accumulate in the tunnel, which clearly is an impossibility. We see that the velocity is changed, but the frequency is unaffected, when the stream of cars enters the tunnel. The spacing between cars, however, *is* decreased in the tunnel (Fig. 13-1). For light entering a denser medium this means unchanged fre-

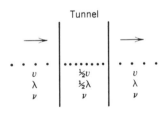

Tunnel

v $\frac{1}{2}v$ v
λ $\frac{1}{2}\lambda$ λ
ν ν ν

Fig. 13-1. Cars entering a tunnel are slowed down to half their original speed, v. The spacing between cars, λ, is also halved. But the number of cars passing a given point per unit time, remains constant. After emerging from the tunnel, the cars regain their original speed and spacing. Inside the tunnel as well as outside, the speed is equal to velocity times frequency.

quency, but decreased wavelength proportional to the velocity: $v = \nu\lambda$, or $\lambda = v/\nu$. Upon the cars' or the light's emergence from the tunnel or the medium, the speed and interval, or the wavelength, are restored to their original values.

Waves can also be characterized by wave number. The *wave number* is simply the number of waves in a given distance interval—so many waves per centimeter. The units, then, are cm^{-1}, called reciprocal centimeters. The symbol used is $\bar{\nu}$.

To summarize, the following relationships exist among frequency, wave number, wavelength, and velocity:

$$c = \nu\lambda \tag{1}$$

$$\bar{\nu} = 1/\lambda \tag{2}$$

$$\bar{\nu} = \nu/c \tag{3}$$

Some attention to units is necessary when using these equations. If the wavelength in Equation (2) is expressed in centimeters, the wave number will come out in reciprocal centimeters. And if the frequency in Equation (3) is expressed in reciprocal seconds and the speed in centimeters per second, the wave number also will come out in reciprocal centimeters.

Waves carry energy; water waves, for example, make corks or ships bob up and down. If they slam into the shore, large waves can do considerable damage. The energy of water waves is a function of their height, or amplitude. We do not, however, think of light and other electromagnetic radiation as having varying amplitude, for reasons which are discussed in Chapter 17. A more intense light beam is simply made up of more waves rather than larger individual waves. The energy, E, of a light wave is proportional to its frequency or, which is equivalent, to its wave number:

$$E \propto \nu \propto \bar{\nu}$$

With the proportionality constant h, the formula becomes,

$$E = h\nu = hc\bar{\nu}$$

We will discuss the significance of this formula in Chapter 17. For the moment it suffices to be aware of the proportionality between energy and frequency or wave number. Since chemists are usually interested in the energies of chemical bonds and chemical reactions, they find frequencies or wave numbers convenient quantities because of their additivity. For example, two light waves of frequency 5000 and 7000 are equivalent in energy to a single light wave of frequency 12,000. Or, doubling the frequency also doubles the energy.

Physicists, on the other hand, especially those interested in optics, prefer to use wavelengths. But wavelength addition is not simply related to energies. Chapter 16 illustrates the advantage of using frequencies: Balmer's original formula for the hydrogen spectrum was expressed in wavelengths; it was only after Rydberg had converted this formula into frequencies that Ritz began to perceive a scheme.

Problem 13-1. Convert the limits of visible light, 4000 and 8000 angstroms, to centimeters (cm); micrometers (μ); and to sec^{-1} (Hz) and cm^{-1}.

INFRARED RADIATION

Electromagnetic radiation extends beyond the visible range in both directions. Some of these radiations were discovered even before the wave character of light had been clearly established by Fresnel and Young in 1818.

One day in 1800 the English astronomer William Herschel (1738–1822) was measuring the heating power of the various colors of the sun's radiation. He used a prism to disperse the light and a thermometer to measure the temperature at the various positions. To his surprise he found heat beyond the red end of the spectrum; therefore, the name *infrared*, "below the red."[1]

The definition of the infrared region is somewhat arbitrary and depends on the detectors used. It is generally considered to extend from the visible region at 0.8 micrometer (8000 A) to about 1000 micrometers.

Problem 13-2. Convert these limits to other wavelength units, and to frequencies and wave numbers.

Even though we may not give the matter much thought, the detection of infrared rays is an everyday occurrence. We feel the warmth of the sun, which comes only in part from visible radiation. We can feel the warmth of an oven, even in the dark and even when convection currents carry most of the heat upward. As we shall see in Chapter 27, infrared radiation is used by spectroscopists to identify and determine the structures of molecules.

ULTRAVIOLET RADIATION

In 1801, only one year after the accidental discovery of infrared radiation, another invisible radiation—this time beyond the violet end of the

---

[1] Friedrich Wilhelm Herschel was born in Hannover, Germany. As a young man he moved to England, where he gained a reputation as a musician. His interest, however, shifted to astronomy. With his homemade Newtonian reflector telescope, he investigated the motions of the planets and discovered the planet Uranus (1781). When King George III appointed him private astronomer to the court, Herschel was able to devote full time to astronomy. Perhaps his most important discovery was that of double stars in 1793.

spectrum (therefore, *ultraviolet*)—was discovered in a similar manner by the German physicist J. W. Ritter (1776–1810). In experiments that proved basic to modern photography, Ritter attempted to determine the efficiency of different colors of light in blackening silver chloride. He found that the greatest effect was produced by invisible radiation beyond the violet end of the visible spectrum.

Today's applications of ultraviolet radiation range from the medical and cosmetic fields to photography and spectroscopy.

Although the principle of the camera had been known for centuries, the development of photography was held back by lack of a process for obtaining permanent pictures rather than mere images on a screen. In the eighteenth century it was found that light produces blackening of silver salts. While investigating the effect of light on silver chloride, Ritter discovered ultraviolet radiation. The first photo-positive using silver chloride was produced in 1816, but it was not until 1822 that the first permanent picture was obtained (not by use of silver salts).

In modern black-and-white photography silver bromide is generally preferred to silver chloride. The process is as follows. Very fine grains of silver bromide, held immobile in a gelatin emulsion on film, are sensitized by light focused through lenses, and an invisible "latent image" is obtained. The development process reduces the sensitized silver ions in the latent image to silver, $Ag^+ \rightarrow Ag^\circ$, while the nonsensitized silver is not affected. Since finely divided silver is black (not silvery), the latent image has now been converted to a visible black image, and we have obtained a negative. Where there was light, there is now a black image.

In order to permanently "fix" the image, the film is treated with a solution that washes off the other chemicals but leaves the black silver grains. This yields the final negative. To obtain a positive, the process is repeated: In effect, a picture is taken of the negative and developed.

Interestingly, two major contributions to photography were made by John Frederick William Herschel (1792–1871), the son of William: the discovery of the solvent power of "hypo," used as fixer (1819); and the invention of photography on sensitized paper, rather than on plates (1839).

RADIO WAVES

Whereas the discovery of infrared and ultraviolet radiation was accidental, that of radio waves was the result of a planned experiment.

The Scottish physicist James Clerk Maxwell (1831–1879) had suggested that light and related radiations are "electromagnetic" radiation. According to this theory (1864) a light beam is an *electromagnetic wave* traveling through space—that is, a varying electric field moving through space with an associated varying magnetic field. The detailed theory explaining electricity, magnetism, and light has been called the greatest scientific publication since Newton's *Principia* of 1687.

On the basis of his theory, Maxwell made a prediction. He remembered several earlier discoveries that linked electric and magnetic phenomena. First, there was Oersted's observation in 1820 that a magnetic needle is deflected by an electric current (p. 124). Conversely, it was found that a magnet exerts a force on a current-carrying wire. Later, in 1831, Faraday used a moving magnet to induce an electric current in a wire loop. Evidently there is an intimate relationship between electric currents and magnetic fields. Maxwell predicted that a varying, or oscillating, current would give off electromagnetic radiations.

In 1887, unfortunately after Maxwell's death, the German physicist Heinrich Hertz (1857–1894) succeeded in showing that rapidly oscillating electric charges do indeed produce electromagnetic radiations, in accord with Maxwell's predictions. These electromagnetic waves were received some distance away, where they were reconverted into detectable electric current (Fig. 13-2). Hertz also showed that these waves had many of the characteristics of light (except visibility): They could be reflected, produced interference effects, and had the same velocity as light.

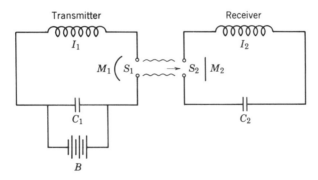

Fig. 13-2. Experimental arrangement used by Hertz. The condenser C_1 is charged by battery B. This causes a spark to jump across gap S_1. The air in the gap has become ionized temporarily and thus acts as a conductor. For a fraction of a second current surges back and forth between capacitor C_1 and induction coil I_1. This oscillating current sends out electromagnetic radiations. These radiations induce a similar, albeit weaker, surging current in the receiver circuit and cause a spark to jump across gap S_2. The radiations can be transmitted through greater distances if an arrangement of concave and plane mirrors, M_1 and M_2, is used to focus the radiation beam, just like visible light.

A decade later, in 1898, Guglielmo Marconi (1874–1937), an Italian inventor and engineer working in London, experimented with the transmission of Hertzian waves over greater distances. In 1901 he succeeded in sending radio signals across the Atlantic.

Problem 13-3. Actually Hertz did not measure the velocity of his waves but determined it indirectly from a measured frequency, 3×10^7 cycles per second, and a measured wavelength, 9.6 meters. Calculate the velocity on the basis of these measurements.

Problem 13-4. Light travels in straight lines. In view of the curvature of the earth, how can radio waves span the ocean? (Apply the principles of reflection and refraction and check your ideas against a reference work, such as an encyclopedia.)

X RAYS

The accidental discovery of X rays by Röntgen in 1895 was described in Chapter 9. These X rays were produced by cathode rays hitting a target in a cathode-ray tube. This process is in accord with Maxwell's electromagnetic theory, which states that electromagnetic radiation should be emitted by a varying electric current (in this case, the electrons being stopped at the target).

Nonetheless, scientists needed another seventeen years to prove that X rays are indeed electromagnetic radiation. They had concluded, from negative results in electric- and magnetic-deflection experiments, that X rays are not beams of either positive or negative particles. But they were unsuccessful, for a long time, in showing that X rays are waves, like light, because they could not obtain diffraction patterns.

The difficulty lay in the extremely short wavelengths of X rays. Whereas the wavelengths of visible light are 4000–8000 angstroms, X rays are several orders of magnitude shorter, from approximately 0.01–100 A. Ordinary grating spaces are too large to produce diffraction patterns from such wavelengths (*see* Problem 12-5).

However, in 1912 the German physicist Max von Laue (1879–1960) had an inspiration. He had just heard of a new theory that crystals are a regular latticework of atoms, with spacings between atoms of the order of 1 A. Why not use these as diffraction gratings? He tried his idea and obtained a diffraction pattern on a photographic plate by directing a narrow beam of X rays through a thin crystal (Fig. 13-3).

Because crystals are not ordinary plane gratings but three-dimensional lattices, the diffraction patterns take on rather complicated, though often very picturesque forms. The specific form depends on the experimental technique (transmission vs. reflection, monochromatic vs. polychromatic

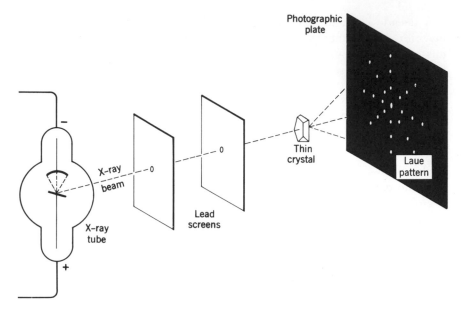

Fig. 13-3. Arrangement for crystal diffraction by the Laue method. The X-ray beam is narrowed by passage through small holes in two sheets of lead.

radiation, single crystals vs. powder, etc.) and the type, arrangement, and spacing of the atoms in the crystals (*see* Chapter 27).

In general, however, the diffraction technique enables us to disperse X rays into their constituent waves and to determine their wavelengths—if we know the grating spacing. Conversely, using known wavelengths, we can apply X-ray diffraction to determine the arrangement of atoms in different crystals and to obtain highly precise measurements of the distances between atomic nuclei (*see* Problem 13-8).

The discovery of similar radiation—gamma rays—was described in Chapter 9. They were discovered by P. Villard in 1900 as part of the radiation from radioactive materials. Gamma rays have still shorter wavelengths and higher frequencies than X rays.

SUMMARY

We have now discussed the most important radiations of the electromagnetic spectrum. It should be emphasized that all these radiations are basically similar, differing mainly in frequency and wavelength. The limits assigned to the various regions are somewhat arbitrary. They are summarized in Figure 13-4.

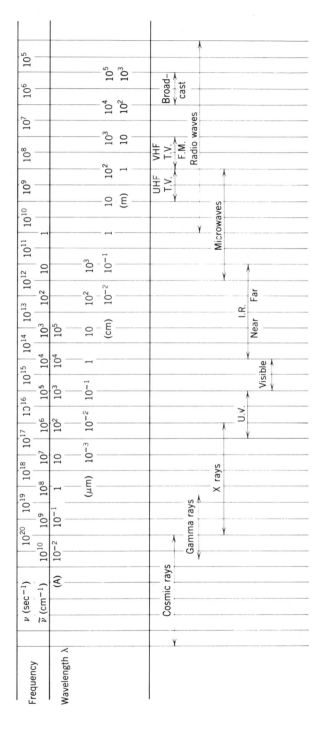

Fig. 13-4. The electromagnetic spectrum.

$\nu\lambda = c$; $\nu = c/\lambda$; $\tilde{\nu} = \nu/c = 1/\lambda$.

$1m = 100cm = 10^3mm = 10^6\mu m = 10^{10}Å$;

$1cm = 10^4\mu m = 10^8Å$.

ADDITIONAL PROBLEMS

Problem 13-5.   Violet light (4000 A) is bent more when entering water from air than is red light. The speed of violet light in water is approximately 2×10^{10} cm/sec.

(a) What, *very* approximately, is the speed of red light in water?

(b) What is the wavelength (in A) of violet light in water?

(c) What is the wavelength (in cm) of violet light in a vacuum?

(d) What is the wave number (cm^{-1}) of violet light in a vacuum?

(e) What is the frequency (sec^{-1}) of violet light in a vacuum?

(f) What is the frequency of violet light in water?

(g) What is the energy (in ergs) of one "wave packet" of violet light in a vacuum? ($h = 6.6 \times 10^{-27}$ erg-sec).

(h) What is the energy, in ergs, of one mole (Avogadro's number) of such wave packets?

(i) How many calories is that? How many kilocalories? (1 kcal = 4.2×10^{10} ergs.)

(j) If a mole of such wave packets is absorbed by a liter of water at room temperature (25°C) and converted into heat, what will be the resultant temperature of the water?

Problem 13-6.   Would green, ultraviolet, or infrared light be bent most by diffraction?

Problem 13-7.   Using the hint given by the previous question, explain why we can "hear" radio transmitters around corners although we cannot see them.

Problem 13-8.   X-ray diffraction patterns reveal that ordinary table salt, sodium chloride (NaCl), consists of a simple cubic lattice in which Na and Cl atoms (or ions) occupy alternating positions (*see* Fig. 13-5.)

From the density of NaCl (2.165 g/cc), its molecular weight, and Avogadro's number, calculate the spacing between atoms in the crystal. From the calculated distance, what can you deduce about the sizes of the sodium and chlorine atoms (or ions)?

(a) Calculate the molecular weight.

(b) How many atoms are contained in one mole of NaCl? (Careful!)

(c) What is the volume of one mole of salt?

(d) What is the number of atoms in 1 cc? (If 100 pencils cost 20 cents, how many pencils for one cent? If $2N$ atoms occupy V cc, how many atoms in 1 cc?)

(e) If we stack atoms in a cubic lattice (say, 1 cc), with n atoms on a side, the total number of atoms in the arrangement will be n^3. Thus

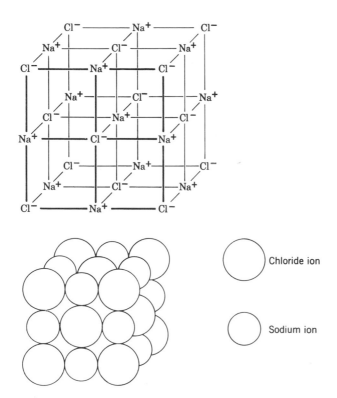

the number of atoms, n, along one edge of a cubic array is the cube root of the total number of atoms per cc. If there are n atoms in length l (namely, 1 cm), what is the spacing, d, between atoms? (Also refer to Problem 4-8 and to p. 144.)

Problem 13-9. Picture the following hypothetical situation. An un-friendly foreign power has managed to construct a gamma-ray gun on Mars. Our intelligence agency, however, has managed to "bug" the gun site with a secret device, which sends out a coded radio signal the instant the gun is fired. How much warning time before the arrival of gamma radiation will the device be able to give us when Mars is 186 million miles from Earth?

ANSWERS TO SELECTED PROBLEMS

13-1. (a) $4000 \text{ A} = 4 \times 10^{-5} \text{ cm} = 0.4 \; \mu$

$\nu = c/\lambda = (3 \times 10^{10} \text{ cm/sec})/(4 \times 10^{-5} \text{ cm}) = 7.5 \times 10^{14} \text{ sec}^{-1}$

$\tilde{\nu} = \nu/c = (7.5 \times 10^{14}\ \text{sec}^{-1})/(3 \times 10^{10}\ \text{cm sec}^{-1}) = 2.5 \times 10^4$ cm^{-1}.

Or, $\tilde{\nu} = 1/\lambda = 1/(4 \times 10^{-5}\ \text{cm}) = 2.5 \times 10^4$ cm^{-1}.

13-5.   (a) More than 2×10^{10} cm/sec (bent less than violet), but less than 3×10^{10} cm/sec (speed of light in vacuum). (Actually, it was experimentally found to be about 2.4×10^{10} cm/sec).

      (b) 2,667 A [2/3 of 4000 A, since $v = (2/3)c$].

      (c) *See* Problem 13-1.

      (f) Same as (e).

      (g) $E = h\nu = 5 \times 10^{-12}$ ergs.

      (h) Multiply by Avogadro's number.

      (i) 70 kcal.

      (j) 95°C.

13-8.   (a) $22.990 + 35.453 = 58.443$.

      (b) $2N_A = 2 \times 6.023 \times 10^{23} = 1.2046 \times 10^{24}$ atoms.

      (c) Density $= \dfrac{\text{Weight}}{\text{Volume}}$; $V = \dfrac{W}{D} \approx \dfrac{58.443\ \text{g}}{2.165\ \text{g/cc}}$ (≈ 27 cc).

      (d) atoms/cc $= \dfrac{2N\ \text{atoms}}{(58.443/2.168)\ \text{cm}^3}$ ($\approx 4 \times 10^{22}$ cm^{-3}).

      (e) (1) Number of atoms per centimeter equals cube root of above value

$$(\approx \sqrt[3]{40 \times 10^{21}\ \text{cm}^{-3}} \approx 3.5 \times 10^7\ \text{cm}^{-1})$$

         (2) spacing d equals one divided by above number

$$(\approx 1/0.35 \times 10^8\ \text{cm}^{-1} \approx 3 \times 10^{-8}\ \text{cm})$$

         (An exact calculation would yield: 2.82 A)

Chapter 14

HISTORY OF CHEMICAL SPECTROSCOPY

Spectroscopy, also known as spectral analysis, is the study of the constituents of light, or radiation in general. More specifically, it is the study of the interaction of radiation with matter. In other words, spectroscopy is a tool in the study of matter.

Spectroscopy is still one of the most advanced and fastest growing branches of chemistry and physics. We will introduce spectroscopy in its relation to the elucidation of atomic structure; but we will also study it in sufficient depth to gain some appreciation of its general applicability and value. Later on, in Chapter 27, we will discuss some specialized fields of spectroscopy in relation to the determination of *molecular* structure.

WHAT NEWTON MISSED

Newton and the other early experimenters with light missed out on an important discovery because of the nature of their experimental setup. They used the most convenient light source available at the time—sunlight, which they admitted through small holes in window shades; on their screens they saw images of these holes. Newton's spectrum, for example, was a superposition of many slightly displaced color images of the original hole. Because of the natural width of the hole, the color images overlapped, and no color appeared pure.[1]

Perhaps a picture will make this important point clearer. If Newton had used a light source containing only red and blue light, he would

[1] *Spectrum* in Latin means "image."

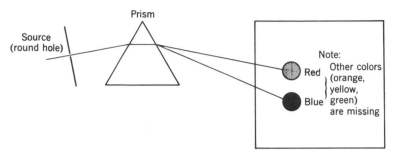

Fig. 14-1. If the original light source contains only red and blue light, two distinct color images of the round hole appear on the screen.

have obtained two clearly separated images, indicating that other colors (orange, yellow, green) were absent from the original light (Fig. 14-1). Had Newton's light source instead contained pure red and yellow light, the two images would have been so close together that they would have overlapped in an area that appeared orange; there would have been no way of telling whether or not the original light contained any pure orange (Fig. 14-2). Had Newton and the others used narrow slits instead of round holes, there would have been less overlap and therefore better definition of the colors in the resultant spectrum.

THE DISCOVERY OF LINES IN THE SOLAR SPECTRUM

Much later, two scientists did just that—they used narrow slits to admit sunlight to their prisms. The two men, working independently, were the English physician William H. Wollaston (1766–1828) and German optician Joseph Fraunhofer (1787–1826), mentioned earlier as the inventor of the diffraction grating. In their better-defined solar spectra, they noticed some dark lines, indicating that certain colors, or wavelengths, were missing from sunlight. Wollaston made his discovery in 1802; Fraunhofer rediscovered the phenomenon in 1814.

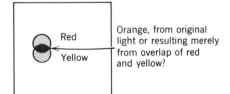

Orange, from original light or resulting merely from overlap of red and yellow?

Fig. 14-2. If the original light source contains red and yellow light, the two images overlap, yielding orange; it then is difficult to determine the presence or absence of orange in the original light source.

Fraunhofer eventually cataloged 600 or 700 of these dark lines. Today we recognize over 15,000 *Fraunhofer lines,* only some of which are in the visible part of the solar spectrum (4000–8000 A).

THE FINGERPRINT METHOD

These curious Fraunhofer lines were explained about a half century later. In 1848 Foucault (who measured the speed of light in water) made an important discovery. When certain materials are vaporized in a colorless or slightly bluish gas flame, they impart color to the flame: sodium, bright yellow; lithium and calcium, red; barium, green, etc. Foucault passed a beam of ordinary white light through a yellow gas flame containing sodium vapor and dispersed the light through a prism; the resultant spectrum had two prominent dark lines in the yellow region. Even more curious, these dark lines were in exactly the same position as two of the Fraunhofer lines in the solar spectrum. Other scientists noted similar correlations between the colors of substances in flames and their spectral lines.

These observations multiplied after the invention of the first practical analytical spectroscope in 1859 by the physicist Gustav Robert Kirchhoff (1824–1887) and the chemist Robert Wilhelm Bunsen (1811–1899). Working together in Heidelberg, Germany, they used an arrangement of lenses and prisms to focus and disperse the light, and thus obtained quite good spectra. (Other spectroscopes subsequently utilized diffraction gratings instead of prisms to yield still better spectra.)[2] Bunsen also devised a special gas burner, which today can be found in every chemistry laboratory. With these two pieces of equipment Kirchhoff and Bunsen systematized the technique of *flame spectroscopy.*

The spectroscope revealed that the colors of flames consist of a set of very sharply defined bright lines, or wavelengths, in the spectrum. Every element has a different and very characteristic *bright-line spectrum.* When several elements are present, the resulting bright-line spectrum is a superposition of the individual bright-line spectra.

The utility of bright-line spectra is easy to see. We can systematically catalog the wavelengths of the important lines belonging to each element. Whenever we have an unknown substance, we can vaporize it in a gas flame, look at its spectrum, and match the observed lines with those in our file, just as criminals are identified by matching a suspect's fingerprints with those in a police file. As a matter of fact, we call this the *fingerprint method* of spectroscopic identification.

[2] Spectroscopes are discussed further in Chapter 27.

Naturally, the fingerprint method is limited to elements in our files. Today these files contain the spectra of all stable elements; in Kirchhoff and Bunsen's time, however, this was not the case. In fact, they were able to use flame spectroscopy to discover new elements, such as cesium and rubidium (1861).

There are at least three common ways of obtaining bright-line spectra: (1) by vaporizing materials and heating them to luminescence in gas flames; (2) by "exciting" gases at low pressures to luminescence by passage of electric currents in cathode-ray tubes (*see* Chapter 8); (3) by vaporizing and exciting solids in electric arcs or with sparks—for example, in iron arcs or carbon arcs (used for searchlight beams). In all these methods, as we shall explain later, the atoms absorb energy from the flame or electric current and are promoted ("excited") to higher energy levels. When the atoms return to lower energy levels, this excess energy is given off in the form of light. The exact energies and wavelengths of this light are different for each atom, resulting in a characteristic bright-line spectrum for each element.

The production of *dark-line spectra* is the reverse process. Atoms in lower energy states, when exposed to light energy, can selectively absorb certain energies, or wavelengths, and be excited to higher energy states. Early spectroscopists observed that the wavelengths absorbed by a given element are generally identical with the wavelengths emitted by them. In other words, bright-line *emission spectra* and dark-line *absorption spectra* yield identical information for our fingerprint files.

Even though old-time spectroscopists did not understand the energy processes involved in the absorption and emission of light, they certainly put the empirical information to use. Extensive spectral files and catalogs were compiled from high-quality spectra obtained with Fraunhofer's diffraction grating. New elements continued to be discovered. Others were identified as old friends; recall, for example, Rutherford's spectroscopic identification of alpha particles as helium ions (Chapter 9).

Spectroscopy, moreover, was not limited to mere earthbound laboratories. Scientists soon realized that Fraunhofer lines are absorption spectra of elements in the sun. All Fraunhofer lines could eventually be matched with spectral lines produced from various elements in the laboratory. Thus it was shown that the elements of the sun and the earth are identical, finally disproving Aristotle's theory that heavenly bodies are somehow different.

The production of a solar absorption spectrum is analogous to the production of a dark-line spectrum obtained by passing light through a gas flame. The surface of the sun—like any hot body, such as a hot filament in an incandescent lamp—emits a *continuous spectrum* (all wavelengths present). When this light passes through the relatively cool atmosphere of the sun, selected wavelengths are absorbed; the energy is used to promote the atoms in the solar atmosphere to still higher energy levels. The absorbed wavelengths are missing from the sunlight that reaches us and appear as dark Fraunhofer lines.

We can also obtain a bright-line emission spectrum from the sun. During solar eclipses, when the moon hides the sun's surface from our view, we can directly view the weaker light emitted by the hot gases in the sun's atmosphere. These gases, like those in our Bunsen burner flame, emit bright-line spectra.

DISCOVERY AND IDENTIFICATION OF THE "INERT" GASES

In Kirchhoff and Bunsen's time, as we have said, not all Fraunhofer lines were matched with the spectra of known elements. In 1868 the French astronomer Pierre Jules Cesar Janssen (1824–1907) observed some unidentified lines in the bright-line spectrum of the sun taken during a solar eclipse. Only two months later Joseph Norman Lockyer (1836–1920), an English astrophysicist, also observed these lines in the solar spectrum and realized that they must belong to a new element, which he named helium (from the Greek *helios,* for sun). Helium was not found on earth until 1895.

The story of the discovery of helium and the other "inert" gases illustrates the close interrelation of the work of astronomers, physicists, and chemists. In the 1880s, the English physicist John William Strutt (1842–1919), better known as Lord Rayleigh, redetermined the atomic weights of oxygen, hydrogen, and nitrogen. He found that the atomic weight of nitrogen seemed to depend on the source of the gas: He obtained a slightly higher weight for nitrogen from the air than for nitrogen from chemicals in the soil.

The Scottish chemist William Ramsay (1852–1916) became interested in this problem. He recalled some earlier work by Henry Cavendish, who about 1785 had tried to combine nitrogen from the air with oxygen and found that there always was a small bubble of unreacted gas left over, "not more than 1/120th part of the whole." Rayleigh and Ramsey in 1894 repeated Cavendish's experiment and applied the tool of spectroscopy. They heated the bubble of leftover gas, examined its emission spectrum, and found new spectral lines that fitted no known element. They named their new element argon (derived from the Greek word meaning "inert"). The argon content of the air is less than 1%, just as Cavendish had found.

The discoverers now were faced with a new problem. Argon has an atomic weight of slightly less than 40. But there was no vacancy in Mendeleev's periodic table at that position. The following elements were known with atomic weights close to 40. (Combining powers are indicated in parentheses.)

Sulfur	34	(2)
Chlorine	35.5	(1)
Potassium	39.1	(1)
Calcium	40.1	(2)
Scandium[3]	45	(3)

If atomic weight were the only criterion for position in the periodic table, the new element argon should be between potassium and calcium; but Mendeleev had previously established the principle that chemical properties and combining power are more important than atomic weight in determining position. Still, there was no vacant position near atomic weight 40 in the periodic table. Now, since argon is very inert (it seemed to have zero combining power), why not call its combining power zero and place the element between chlorine and potassium? (Note that this represents one of the reversals of atomic weights.)

If the periodic table is any guide, then one would expect argon not to be the only element of combining power zero. Where were the other members of this family?

Ramsay began a systematic search. He had learned of a strange gas obtained in the United States in 1891 from uranium minerals by W. F. Hillebrand (1853–1925). This gas consisted mainly of nitrogen, but its spectrum contained additional lines not identifiable with those of any element known to exist on earth. Hillebrand attributed these lines to pressure variations and did not investigate further; he actually rejected a suggestion by one of his colleagues that a new element might be present. Ramsay repeated Hillebrand's work and also found lines in the spectrum' that belonged neither to nitrogen nor, to his surprise, to argon. He asked Sir William Crookes to measure the wavelengths of these lines, and they turned out to match those observed by the French astronomer Janssen in the solar eclipse of 1868. In other words, the new gas was identical with Janssen's and Lockyer's helium. Thus, although helium was the first of the inert gases to be discovered, it was the second to be found on earth.

---

[3] Discovered in 1879 and later shown to be identical with "cka-boron," one of the three elements predicted by Mendeleev (see Chapter 5).

That same year (1895) Ramsay also found helium in meteoric iron. In 1905 it was found in natural gas in Kansas (almost 2%). In 1899 Rutherford discovered alpha particles from radioactivity; he suggested that they were helium ions in 1906 and proved his case through fingerprint spectroscopy in 1909 (*see* Chapter 9).

Ramsay continued his search for other members of the inert-gas family. In 1898 he boiled some liquefied air and identified the gases as they appeared in the vapor. In addition to argon and traces of helium, he found three new inert gases which he named neon (from the Greek word meaning "new"), krypton ("hidden," "cryptic"), and xenon ("stranger," as in "xenophobe"). In 1900 Rutherford and the German physicist F. E. Dorn (1848–1916) discovered radon, the last of the inert gases.

At first these gases were considered mere curiosities. In 1910, however, the French chemist Georges Claude (1870–1960) showed that in a type of cathode-ray tube, an electric current will cause these and other gases to give off a gentle glow, whose color depends on the gas in the tube. These tubes were the forerunners of the modern neon tubes used in advertising.

In closing it should be mentioned that the so-called inert gases are not totally inert. In 1962 N. Bartlett (1932–   ) succeeded in preparing compounds of xenon and fluorine. Other inert-gas compounds have been made since. We now refer to these gases as *noble gases*.

Chapter 15

ATOMIC SPECTRA

Once Kirchhoff and Bunsen had built the first practical spectroscope and fully interpreted the Fraunhofer lines (1859), spectroscopy branched out. Chemists used spectroscopy for chemical analysis and the identification of new elements. Physicists were more interested in developing a theory of spectroscopy, especially in understanding the origin of spectral lines at certain wavelengths.

In 1882 the American physicist Henry Rowland (1848–1901) began a new era in spectroanalysis with the construction of a new ruling engine for diffraction gratings and his invention of a new kind of reflection grating. The new gratings yielded increasingly precise spectroscopic data, and physicists began looking for some sort of relationship among the wavelengths of the lines in each spectrum.

THE SEARCH FOR REGULARITIES

The spectrum of the hydrogen atom (Fig. 15-1) was particularly interesting because of its simplicity and because of its prominence in the spectra of stars. (Hydrogen is a very common element in stellar atmospheres.) Important features of the spectrum of atomic hydrogen are the presence of discrete sharp lines and a regular decrease in the spacing of some of these lines with diminishing wavelength.

From about 1870 on there were many attempts to define and explain the regularity in the hydrogen spectrum. The first breakthrough was scored in 1885 by Johann Balmer (1825–1898), a Swiss school teacher. Balmer had been studying the only four lines of the hydrogen spectrum known at that time and discovered a relationship among the numbers representing the wavelengths. No theory was involved; Balmer simply hit

210

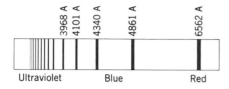

4101 A
4340 A
4861 A
6562 A

Ultraviolet   Blue   Red

Fig. 15-1. The Balmer series in the emission spectrum of atomic hydrogen. Note that the lines are closer together at shorter wavelengths. Only the lines at 6562, 4861, 4340, and 4101 A are in the visible part of the spectrum and were known to Balmer in 1885. Similar series of lines were discovered later in the infrared and ultraviolet regions of the spectrum.

upon the relationship and could give no reason for its existence. His formula for the wavelength of any line in the spectral series was:

$$\lambda_n = b \, \frac{n^2}{n^2 - 4}$$

where $b = 3646$ angstroms and $n = 3, 4, 5, \ldots$. For example, for $n = 3$, Balmer's formula yields the wavelength,

$$\lambda_3 = 3646 \, \frac{3^2}{3^2 - 4} = 6563 \text{ A}$$

in good agreement with observation (*see* Table 15-1).

Problem 15-1. (a) Use Balmer's formula to calculate values of λ_n when $n = 5, 7, 9, 11, 100, 1000,$ and ∞ (infinitely large). Compare your values with those in Table 15-1.

(b) In what regions of the electromagnetic spectrum are these wavelengths? What are the colors of the visible lines?

Although Balmer's formula was fairly simple and was devised for only four wavelengths, it fit nine other spectral lines discovered shortly thereafter. Spectroscopists have since found more than thirty spectral lines that tally with this formula. Perhaps no other empirically obtained formula in all of physics fits observed data so precisely. Such precision suggested that the formula was not just a mathematical trick or coincidence but might have some real physical meaning.

We can better interpret the formula if we convert wavelengths to wave numbers. Since the wave number, $\tilde{\nu}$, is the reciprocal of wavelength, the Balmer formula becomes:

$$\tilde{\nu}_n = R\left(\frac{1}{4} - \frac{1}{n^2}\right), \quad \text{or,} \quad \tilde{\nu}_n = \frac{R}{4} - \frac{R}{n^2}$$

where $R = 109{,}677.58$ cm^{-1}, and $n = 3, 4, 5, \ldots$. R is known as the *Rydberg constant,* in honor of the Swedish spectroscopist J. R. Rydberg

Table 15-1. Selected Lines from the Hydrogen Spectrum

Series	Quantum number[a]		Wavelength (A)	Wave number (cm^{-1})
	n_1	n_2		
Lyman	1	2	1216	82,260
	1	3	1026	97,490
	1	4	973	102,800
	1	5	950	105,300
	1	6	940	106,600
	1	∞	912	109,700
Balmer	2	3	6562[b]	15,230
	2	4	4861[b]	20,560
	2	5	4340[b]	23,030
	2	6	4101[b]	24,370
	2	7	3968	25,180
	2	∞	3650	27,420
Paschen	3	4	18,760	5330
	3	5	12,820	7799
	3	6	10,940	9139
	3	7	10,050	9948
	3	8	9540	10,470
	3	∞	8220	12,190

[a] Integers n_1 and n_2 for the general formula,

$$\lambda = \frac{n_2{}^2}{n_2{}^2 - n_1{}^2}, \quad \text{or} \quad \bar{\nu} = \frac{R}{n_1{}^2} - \frac{R}{n_2{}^2}$$

[b] The four lines known to Balmer

(1854–1919), who worked on subsequent phases of this problem. It is one of the most precisely determined constants in all science, having eight significant figures. This formula can also be used for calculating frequencies; in that case, the Rydberg constant becomes 3.288 . . . × 10^{15} sec^{-1}.

Problem 15-2. (a) Calculate $\bar{\nu}_n$ (in cm^{-1}) for $n = 3, 5, 9, 100, 1000, \infty$. (b) Convert $\bar{\nu}_3$, $\bar{\nu}_5$, and $\bar{\nu}_\infty$ to λ (in angstroms) and compare these with the λ's calculated directly from Balmer's formula in Problem 15-1.

From the wave-number or frequency version of Balmer's formula we can see two things:

1. As n becomes larger, the second term becomes smaller (1/9, 1/16, 1/25, etc.), so that successive wave numbers or frequencies differ by less

and less; that is, they are spaced more closely (as observed) and approach the limiting value of $R/4$.

2. Each wave number or frequency (but *not* wavelength) represents a difference between two arithmetic terms—$R/4$ and R/n^2.

Balmer speculated about the existence of other series. In 1889 Rydberg set out deliberately to find a more general formula. The following year he devised a formula, which fit other spectral series found later. Rydberg simply replaced the first term in Balmer's formula (which was equal to $R/2^2$) with a more general term (R/n_1^2), so that the formula became

$$\nu = \frac{R}{n_1{}^2} - \frac{R}{n_2{}^2}$$

where both n_1 and n_2 are integers but n_2 is always greater than n_1. When n_1 equals 2, we obtain the Balmer series; that is, the Balmer formula is a special case of Rydberg's more general formula. Note that when n_2 becomes very large, the second term approaches zero and the wave number becomes R/n_1^2. This means that the separation of consecutive lines in each series decreases progressively and approaches the series limit R/n_1^2. Table 15-2 lists the spectral series that have been observed to date.

The dates of discovery hint at some of the experimental problems associated with the various spectral regions. The Balmer series in the visible and near-ultraviolet region was easiest to detect and was discovered first. The Lyman series, spotted in 1906 but not fully identified until 1914, occupies the far-ultraviolet region (1216-912 A; that is, far from the visible region). Unfortunately, glass is transparent only to visible and near-ultraviolet radiation. Even quartz transmits light only to about 1850 angstroms, and fluorite (CaF_2) or lithium fluoride (LiF) to about 1200 A. No transparent materials were found for the Lyman series in the far-ultraviolet region, and reflection gratings had to be used. Special photographic plates had to be used, because gelatin in the emulsion of ordinary plates absorbs ultraviolet light so strongly that the radiation could not penetrate to the light-sensitive material. Also, since air is opaque in this region, the entire apparatus had to be evacuated—hence the expression "vacuum ultraviolet." After these problems had been solved, it was possible to detect the entire Lyman series.

Similar difficulties arose for the other series—all in the infrared region, where glass again is opaque. Here, prisms and lenses of special materials (e.g., rock salt, KBr, CsBr) were needed, or reflection gratings had to be used. Detection also posed a problem. Photographic detection was possible only in the very-near-infrared region; beyond that, heat detectors of various sorts were used. Because of experimental difficulties, the first— and so far only—member of the Humphreys series was not observed until 1952.

Table 15-2. The Series of Lines in the Hydrogen Emission Spectrum

Series (named after discoverer) and year of discovery	Spectral region		Formula[a] for ν_n
	Wavelength (A)	Wave number (cm^{-1})	
Lyman (1906)[b]	far UV 1216–913	82,620–109,700	$\dfrac{R}{1^2} - \dfrac{R}{n^2}$ $n = 2,3,4, \ldots$
Balmer (1885)	visible and near UV 6562–3646	15,230–27,420	$\dfrac{R}{2^2} - \dfrac{R}{n^2}$ $n = 3,4,5, \ldots$
Paschen (1908)	IR 18,760–8220	5331–12,190	$\dfrac{R}{3^2} - \dfrac{R}{n^2}$ $n = 4,5,6, \ldots$
Brackett (1922)	IR 40,500–14,600	2468–6855	$\dfrac{R}{4^2} - \dfrac{R}{n^2}$ $n = 5,6,7, \ldots$
Pfund (1924)	IR 75,000–22,800	1300–4400	$\dfrac{R}{5^2} - \dfrac{R}{n^2}$ $n = 6,7,8, \ldots$
Humphreys (1952)	IR 125,000–33,000[c]	800–3000	$\dfrac{R}{6^2} - \dfrac{R}{n^2}$ $n = 7,8,9, \ldots$

[a] $R = 109,700$ cm^{-1}, or 3.288×10^{15} sec^{-1}.
[b] Fully identified only in 1914.
[c] Only one line, 125,000 A, observed to date.

Let us now return to Rydberg's general formula and note that this formula expresses all observed wave numbers or frequencies as the difference between two arithmetic terms. In 1895, when only the Balmer series and some other individual frequencies were known, the Swiss physicist W. Ritz (1878–1909) speculated that the difference between *any* two terms would lead to a spectral line. An example is

$$\frac{R}{3^2} - \frac{R}{7^2}$$

Table 15-2 shows that this frequency was found later as the fourth line of the Paschen series. Ritz's *Combination Principle* has been found to hold true, with some limitations, for *all* atoms—not only for hydrogen, for which it was originally devised.

To summarize, we have the following general picture of atomic spectra:

1. The wave number or frequency of each observed spectral line can be expressed as the difference between two numbers, called *terms*. (By itself, this statement would say nothing, since we can express *any* number as the difference between two other numbers—e.g., $67 = 84 - 17$.)

2. All terms can be expressed as fractions of the type of R/n^2, where R is a constant and n can be any *integer*.

3. These terms can be subtracted in various ways to yield wave numbers or frequencies of other spectral lines. It should be emphasized that the Rydberg and Ritz formulas *predicted* spectral lines other than those observed by Balmer.

The fact that a fairly simple scheme appears to be generally applicable to all atoms reconfirmed scientists' belief in the orderliness and simplicity of nature and sent them looking for a single, simple, and universal mechanism by which such spectra are emitted.

With hindsight, we can now direct attention to the apparent importance of the terms (R/n^2)—instead of the spectral lines, which are differences between terms. It seems the terms themselves have intrinsic physical meaning. Could these terms represent energy states of the atom before and after emission of radiation? But why, if this is true, are integral (n^2) relationships involved?

THE FAILURE OF EXISTING ATOM MODELS

It was difficult to imagine a mechanism that would account for the observed features of spectra. What spectra could be expected from the two then-existing models of the atom—the Thomson raisin-pudding model and the Rutherford nuclear model? (*See* Chapter 10.)

The Raisin-Pudding Model

The raisin-pudding model, suggested at the turn of this century, was shown to be incompatible with Rutherford's alpha-ray scattering experiments (1911). But even before that it had been in conflict with evidence from atomic spectra. In Thomson's picture the positive charge was uniformly distributed in a sphere (the pudding); electrons (the raisins) were arranged inside the sphere. The lone electron of the hydrogen atom, when excited, would oscillate about its equilibrium position with a certain frequency, depending on its mass, its charge, and the atom's diameter.[1] According to Maxwell's electromagnetic theory, such an oscillating charge would radiate energy with a frequency equal to its frequency of oscillation. Therefore, Thomson's model would allow only a *single* spectral line for hydrogen, in conflict with observation. Evidently the Thomson model does not represent the hydrogen atom very well.[2]

The Rutherford Model

We have already mentioned some difficulties of Rutherford's model, which had negative electrons and a positive nucleus. According to Rutherford, the electrons had to orbit to keep from being drawn into the nucleus by electrostatic forces. At this point the model came into conflict with Maxwell's electromagnetic theory, which postulates that a constantly accelerated charged particle, such as an orbiting electron, should radiate energy continuously. In fact, this prediction has been proved correct: Electrons traveling in circular paths in the modern betatron *do* radiate and lose energy. If electrons in *atoms* were to radiate in this manner—as they do not!—this radiation would be at the expense of their kinetic energy; the electrons would slow down and eventually fall into the nucleus. Thus, paradoxically, the very feature Rutherford introduced to stabilize the atom would instead cause its collapse.

What is more, ordinary atoms do not radiate at all. Only when atoms are "excited" (e.g., by heat, by electric discharge in a cathode-ray tube, or by electron impact in an X-ray tube) do they radiate, and then we usually get not "continuous radiation" (all frequencies) but emission of selected discrete frequencies. It would be difficult to explain these on the basis of Rutherford's model.

[1] Various charge-to-mass determinations for hydrogen ions, always yielding only one faraday per gram—never more—supported the belief that the hydrogen atom contains only one electron.

[2] In analogy to sound waves, we might think of accompanying overtones, or harmonics, of the single frequency; but these harmonics would be at evenly spaced frequencies, not getting closer and not approaching a limit.

From the evidence we must conclude that either Rutherford's model or Maxwell's electromagnetic theory is wrong or inadequate; or perhaps Maxwell's theory does not apply to atoms.

There is yet another difficulty with the Rutherford model. Some materials in the gaseous state, such as sodium vapor, emit characteristic discrete frequencies when excited; then, when condensed to the solid state, these materials emit and absorb *all* frequencies. The latter observation could be explained somehow by the interaction of electron orbits in the solid state. But when we revaporize the materials, they behave exactly as before, absorbing and emitting the same sharp spectral frequencies as before. How could we explain this on the basis of a simple planetary model? How could the electrons have found their way back exactly into their original orbits? This is equivalent to an earth satellite's unaided return to its original orbit after perturbation by some other body. What would be so special about these particular orbits? And why do all atoms of one element behave identically—emit the same radiations and presumably have the same orbital configurations?

The next chapters explain how Niels Bohr attacked these very fundamental problems and how he tried to answer these questions.

Chapter 16

BOHR'S ATOM MODEL

Efforts to learn more about the structure of the atom had reached an impasse by 1913. Rutherford's scattering experiments clearly showed that the atom consists of a tiny positive nucleus, containing most of the atom's mass, and enough lightweight external electrons to preserve the atom's neutrality. The dilemma was that these electrons presumably must orbit or else fall into the nucleus; but circling electrons should radiate energy, according to Maxwell's electromagnetic theory, and thus be slowed down by loss of energy and fall into the nucleus after all. It appeared that the orbiting-electron model must be self-destructive.

Additionally, an orbiting electron would radiate *one* frequency of energy if it was postulated to be in a stable orbit, and thereby would violate the Law of Conservation of Energy (a *stable* orbit means *no* loss of energy); or else the electron would radiate a continuously changing range of energies as it spiraled inward toward the nucleus. Neither of these possibilities corresponded to observed facts: (1) The atom ordinarily does not radiate at all; (2) when it does, upon excitation, the radiations consist of several discrete frequencies.

Niels Bohr (1885–1962) addressed himself to this problem and designed an atom model that specifically accounts for the observed spectra.[1]

---

[1] Bohr was born in Denmark the year that Balmer discovered his spectral formula. After obtaining his Ph.D. in Copenhagen in 1911, the year of Rutherford's nuclear atom, Bohr went to England to work with J. J. Thomson and with Rutherford. Hence Bohr's interest in the nuclear atom. He was only twenty-eight when he made his bold proposal regarding atomic structure. Later he became professor of theoretical physics at Copenhagen. On the eve of World War II, in January 1939, he attended a conference on theoretical physics in Washington and reported some ideas about uranium fission suggested by Otto Frisch and Lise Meitner, two Austrian refugees working in Copenhagen and Stockholm. These ideas turned out to be the

That this model, so restrictedly devised, could be modified and broadened to explain many more observations of an entirely different nature bears witness to the apparent validity of Bohr's concepts. These other observations include the combining power of atoms, the periodicity of chemical properties, ionization energies of atoms, the photoelectric effect, and information from collisions between atoms and electrons. Some of these topics have already been broached, others are discussed in later chapters.

BOHR'S BASIC ASSUMPTIONS

The emission and absorption of radiation by an atom imply, according to the Law of Conservation of Energy, that the atom itself has changed its energy content. Perhaps one term (R/n^2) in the Rydberg-Ritz equation, $\nu = R/n_1^2 - R/n_2^2$, represents the initial energy state and the other term the final energy state of the atom. For emission, the difference between the two would represent the energy released by the atom; for absorption, the difference would represent the energy taken in by the atom. Since discrete frequencies are observed, conceivably only certain energy states of the atom are possible, and these energies would correspond to R/n^2, n being limited to integers. Also, if these different energy states of the atom were associated with different electron orbits, this would suggest that only certain orbits are possible.

These ideas led Bohr to the following postulates:

1. Of the infinite number of possible circular orbits of an electron about an atomic nucleus, only *certain discrete orbits* actually occur. Each of these orbits is associated with a particular energy state of the atom (potential plus kinetic energy).

2. Contrary to Maxwellian theory, the electron, in spite of being in accelerated motion, does *not* emit electromagnetic radiation while it is in one of these orbits. (This was simply an ad-hoc postulate to account for the observation that atoms do not radiate unless excited.)

3. Radiation is emitted or absorbed only when an electron goes from one allowed orbit to another. In that case the energy is not dribbled out gradually but is emitted or absorbed suddenly in a single package called a *quantum*. The energy in this package is equal to the difference between the initial and final energy states of the atom:

$$E_{radiation} = E_{initial} - E_{final}$$

seed of the Allied atomic bomb. When the Germans occupied Denmark, Bohr escaped to England in a small boat and later went on to the United States to participate in the Manhattan Project, devoted to the top-secret development of the atomic bomb.

Einstein had related the energy of a radiation package to its frequency by the formula

$$E = h\nu$$

where h is a proportionality constant, called Planck's constant.[2] This relationship together with Bohr's third assumption would explain why the frequencies of emitted radiation can always be expressed as the difference between two arithmetic terms,

$$\nu_{\text{radiation}} = \frac{E_{\text{radiation}}}{h} = \frac{E_{\text{initial}}}{h} - \frac{E_{\text{final}}}{h}$$

The terms E/h are equal to the terms R/n^2 in the Rydberg-Ritz formula; the energy states are therefore equal to hR/n^2.

THE ENERGY-LEVEL DIAGRAM

On the basis of these assumptions and the experimentally observed frequencies, Bohr drew up an *energy-level diagram* of the hydrogen atom. For each value of the integer n he calculated an energy state, or *energy level*,

$$E_n = \frac{hR}{n^2}$$

When he inserted numerical values for the Rydberg constant (3.288×10^{15} sec^{-1}) and for Planck's constant (6.626×10^{-27} erg-sec), he obtained these energy levels:

$$E_n = 217 \times \frac{10^{-13}}{n^2} \text{ ergs}$$

The energy level for $n = 1$ accordingly is 217×10^{-13} erg; for $n = 2$, is 54.3×10^{-13} erg (remember to divide by n^2); and for $n = \infty$, is zero erg (Table 16-1).

Sometimes this energy is expressed in a smaller energy unit, the *electron volt, ev*, which is equal to 1.60×10^{-12} erg. The equivalent form of the previous formula then is

$$E_n = \frac{13.6}{n^2} \text{ ev}$$

The first level is 13.6 *ev*; the second is 13.6/4, or 3.40 *ev*; the third is 13.6/9, or 1.51 *ev*; the level for $n = \infty$ again is zero (Fig. 16-1).

[2] This formula was introduced in Chapter 13. Its origin is discussed in the next chapter.

Table 16-1. Energy Levels of the Hydrogen Atom

	Energy		Equivalent
n	Ergs	Electron volts	wave number[a] (cm^{-1})
1	-217.3×10^{-13}	-13.60	$-109,678$
2	-54.3	-3.40	$-27,420$
3	-24.2	-1.51	$-12,186$
4	-13.6	-0.85	-6855
5	-8.7	-0.54	-4387
6	-6.0	-0.38	-3047
7	-4.4	-0.28	-2238
8	-3.4	-0.21	-1714

[a] $\tilde{\nu}_{cm^{-1}} = \dfrac{\nu_{sec^{-1}}}{c} = \dfrac{E/h}{c} = \dfrac{E}{hc}$

The reader may be puzzled by the negative energy values in Figure 16-1 and Table 16-1. Negative values are possible because the choice of a reference point for such an energy-level diagram is arbitrary.[3] After all, the only experimental data—the observed frequencies—are *differences* between energy levels. It is mathematically more convenient to have energies run from 13.6 *ev* at the bottom to zero at the top, rather than, say, from zero to $+13.6$ *ev*. The advantage of this scheme will become evident later on, when we present a more detailed physical model of Bohr's atom in Chapter 18.

TRANSITIONS BETWEEN ENERGY LEVELS

With this scheme of energy levels, emission of radiation is associated with transitions from higher-n levels (higher energies, *less*-negative values) to lower-n levels (lower energies, *more*-negative values). For instance, emission is associated with the transition from -1.51 *ev* ($n = 3$) to -3.40 *ev* ($n = 2$):

$$E_{radiation} = E_{initial} - E_{final}$$

$$= (-1.57) - (-3.40) = 1.89 \ ev$$

Or, in terms of wave numbers (*see* Table 16-1, last column):

$$\tilde{\nu}_{radiation} = (-12,186) - (-27,420) = 15,234 \ cm^{-1}$$

[3] An analogous arbitrary reference point is our choice of sea level as "zero" elevation." On that basis we have positive elevations on mountains, and negative elevations in places like Death Valley.

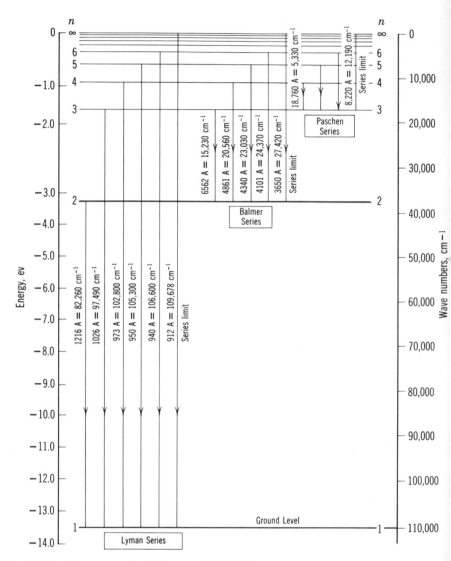

Fig. 16-1. The energy-level diagram of the hydrogen atom.

We can convert this wave number to wavelength:

$$\lambda = 1/\nu = 1/15{,}234 \text{ cm}^{-1} = 6.562 \times 10^{-5} \text{ cm} = 6562 \text{ A}$$

Table 15-1 (p. 212) lists this wavelength as the first line of the Balmer series ($n_1 = 2$, $n_2 = 3$).

Absorption of radiation is associated with the reverse process—that is, transitions from lower to higher energy levels. It should be evident that the same energy differences—and therefore the same observed frequencies, wave numbers, and wavelengths—are involved in absorption and in emission of radiation.

We can now explain the various spectral series in the emission spectrum of hydrogen. The Lyman series, $\nu = R(1/1^2 - 1/n^2)$, is the result of transitions from the various levels ($n = 2, 3, 4, \ldots$) to the *ground level* ($n = 1$). The energy-level diagram shows that these are all relatively high-energy transitions, because the first of them (from $n = 2$ to $n = 1$) alone involves $13.60 - 3.40 = 10.20$ *ev*, and all the other transitions of that series are greater. High energy means high-frequency radiation; hence the Lyman series is found in the ultraviolet region of the spectrum. Note that the lengths of the transition arrows are proportional to the energy differences, and therefore to radiation energies and frequencies. Because the upper energy levels are more closely spaced, transitions from successive levels to the same lower level have smaller frequency differences. This accounts for the convergence of each spectral series.

The Balmer series, $\nu = R(1/2^2 - 1/n^2)$, represents transitions from $n = 3, 4, 5, \ldots$ to $n = 2$. The lowest transition ($3{\rightarrow}2$) involves $3.40 - 1.51$ *ev*, or 1.89 electron volts. Radiation of this energy has a wavelength of 6562 A, which is the first member of the Balmer series, in the red part of the visible spectrum. The largest of the Balmer transitions, from $n = \infty$ to $n = 2$, releases 3.40 *ev* of energy, or a wavelength of 3650 A, which is the series limit. This radiation is in the near ultraviolet. Thus almost the entire Balmer series is in the visible part of the spectrum. The other series all involve smaller energy transitions, so that they appear in the infrared region.

At this point we wish to call attention to the differences between the contributions of Balmer and Bohr. Balmer's method involved no theory and only a little logic; it was essentially a trial-and-error procedure of fitting numbers to a formula. Balmer did not have an explanation for the spectral lines or their mathematical regularity. Bohr, on the other hand, formulated a theory that offered a logical explanation for the Balmer formula. Bohr introduced these novel features:

1. The postulation of only certain allowed energy levels.

2. The postulation of their stability, in violation of classical principles.

3. The assumption that, when a transition occurs between energy levels, energy will be emitted or absorbed suddenly in a single package (a quantum) rather than spread out in a continuous spectrum over a period of time.

The background and basis for these ideas is examined in the next chapter.

BRIEF DESCRIPTION OF BOHR'S ATOM MODEL

Bohr combined his assumptions with the existing theory to build a model of the hydrogen atom. This model is discussed in detail in Chapter 18. It has circular electron orbits, with diameters proportional to n^2. The smallest orbit corresponds to $n = 1$; the next orbit is four times (2^2) as large, and so on. In terms of this model, energy transitions can be pictured as sudden jumps of electrons from one orbit to another. Emission means jumps from outer to more stable inner orbits; absorption of radiation means promotion of electrons from inner to outer orbits.

With this model Bohr predicted a hydrogen spectrum that agreed excellently with the observed spectrum. This close agreement lent support to his model but did not prove it—after all, wasn't the model devised specifically to account for the spectrum? But when *other* observations, for which the model was not expressly engineered, could be predicted correctly, scientists began to place more trust in the model (*see* Chapters 18 and 19).

Before we end this chapter, we should point out that the energy-level concept we have developed so far is independent of any physical pictures, or models, of the atom. A physical picture may help us visualize the various processes; furthermore—and this is very important—a physical picture may allow us to predict *other* properties of the atom. Nevertheless, if the physical picture proves to be false or in need of amendment, its incorrectness will *not* affect the energy-level diagram.

Chapter 17

THE BACKGROUND OF
BOHR'S IDEAS

> *"Certain ideas at certain times are in the air; if one man does not enunciate them, another will do so soon afterwards."*
>
> F.A. Kekulé (1829–1896)

Bohr's quantum theory of atomic structure, proposed in 1913, arose out of the inability of classical physics to explain the experimentally observed spectrum of atomic hydrogen.

As usual, however, Bohr's unorthodox ideas did not evolve in a scientific vacuum. The way had been paved by two earlier quantum theories: Planck's quantum theory, proposed in 1900 to explain the experimentally observed distribution of energy in the spectrum of a black body; and Einstein's quantum theory of light, proposed in 1905 to explain the photoelectric effect. We shall discuss both of these in the present chapter, and then return to a more detailed examination of Bohr's atom model.

PLANCK'S QUANTUM THEORY FOR BLACKBODY RADIATION

Planck's quantum theory arose out of his attempt to explain radiation from solid surfaces, in particular from the surfaces of "black bodies." An ideal *black body* is defined as a body that absorbs all radiation falling upon it and reflects none. Although such a body does not reflect, it does emit radiation—mostly infrared (heat) radiation if it is at room temperature, and mostly visible or ultraviolet radiation at higher temperatures. In fact, a black body is a better emitter of radiation than a nonblack body, which reflects much of the light incident upon it.

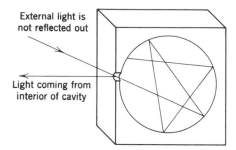

External light is
not reflected out

Light coming from
interior of cavity

Fig. 17-1. The ideal blackbody radiator—a cavity with a small entrance/exit hole. The whole apparatus is kept at constant temperature in an oven. Entering radiation is eventually absorbed after multiple reflections inside the cavity. Escaping radiation has the properties of black-surface radiation.

However, there is no such thing as a perfectly black surface. Even the blackest surface reflects perhaps one percent of the radiation falling upon it. But we can obtain the equivalent effect of a totally black surface by means of a specially designed cavity. Picture a hollow ball or cylinder with a single small hole in it (Fig. 17-1). All radiation entering the hole is eventually absorbed inside after multiple reflections. The hole thus acts as a perfect absorber, or ideal black surface. If we maintain this device at constant temperature, for instance in an oven, its inside wall will radiate into the cavity. The radiation escaping through the small hole has the properties of black-surface radiation. Such radiation is called cavity radiation or, more commonly, *blackbody radiation.*

Radiation from black bodies was first studied by G. R. Kirchhoff in 1859. In the next forty years various attempts were made to find a formula describing the spectrum of blackbody radiation; that is, to find a formula defining the intensity of each wavelength, or frequency, of radiation emitted by a black body. Two formulas in particular were partially successful. But one of them failed to predict correct intensities at the ultraviolet end of the spectrum; the other failed at the infrared end, as shown by measurements obtained in 1900. Later that same year the German physicist Max Planck (1858–1947) came up with a formula that predicted correct radiation intensities over the whole spectral range. He combined the best features of the two earlier formulas and modified them until the predictions of his formula were in accord with observed intensities.

Devising a workable formula was in itself a major achievement. But Planck was not satisfied with merely creating a formula. After all, *any* set of data can be fitted by a formula if one is sufficiently persistent. Up to this point Planck's accomplishment was akin to Balmer's, who found an empirical formula to fit the spectrum of atomic hydrogen. Planck wanted more than that. He wanted a theoretical explanation for his formula.

Heinrich Hertz's discovery in 1887 of radio waves from oscillating electric currents had convinced Planck that the key to the blackbody spectrum would be found in the laws governing absorption and emission of radiation by electric oscillators. These laws were embodied in Maxwell's electromagnetic theory (Chapter 13). Planck pictured a black body as containing many tiny electric charges in constant oscillation. Some of these charges might be electrons, which had been discovered only three years earlier by J. J. Thomson. Each little oscillator would vibrate at a definite frequency—analogous to the characteristic frequency of a given spring-and-weight system or of a pendulum with a given length. Planck believed that any given oscillator would absorb or emit radiation only of the same frequency as that of the oscillator itself. Thus for a black body to absorb and emit radiation of all frequencies—from the ultraviolet to the infrared—it would have to contain very many oscillators, having the whole range of frequencies. But any one oscillator would vibrate at only one characteristic frequency.

The energy of a given oscillator would depend on its characteristic frequency and would vary with the amplitude of the vibration, just as a pendulum's energy depends on the amplitude, or height, of its swing. Absorption or emission of radiation by an oscillator would affect its energy and thereby its amplitude of oscillation, but not its frequency.

Up to this point Planck was in accord with accepted, or "classical," theory. But on that basis Planck, like others before him, was unable to explain the blackbody spectrum. To do this he had to break drastically with traditional theory. Planck did so, with great reluctance, by introducing two novel assumptions.

According to classical theory, a given oscillator could have *any* energy. In Planck's picture this would have meant that a given oscillator could oscillate with *any* amplitude, just as a pendulum can swing with any amplitude. Not so, according to Planck. He made the novel assumption that a given oscillator can have *only certain discrete energy states* and that these states are all *whole-number multiples* of the lowest positive energy state. Thus, if we designate the value of the lowest positive energy state of a given oscillator by the symbol E_0, the allowed energy states of this oscillator might be 0, E_0, $2E_0$, $3E_0$, $4E_0$, etc. Classical theory would have allowed *any* energy, such as $1.19E_0$ or $2.64E_0$. But Planck restricted the allowed energies to *integral* multiples of E_0. It is often useful to picture these allowed *energy levels* as evenly spaced rungs on a ladder (Fig. 17-2). Note that the spacing between adjacent energy levels of a given oscillator is a constant E_0.

Planck's other novel assumption involved the emission of radiation by the oscillators. According to classical electromagnetic theory, an electric

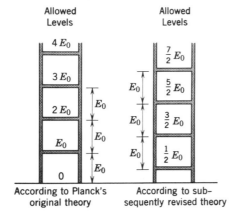

Fig. 17-2. Allowed energy levels of a given oscillator. In Planck's original theory, the allowed levels were 0, E_0, $2E_0$, $3E_0$, $4E_0$, etc. In the subsequently revised theory, zero was *not* an allowed level. Instead, the first level was half a rung-interval above zero; we have accordingly designated it as $1/2\ E_0$, and the following levels as $3/2\ E_0$, $5/2\ E_0$, $7/2\ E_0$, etc. The reason for the revision will become clear during our discussion of quantum mechanics in Part Four. The distinction between the two versions is of no importance if we concern ourselves only with blackbody radiation, because the spacing between levels is identical in the two versions, namely E_0. Other oscillators vibrating at other frequencies would have different values of E_0 and therefore different spacing between levels.

oscillator radiates all the time it is vibrating. But Planck postulated that radiation is emitted only when an oscillator jumps down to a lower energy level. If it jumps to the next-lower level the energy emitted will be E_0. Planck postulated further that this energy is given off suddenly in a short pulse of radiation, characterized by a single frequency or wavelength. Such a pulse, or packet, of radiation energy is called a *quantum*.

To understand how various frequencies of radiation are obtained from different oscillators, we first need to examine the quantitative relationship of the energy E_0 to an oscillator's frequency, f. We stated previously that, in Planck's picture, the energy of an oscillator depends on the frequency and amplitude of its vibrations. In a series of different oscillators all vibrating at their lowest amplitudes, the different energy values, E_0, are proportional to the frequencies, f. That is, those oscillators having, say, twice the frequency of vibration as others will have twice the energy; and so on. We can express this mathematically by the expression

$$E_0 \propto f$$

(where the symbol $\propto$ means "is proportional to") or by the equation

$$E_0 = hf$$

where the constant h is a "proportionality constant," subject to experimental determination.

Planck also assumed that oscillators absorb and emit only radiation of the same frequency, ν, as that of the oscillator itself:

$$\nu = f$$

$$\text{and} \quad h\nu = hf = E_0$$

Accordingly, different oscillators having different frequencies—and different spacings, E_0, on their energy ladders—are responsible for emitting different frequencies of radiation.

With his model of little oscillators having definite energy levels and emitting quanta of energy only when making transitions from one level to another, Planck satisfactorily explained the observed features of blackbody radiation and could *derive* the formula that he had already obtained empirically.

There remained one major difficulty. His model satisfactorily explained *emission* of radiation from black bodies, but it did not do so well for *absorption* of radiation. In emission, when an oscillator decreased its energy by dropping to a lower level, it radiated energy in a single short pulse, or quantum. Planck did not concern himself further with the nature of the radiation. Presumably, in accordance with Maxwell's electromagnetic theory, this packet of energy would spread out in all directions in a wavelike manner. Fine, so far. But, in the reverse process, in absorption, how could an oscillator suddenly gather up enough energy from dissipated waves to promote itself all the way up to the next-higher energy level, all in one jump? Planck did not satisfactorily answer that question.

EINSTEIN'S QUANTUM THEORY OF LIGHT

Planck could have solved the problem of blackbody absorption of radiation by assuming that energy not only is emitted in packets called quanta, but also is absorbed in quanta. This would have implied that radiation travels around as quanta of energy. But Planck was too strong a believer in the electromagnetic wave theory and was too conservative to make more than one radical proposal at a time. Thus it was left to the German-born scientist Albert Einstein (1879–1955) to make such a proposal five years later.

Actually, Einstein originally proposed his quantum theory of radiation in order to explain the photoelectric effect rather than blackbody absorption. The photoelectric effect had been discovered in 1887 by Heinrich Hertz but was not clarified until further experiments had been per-

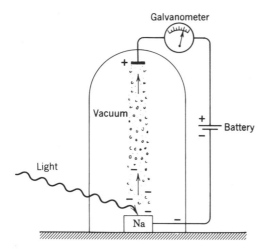

Fig. 17-3. The photoelectric effect.

formed in 1900 by Philipp Lenard; thus Planck was probably not aware of this phenomenon or did not realize its significance.

The Photoelectric Effect

In 1887, during the experiments that resulted in his discovery of radio waves (*see* p. 196) Heinrich Hertz observed that electric sparks jumped across a gap between two wires more readily when light fell upon the wires. Further investigation revealed that violet and ultraviolet light caused negative charges to be emitted by some metals, and that it was these charges that helped maintain the electric current in Hertz's experiment. Occurrence of this phenomenon in a vacuum demonstrated that the presence of air was not necessary.

A schematic experimental arrangement is shown in Fig. 17-3. A block of sodium metal is connected to the negative end of a battery. Above the block, a copper plate is connected to the positive end of the battery. Initially, no current flows through the vacuum inside the sealed glass jar. But when light shines onto the sodium surface, current immediately flows and registers on the galvanometer. The light has knocked negative charges out of the surface of the sodium block. These charges are repelled by the negative sodium surface and are attracted by the positive copper plate; thus current flows.[1]

---

[1] The modern "electric eye" makes practical use of this phenomenon: Light falling upon an electric eye causes electric current to flow; the current can then be used to open a door, activate a counter, turn on an alarm, etc.

In 1900 Philipp Lenard proved that the emitted charges were ordinary electrons. His method was similar to the magnetic-deflection method used by J. J. Thomson three years earlier for measuring the mass-to-charge ratio of the electron (Chapter 8).[2] In other words, certain frequencies of light can knock electrons out of the surfaces of some metals. The phenomenon accordingly is called the *photoelectric effect*. Lenard's identification of the emitted charges as electrons further strengthened the idea that electrons are contained in the atoms of many, if not all, elements (*see* Chapter 8).

The photoelectric effect cannot be explained on the basis of the electromagnetic wave theory of light. Spread-out light waves would not be able to concentrate enough energy in one place to knock electrons out of a surface. Nor can the explanation be found in an energy-storage mechanism. Computations have shown that it would take several months before an electron could gather the energy necessary to tear itself loose from the metal surface, but experiments revealed that current starts to flow immediately when light falls upon the surface.

Photons

The photoelectric effect remained unexplained till 1905. In that year, five years after Lenard's identification of photoelectrons (and Planck's quantum theory of blackbody radiation), Einstein proposed the quantum theory of light. He suggested that light energy is concentrated in packages, or quanta. These quanta of radiation later were called *photons,* the name most commonly used today. These photons would have enough concentrated energy, so that an electron absorbing this energy in a collision would gain the energy needed to escape from the metal surface.

The idea of the existence of photons is supported by additional experiments. If we use a beam of monochromatic light (only one frequency; say, violet), a certain number of electrons will be knocked out of the metal, and these electrons will possess a certain average velocity and thus a certain average kinetic energy, since $KE = \frac{1}{2}mv^2$ (*see* Chapter 6). Increasing the intensity—that is, brightness—of the violet beam increases the number of electrons, but not their velocity, or energy. This result must mean that each photon of violet light has a certain amount of energy, which in a collision is transferred to an electron. A more in-

---

[2] The reader will recall that Thomson, the "discoverer" of the electron, proved that cathode rays consist of particles with definite mass and charge. Lenard has been previously mentioned on p. 146 for his pioneering experiments on electrons in beta rays: He found in 1903 that these high-powered electrons from radioactive sources passed easily through thin metal foils, but sometimes were deflected quite sharply. Rutherford, repeating these experiments with heavier alpha particles, made similar observations and subsequently proposed his nuclear, or solar system-type, atom model.

tense beam merely means more photons, which can then eject more electrons, but the amount of energy per photon is the same as before.

Experiments show that we can increase the velocity of the photoelectrons by using ultraviolet rather than violet light. Evidently, the higher frequency of ultraviolet light means more energy per photon.

From a series of measurements of the velocities, or energies, of ejected photoelectrons, we can show that the energy, E, of a single photon, is proportional to the frequency, ν, of the radiation. Einstein expressed this relationship in the equation

$$E = h\nu$$

where h is a proportionality constant that can be determined by experiment.

The value of h was determined eleven years later by R. A. Millikan (who had measured the charge of the electron in his famous oil-drop experiment—see Chapter 8). The value of h found by Millikan from his photoelectric experiments is identical to the value of the constant h determined previously from measurements on blackbody radiation (p. 229). This constant has become known universally as *Planck's constant*. Its modern value is

$$6.626 \times 10^{-7} \text{ erg-sec}$$

It appears again in Bohr's theory of atomic structure (Chapter 18) and in several other seemingly unrelated contexts; consequently it is considered one of the fundamental constants of nature, like the velocity of light.

The dependence of photon energy on wave frequency, $E = h\nu$, explains why violet light will cause the ejection of electrons from sodium metal whereas red light will not. Red light has double the wavelength (8000 A) and therefore half the frequency of violet light ($\lambda = 4000$ A); thus red light has only half the energy of violet light—not enough to knock out photoelectrons. On the same basis the relationship $E = h\nu$ explains the great penetrating power of X rays and gamma rays.

Waves or Particles?

It is ironic that Hertz not only discovered radio waves and thus helped confirm Maxwell's electromagnetic wave theory, but in the same series of experiments also discovered the photoelectric effect, which turned out to be incompatible with Maxwell's theory.

But Einstein's new theory was not easily accepted. Even Max Planck, the inventor of quantum theory, at first resisted Einstein's photon theory, feeling that it would be a reversion to Newton's discredited corpuscular theory.

Actually, Einstein's photons are not corpuscles, or particles, in the usual meaning of the word. In contrast to true corpuscles, photons are merely bundles of energy that have no mass. And photons exist only in motion; they cease to exist once they are stopped.

Beyond having to overcome the reluctance of scientific conservatives, Einstein's theory had a built-in ambiguity. Einstein speaks of *packages* of light, but he described the energy of these light packages in terms of their *frequency,* and frequency definitely is a *wave* property. The frequency of monochromatic light is computed through the equation $\nu = c/\lambda$, where c is the velocity of light and λ is the wavelength. The latter is determined by measuring diffraction angles (*see* p. 182). But diffraction and interference can be explained only on the basis of a wave theory of light (*see* p. 180). Thus Einstein used the results of wave theory to describe the energy of his photons.

Moreover, although some phenomena could be explained through either a wave theory or a corpuscular theory, a wave theory was definitely needed to explain phenomena like diffraction and interference; and some sort of "corpuscular" theory was definitely needed to explain phenomena like blackbody radiation and the photoelectric effect. Two seemingly contradictory theories were needed to explain the observed properties of light! Scientists eventually learned to live with this dilemma and used whichever theory provided the best answers. We shall see in Chapter 21 how, two decades later, the new "quantum mechanics" finally resolved the dilemma of particle-wave duality by unifying the two theories.

CONCLUSION

We have now seen where Bohr in 1913 obtained the radical ideas for his model of the hydrogen atom:

1. discrete orbits, each associated with definite energy levels;
2. nonemission of radiation while the electron is in one of these orbits;
3. absorption and emission of radiation only when the electron jumps to another orbit, or energy level; and
4. absorption and emission of radiation only in quanta.

The first three are direct adoptions from Planck's model of blackbody radiators, and the fourth has its origin in both the blackbody model and Einstein's theory of radiation.

Planck's and Einstein's success resulted from persistent efforts to make theory fit the facts, according to the motto *Theory Guides, Experiment Decides*. When possible, they adapted traditional theory. When this procedure proved inadequate to explain observed fact, they did not hesitate to break with classical theory and introduce completely new ideas. Bohr, as we have seen, likewise amalgamated concepts from classical theory with unconventional new ideas. In the next chapter we examine his assumptions in more detail and show just how he proceded to "build" his atom model.

Chapter 18

BOHR'S HYDROGEN ATOM IN MORE DETAIL

In Chapter 16 we devised an energy-level diagram for the hydrogen atom to account for the observed spectrum. No particuar physical picture of the atom was involved in establishing relative energy levels (although we did mention Bohr's idea of circular orbits of the electron). In our approach, the spacing between energy levels was derived simply from the frequencies of observed spectra.

This energy-level diagram enables us to understand the salient features of the hydrogen-atom spectrum: (1) the presence of discrete lines; (2) the appearance of these lines in groups, or series, in more or less separated regions of the spectrum; (3) the decreasing separation of lines in each series, with convergence toward a limiting frequency; and (4) the fact that various arithmetic terms of the Rydberg-Ritz formula combine to yield other observable frequencies. The energy-level diagram not only systematizes knowledge acquired experimentally; it also allows us to predict the presence of other hydrogen spectral lines. In fact, Humphreys was guided by Ritz's Combination Principle and the energy-level diagram in his successful search for a new spectral line in 1952.

But this approach to the energy-level diagram leaves us without a tangible physical picture of the atom. Fortunately, Bohr did not proceed in this manner. As we pointed out in Chapter 16, Bohr postulated a physical atom with definite electron orbits. He then added two more postulates—(1) no radiation is emitted while the electron remains in any orbit, and (2) a single quantum of radiation is emitted when the electron jumps from one orbit to another. On the basis of these three postulates Bohr *derived* the energy levels. In other words, he started with

235

basic postulates, calculated energy levels, and then showed that his derived spectral frequencies were in accord with observation.

In the first part of this chapter we trace Bohr's procedure in some detail. In the later sections and the next chapter we discuss the achievements and shortcomings of Bohr's model. Correct prediction of observed frequencies on the basis of fundamental postulates was a major achievement in itself. But Bohr's model accomplished much more than that. Using it, scientists correctly predicted, or at least explained, various additional observations, some of which were quite independent of spectroscopy.

Today we know that Bohr's model is not wholly accurate. Nevertheless, scientists frequently still use it to grasp some chemical problem quickly. Because of this and because Bohr's ideas were an important milestone along the path of understanding atomic structure, we devote this chapter and the next to closer scrutiny of Bohr's ideas and to interpretation of various observations in terms of Bohr's atom model.

ANGULAR MOMENTUM

Bohr's assumption of definite electron orbits was cast in the form of his postulating definite ("quantized") values of *angular momentum* of the circling electron. Just what is angular momentum?

Chapter 6 introduced momentum as the property that tends to keep a body moving with unchanged velocity—that is, with unchanged speed and in the same direction. To change the momentum of a body or a system requires an outside force (Newton's Second Law—$F = ma$; $F\Delta t = \Delta p$); in the absence of such force, momentum remains constant (Newton's Third Law—Conservation of Momentum).

We can say similar things about circular motion. In the absence of friction, a rotating wheel continues to rotate indefinitely. To increase or decrease its rate of rotation requires an outside force. In the absence of such force, its rotation remains constant; we say, its "angular momentum" is conserved.

Angular momentum, like linear momentum (which we have heretofore simply called "momentum"), is a vector quantity; it has both magnitude and direction. Force is required not only to change a wheel's rate of rotation but also to change its direction—for instance, to tilt it. We can demonstrate this principle with a bicycle. A very slowly moving bicycle rider cannot lean sideways far without tipping over; but a rapidly moving rider can lean quite far from the vertical without falling. Rapidly spinning bicycle wheels have more angular momentum, which resists change in orientation—i.e., resists tipping over. Likewise, it is angular

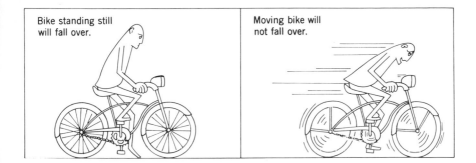

momentum that gives gyroscopes their peculiar stabiltiy, which makes them so useful in ships' compasses and stabilization systems and in inertial guidance systems of missiles.

To deal quantitatively with angular momentum, we introduce *angular velocity*. Whereas linear velocity is linear distance traveled divided by time, angular velocity is *angular* distance divided by time. For example, the earth rotates once on its axis (360°, or 2π radians) every 24 hours.[1] Its angular velocity (the Greek letter omega, ω, is its usual symbol) is:

$$\omega = \frac{360°}{24 \text{ hours}} = 15° \text{ per hour}^2$$

or

$$\omega = \frac{2\pi \text{ radians}}{24 \text{ hours}} \approx \frac{2 \times 3.14 \text{ radians}}{24 \text{ hours}} = 0.262 \text{ radians/hour}$$

Problem 18-1. An LP record makes 33 1/3 revolutions per minute (rpm). What is its angular velocity in degrees and in radians per mintue? In degrees and in radians per second?

Linear velocity v and angular velocity ω are related as follows.

$$v = \omega r \tag{1}$$

or

$$\omega = v/r \tag{2}$$

[1] Radians are angular measure; a circle contains 2π radians (approximately 6.28 radians), making a radian equal to about 57°. Radians are defined as arc length divided by radius—i.e., length divided by length. Thus radians are dimensionless.

[2] Accordingly, time zones (1 hour's time difference) span 15° longitude on the average. For example, the difference in standard time between New York (74° longtitude) and Los Angeles (118°) is 3 hours.

where r is distance from the axis of rotation. Linear velocity increases with r. For instance, on a merry-go-round, children near the rim travel faster than children closer to the center, but both have the same angular velocity.

Problem 18-2. (a) What is the approximate linear velocity, in miles per hour, of a body on the surface of the earth? Calculate this in two different ways: via the earth's circumference (25,000 miles), and via the angular speed (in radians) and radius (4000 miles).
(b) What is the approximate speed in miles per hour of a satellite circling the earth every 1 1/2 hours at an altitude of 200 miles? (Its orbital radius is *not* 200 miles.)

In dealing with circular motion it is convenient to introduce another concept—*moment of inertia, I.* For an object of mass m rotating about an axis at a distance r, we define the moment of inertia as:

$$I = mr^2 \tag{3}$$

Moment of inertia plays a role in rotational motion analogous to the role of mass in translational motion. Thus linear momentum, p, is defined as the product of mass times linear velocity,

$$p = mv \tag{4}$$

whereas *angular momentum, L,* is defined as the product of moment of inertia times angular velocity,

$$L = I\omega$$

Alternatively, using Equations (2) and (3), we can express angular momentum as follows.[3]

$$L = I\omega = (mr^2)\left(\frac{v}{r}\right) = mvrv \tag{5}$$

$$\frac{1}{2}I\omega^2 = \frac{1}{2}(mr^2)\left(\frac{v^2}{r^2}\right) = \frac{1}{2}mv^2$$

Problem 18-3. The dimensions of angular momentum are ml^2t^{-1} (mass times length^2/time, from $mvr = m\frac{l}{t}l$). Show that Planck's constant, h, has the same dimensions. (Perhaps use the following steps: $E = h\nu$; $h = E/\nu$; find the dimensions of energy from $1/2\ mv^2$; recall the dimensions of frequency.)

---

[3] The kinetic energy of any orbiting body is $1/2\ mv^2$. In terms of rotational quantities it is $1/2\ I\omega^2$. We can show that these expressions are equivalent:

The *Law of Conservation of Angular Momentum* is one of nature's basic conservation laws, along with the Laws of Conservation of Mass, Energy, Linear Momentum, and Charge. Conservation of angular momentum was discovered in 1746 by the Swiss mathematician Leonhard Euler (1707–1783) and the Dutch-Swiss scientist Daniel Bernoulli (1700–1782) (although Newton had practically derived this law from Kepler's Second Law of Planetary Motion—*see* following paragraph).

A whirling ballerina applies the Law of Conservation of Angular Momentum. First she whirls with outstretched arms. Then, putting her arms closer to her body, she reduces her effective radius; to maintain constant angular momentum her rotational rate is increased:

$$L = I\omega = (mr^2)\,\omega$$

(ω increases as r^2 decreases)

Conservation of angular momentum also explains why a planet or a comet speeds up as it moves closer to the sun in an elliptical orbit. This is expressed in Kepler's Second Law of Planetary Motion, also called the Law of Equal Areas (1609).

Problem 18-4. Attach a small object to a piece of string. Whirl it in a circular orbit, allowing the string to wind up on your finger or your arm.

(a) Describe the object's rate of orbiting (i.e., its angular speed) as it spirals inward.

(b) How do you account for your observation?

(c) As it spirals inward, has the object maintained the same *linear* speed (v)?

(d) When the radius is halved, how are the angular and the linear velocities affected?

DERIVATION OF BOHR'S ENERGY-LEVEL FORMULA

Orbit Radii and Velocities

Bohr's hydrogen atom has an electron circling about a nucleus. This electron is held in orbit by an electrostatic force of attraction between the negative electron of charge $-e$ and the positive nucleus of charge $+Ze$, where Z is the number of protons (the atomic number, namely, one). According to Coulomb's Law of Electrostatic Forces (Chapter 7), attractive force is inversely proportional to the square of distance, which in this case is the radius:

$$F = \frac{(Ze)(-e)}{r^2} = \frac{-Ze^2}{r^2} \tag{6}$$

An object in circular motion at constant speed v has an average acceleration of v^2/r, where r is the radius. Newton's Second Law, $F = ma$, then takes this form for uniform circular motion:[4]

$$F = \frac{mv^2}{r} \tag{7}$$

Combining Equations (6) and (7), and remembering from Chapter 6 that acceleration is a vector quantity and has the same direction and therefore same sign as force, we obtain

$$\frac{mv^2}{r} = \frac{Ze^2}{r^2}$$

Canceling one r and cross-multiplying yields

$$mv^2r = Ze^2 \tag{8}$$

At this point, Bohr introduced a "quantum condition." He postulated that only certain values of angular momentum (mvr) are allowed. We say that momentum is *quantized*. According to Bohr, it is packaged in multiples of $h/2\pi$:

$$mvr = n(h/2\pi), \qquad \text{where } n = 1, 2, 3, \ldots \tag{9}$$

How did Bohr arrive at this postulate? As we discussed in Chapter 16, the discreteness of spectral lines indicated discrete ("quantized") energy levels. In both previous works on quantization—Planck's quantum theory for blackbody radiation and Einstein's quantum theory for light—the proportionality constant h appears and has the dimensions of angular momentum (*see* Problem 18-3). Bohr may at first have guessed that orbit sizes are fixed by the condition that the angular momentum of the revolving electron be a multiple of h; but he found that predicted and observed spectra actually agree when the angular momentum is assumed to be an integral multiple of $h/2\pi$, rather than of h itself.[5]

---

[4] Although we do not wish to derive this formula here, it is reasonable: (1) It takes more force ("centripetal force") to whirl an object in a small circle (small r) than in a large one; hence there is an inverse relationship between F and r. (2) It takes more force to hold a heavy and rapidly moving object in a circular path than a light and slowly moving object; hence there is a direct relationship between F and mv (actually, v²). (3) The dimensions of v^2/r are lt^{-2}, where l is length and t is time; these also are the dimensions of acceleration; hence the dimensions of mv^2/r are the same as those of force ($F = ma$).

[5] Bohr's reasoning was not quite as arbitrary as the text implies. Two principles guided him. One was the Correspondence Principle, which in this case relates the frequency of the emitted radiation to the frequency of revolution of the electron in its orbit. The other was an argument involving integration of the angular momentum over one revolution; the factor $1/2\pi$ appears in the result.

In Bohr's time (1913) four quantities in Equations (8) and (9) were known: Planck's constant, h (1900); the mass, m, and charge, e, of the electron (Thomson and Millikan, 1897–1910); and the atomic number, Z (Moseley, 1913). Hence Bohr could solve the two equations for the two unknowns: the electron's orbital radius, r, and velocity, v.

The velocity can be obtained by substituting Equation (9) into Equation (8):

$$mv^2r = (mvr)v = Ze^2$$

$$\frac{nh}{2\pi} v = Ze^2$$

$$v_n = \frac{2\pi Ze^2}{nh} \tag{10}$$

The subscript n emphasizes that the velocity depends on a particular value of n, which may be 1, 2, 3, etc. This velocity is proportional to $1/n$; that is, the electron travels more slowly in an outer orbit, as do planets in the solar system.

The radius can be obtained by first solving Equation (9) for r:

$$mvr = \frac{nh}{2\pi} \tag{9}$$

$$r_n = \frac{nh}{2\pi mv}$$

and then substituting Equation (10) for v:

$$r_n = \frac{nh}{2\pi m} \frac{nh}{2\pi Ze^2}$$

$$r_n = \frac{n^2h^2}{4\pi^2 mZe^2} \tag{11}$$

The subscript n again emphasizes that r depends on the value of n. Note that the radius is proportional to n^2. A larger quantum number n means a larger radius; the orbit radius increases as the square of n:

$$r_n = n^2r_1$$

where r_1 is the smallest allowed radius (when $n = 1$). Fig. 18-1 shows the first few orbits of the hydrogen atom, drawn to scale.

From Equation (11) Bohr calculated a value for r_1, the radius of the first orbit. The result, 0.53 angstrom, was in good agreement with values obtained from experiments on collision cross-sections (*see* Chapter 10)— a great accomplishment of Bohr's theory.

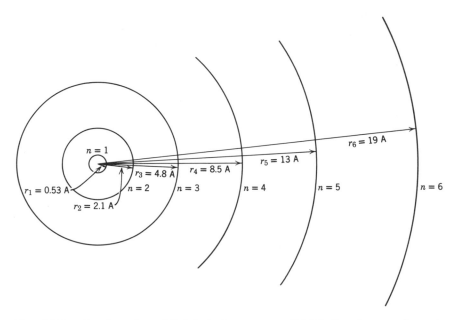

Fig. 18-1. Circular orbits of Bohr's hydrogen atom. Orbital diameters are drawn to scale.

Problem 18-5. Given the following experimentally determined parameters (all in cgs units, to simplify the calculations):

mass of electron, $m = 9.11 \times 10^{-28}$ g;

charge of the electron, $e = -4.80 \times 10^{-10}$ statcoul;

Planck's constant, $h = 6.63 \times 10^{-27}$ erg-sec;

atomic number of hydrogen, $Z = 1$.

(a) Calculate the radius, r_1, of the first Bohr orbit ($n = 1$) of the hydrogen atom, in centimeters. (The answer can be obtained directly by simple substitution in Equation (11), because all quantities are given in cgs units.)

(b) What is the diameter of this orbit in angstroms?

(c) From the answer to (a), without again resorting to Equation (11) determine the radius of hydrogen orbits 2, 3, 4, and 5; compare your answers with the scale drawing of Fig. 18-1. What is the relation of r_2, r_3, r_4, and r_5 to r_1?

Problem 18-6. (a) From Equation (14) calculate the velocity of the electron in the first orbit; in the second and third orbits.

(b) What percent of the speed of light is the electron's speed in the first orbit?

Energy Levels

The total energy of an electron in any orbit consists of kinetic and potential energies. For the one electron in a hydrogen atom, the kinetic energy can be expressed with the help of Equation (8) as a function of charge and radius:

$$mv^2r = Ze^2$$

$$mv^2 = \frac{Ze^2}{r} \tag{8}$$

$$KE = \frac{1}{2}mv^2 = \frac{1}{2}\frac{Ze^2}{r} \tag{12}$$

The potential energy is[6]

$$PE = \frac{-Ze^2}{r} \tag{13}$$

Note that PE is twice as large as KE for any given r, but has the opposite sign. PE thus is the dominant factor in determining the total energy, E:

$$E = KE + PE$$

$$= \frac{Ze^2}{2r} + \frac{-Ze^2}{r}$$

$$= \frac{Ze^2}{2r} - \frac{2Ze^2}{2r}$$

$$E = \frac{-Ze^2}{2r} \tag{14}$$

Total energy here is always negative. With increasing radius the energy becomes less negative; that is, the energy increases with increasing radius. When the radius becomes very large ($r \to \infty$), the energy of the electron approaches zero. In other words, the electron's energy can vary from various negative values in inner orbits to near zero in far-out orbits.

[6] The potential-energy expression is derived by multiplying force by distance. Since the force is variable, decreasing with distance, we must use integration:

$$PE = \int_{\infty}^{r} Fdr = -\int_{\infty}^{r}\frac{Ze^2}{r^2}dr = -\left(0 - \frac{-Ze^2}{r}\right) = \frac{-Ze^2}{r}$$

The minus sign indicates that energy is released from the atom when its electron moves toward the nucleus. As the electron moves in closer to more stable levels, in response to the attractive force, the potential energy decreases and so becomes more negative.

According to Bohr, not all radii are allowed, only those in which

$$r_n = \frac{n^2 h^2}{4\pi^2 m Z e^2} \tag{11}$$

Substituting this expression into Equation (14) yields the allowed (quantized) energy values for the electron in the hydrogen atom:

$$E = \frac{-Z e^2}{2r} \tag{14}$$

$$= \frac{-Z e^2}{2} \frac{4\pi^2 m Z e^2}{n^2 h^2}$$

$$E_n = -\frac{1}{n^2} \frac{2\pi^2 m Z^2 e^4}{h^2} \tag{15}$$

Note that the values of allowed energy levels are inversely proportional to n^2:

$$E_1 = \frac{2\pi^2 m Z^2 e^4}{h^2} \tag{16}$$

$$E_n = \frac{E_1}{n^2} \tag{17}$$

where E_1 is the energy of the electron when it is in the smallest ($n = 1$) orbit. Numerically, E_1 is 217.3×10^{-13} erg, or 13.60 electron volts.

Problem 18-7. (a) Calculate E_1 (in ergs) from Equation (16) and the known values of m, Z, e, and h.
(b) What is the energy, in ergs and in electron volts, of the electron in the first Bohr orbit?
(c) What is the energy of the electron when it is in the second orbit? Express E_2 in terms of E_1 and also in electron volts.
(d) What are the values of E_3, E_4, and E_5, in terms of E_1 and in ev? Compare your answers with Table 16-1.
(e) Prepare an energy-level diagram from these values and compare your diagram with Figure 16-1.

THE RYDBERG CONSTANT

Chapter 16 gave the relation between energy levels and the experimentally observed Rydberg constant:

$$E_n = \frac{hR}{n^2}$$

With this formula Bohr could *calculate* the value of the Rydberg constant from his theoretically derived energy levels:

$$R = \frac{E_1}{h} = \frac{217 \times 10^{-13} \text{ erg}}{6.63 \times 10^{-27} \text{ erg-sec}}$$

$$R = 3.29 \times 10^{15} \text{ sec}^{-1}$$

This value agrees with the experimentally determined R value—another triumph for Bohr's model.

TRANSITIONS BETWEEN LEVELS

When an electron makes a transition from a higher energy level to a lower one, the radiated energy is the difference (ΔE) between the initial and final energy levels:

$$E_{\text{radiation}} = \Delta E = E_{\text{initial}} - E_{\text{final}}$$

Together with Equation (17) [$E_n = E_1/n^2$], the foregoing formula yields:

$$E_{\text{radiation}} = \Delta E = \frac{E_1}{n_i^2} - \frac{E_1}{n_f^2}$$

or, $$\Delta E = \left(\frac{1}{n_i^2} - \frac{1}{n_p^2}\right)E_1 \qquad (18)$$

where n_i and n_f are the n-values, respectively, of the initial and final states. For example, when an electron makes a transition from the $n = 3$ level to $n = 2$, the radiated energy is

$$\Delta E = E_3 - E_2 = \left(\frac{1}{3^2} - \frac{1}{2^2}\right)E_1 = \left(\frac{1}{9} - \frac{1}{4}\right)E_1$$

$$= (-0.139)(-13.6 \ ev) = 1.89 \ ev$$

We could have obtained this radiated energy directly from Table 16-1 by subtraction:

$$\Delta E = E_3 - E_2 = (-1.51) - (-3.40) = 1.89 \ ev$$

Problem 18-8. With the help of Table 16-1,
(a) determine the wave number (cm^{-1}) of the radiation for the $3 \rightarrow 2$ transition;
(b) convert this value to wavelength in angstroms. Use the formula

$$\lambda = 10^8 \lambda_{\ cm} = 10^8/\nu_{\ cm^{-1}}$$

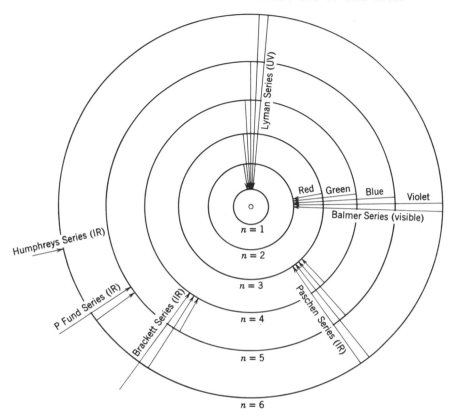

Fig. 18-2. Spectral series in relation to circular orbits of Bohr's hydrogen atom (not drawn to scale).

Compare your value with that listed for the first line of the Balmer series (*see*, for instance, the energy-level diagram in Fig. 16-1).

In Chapter 16 we used the energy-level diagram to explain the various series of the observed spectrum. Fig. 18-2 presents an analogous explanation in terms of electron orbits. A spectral line of the Lyman series results when an electron jumps from an outer orbit to the innermost ($n = 1$) orbit. A Balmer series line occurs when the electron jumps from an outer orbit to the $n = 2$ orbit; and so on for the other series.

Fig. 18-2 may be more graphic than the energy-level diagram of Chapter 16. From the latter, however, we can see more clearly why some transitions result in ultraviolet light while others yield infrared. More important, the energy-level diagram remains valid even though scientists later had to replace Bohr's orbits with another model.

Problem 18-9. Suppose that a Bohr hydrogen atom has been sufficiently excited so that its electron is in the third level. The electron can now return to the ground state—the normal ($n = 1$) state—by two paths. It can go directly from $3 \rightarrow 1$ (Case A), or it can go first from $3 \rightarrow 2$ and then later from $2 \rightarrow 1$ (Case B).

(a) From Table 16-1, determine the wave number of the emitted radiations in Cases A and B. Identify these as members of spectral series.

(b) Show that $\nu_{3 \rightarrow 2} + \nu_{2 \rightarrow 1} = \nu_{3 \rightarrow 1}$

(c) Convert the wave numbers to wavelengths and show that $\lambda_{3 \rightarrow 2} + \lambda_{2 \rightarrow 1}$ does *not* equal $\lambda_{3 \rightarrow 1}$ (*see* p. 193).

Problem 18-9 points up a major limitation of Bohr's theory. Although an electron promoted to the $n = 3$ level has two alternative ways of returning to the ground state, the theory fails to predict the relative probabilities of the two processes. These probabilities in turn determine the relative intensities of various lines in the spectrum. From still higher levels there are many more ways of returning to the ground level. Bohr's theory predicts only frequencies, and not intensities, of spectral lines.

Problem 18-10. To anwer the following questions, refer to Figure 18-2 or the energy-level diagram of Chapter 16, and to the table of energy levels in Chapter 16. The Bohr hydrogen atom at room temperature has its electron in the smallest orbit, $n = 1$, of energy E_1. The atom can absorb only energy that will promote the electron to another allowed orbit. The electron cannot exist between allowed orbits or between energy levels; hence the atom can absorb and emit energy only in definite quanta. The minimum energy that the atom can absorb to be promoted to the next level—in this case to $n = 2$—is called the *excitation energy*.

(a) What is the excitation energy, in electron volts, for a ground-state hydrogen atom?

(b) If the electron in (a) returns to the ground state ($n = 1$), what is the wave number of the emitted radiation?

(c) How many electron volts of energy are necessary to remove the electron completely from a ground-state hydrogen atom (that is, to promote the electron from $n = 1$ to $n = \infty$)? This energy is known as the *ionization energy*.

SUPPORTING EVIDENCE FROM ELECTRICAL EXPERIMENTS

In Problem 18-10 we derived the excitation and ionization energies of the hydrogen atom. These energies were deduced from the energy-level diagram, which was constructed solely on the basis of spectroscopic observations. In 1914 James Franck (1882–1964) and Gustav Hertz (1887–) actually measured excitation and ionization energies in several experiments.[7] Some experiments involved a combination of spectroscopic and electrical observations, and some involved only electrical observations. These independent electrical experiments lent further support to Bohr's idea of energy levels.

In their experiments on excitation energies, performed in modified cathode-ray tubes, Franck and Hertz bombarded gaseous atoms with electrons of known energies. Their electrical measurements revealed the following: (1) When the energy of the bombarding electrons was less than the excitation energy of the atoms, no energy transfer took place; the electrons retained their initial kinetic energies. (2) When the energy of the bombarding electrons was higher than the energy necessary to excite the atoms, the electrons lost energy only in amounts (quanta) equal to the atoms' excitation energy. In other words, atoms can accept energy only in definite-sized packages equal to the transition energies between levels.

Although Franck and Hertz used mercury vapor as the gas, we shall describe their experiments schematically as if they had used hydrogen atoms. In Problem 18-10 we deduced from the energy-level diagram that the (minimum) excitation energy of hydrogen ($n = 1 \rightarrow n = 2$) is 10.20 ev; when the electron returns to the ground state, this transition ($2 \rightarrow 1$) results in emission of the first Lyman series line—1216 A, or 82,258 cm^{-1}.

In the first experiment, hydrogen is bombarded with increasingly energetic electrons. At low accelerating voltages the hydrogen gas remains "dark" (no emission of radiation). But at 10.20 volts or more, when the electrons have an energy of 10.20 ev or more, the gas suddenly emits ultraviolet light with a wavelength of 1216 A (and an energy of 10.20 ev).

We explain this observation as follows. At lower energies the cathode-ray electrons collide *elastically* with the atoms—the electrons bounce off, retaining their original speeds and kinetic energies. No energy transfer occurs; the atoms do not accept energy from the electrons. But when the cathode-ray electrons have a critical energy of 10.20 ev or more, some

[7] The German physicist Gustav Hertz was born the same year (1887) in which Heinrich Hertz (1857–1894) discovered radio waves and the photoelectric effect. James Franck (1882–1964) was born in Germany, later became a U. S. citizen. In 1925 they were awarded the Nobel Prize for their work.

collisions are *inelastic*—precisely 10.20 *ev* of energy is transferred to the atom, and the colliding electron loses that much kinetic energy. In the atom, an orbital electron is promoted to the next-higher energy orbit. Radiation of 1216 A (10.20 *ev*) is emitted when that electron returns to its original level. The experiment thus indicates that 10.20 *ev* is the minimum electrical energy necessary to excite hydrogen atoms. This value agrees with that deduced from the spectrum of hydrogen.

In a similar experiment, Franck and Hertz measured the cathode-ray voltage needed to completely remove an electron from an atom ($n = 1 \rightarrow n = \infty$); that is, the ionization energy. For hydrogen, that energy is 13.60 *ev*. Again, electrical measurements were in complete accord with values from spectroscopically derived energy-level diagrams. We shall say more about ionization energies in the next chapter.[8]

EXTENSION OF BOHR'S THEORY TO OTHER ONE-ELECTRON "ATOMS"

Ordinary helium gas has a complex spectrum. But under very-high-energy excitation, a second spectrum appears with a simple structure that is very similar to the hydrogen spectrum. Similar series of lines appear but with larger spacing and with frequencies shifted to higher values. The frequencies can be described by the formula

$$\nu = 4R\left(\frac{1}{n_1{}^2} - \frac{1}{n_2{}^2}\right) \tag{19}$$

which differs from the Rydberg-Ritz formula for hydrogen only by the factor 4.

We can explain this spectrum as resulting from singly ionized helium, He^+; in other words, a doubly charged helium nucleus with a single electron. Because this ion, like a hydrogen atom, is composed of a nucleus and a single orbiting electron, Bohr's theory and its formulas should apply here also:

$$r_n = \frac{n^2h^2}{4\pi^2mZe^2} \tag{11}$$

and

$$E_n = -\frac{2\pi^2mZ^2e^4}{n^2h^2} \tag{15}$$

[8] This ionization process must not be confused with the photoelectric effect. Ionization energies refer to the removal of electrons from individual gaseous atoms or ions; the photoelectric effect deals with the ejection of electrons from solid surfaces, containing very many tightly packed atoms or ions.

For helium the nuclear charge, Z, is 2 (instead of 1 as for hydrogen). Accordingly, the radii of the various orbits of He^+ are only half as large as comparable orbits of hydrogen. Also, He^+ energy levels have four (2^2) times the values of the corresponding H levels. We can therefore use the same energy-level diagram, only we must multiply the scale by four. Now the lowest level, E_1, is 4 × 13.60, or 54.4 ev; E_2 is 4 × 3.40, or 13.6 ev; and so on. The minimum excitation energy ($1 \rightarrow 2$) is 54.4 − 13.6, or about 41 ev; the ionization energy ($1 \rightarrow \infty$) of He^+ is 54.4. These last two values also are four times the corresponding values of the hydrogen atom. Note that the 54.4-ev ionization energy refers to He^+, not He; that is, it applies to the removal of the *last* electron from helium.

The spectra of other single-electron ions such as Li^{++}, Be^{+3}, B^{+4}, C^{+5}, etc., can be explained similarly. Since $Z = 3$ for lithium, its energy levels lie nine times as low; and so on.

These atomic numbers, Z, were themselves just being determined in 1913–1914 by Moseley in his X-ray-scattering experiments (Chapter 11). Their successful use by Bohr in predicting spectra of single-electron ions provided an independent verification of their values.

So far, so good. However, spectroscopy allowed very precise measurements. These revealed that the Rydberg constant for He^+ is 109,722 cm^{-1}; but it is only 109,678 cm^{-1} for H, a difference of 44 cm^{-1}, or 0.04%. Not a large discrepancy, but definitely beyond experimental error. Even larger discrepancies were observed for Li^{++} and still-heavier one-electron ions.

Does the Bohr theory predict such a shift? Yes, if it is applied properly. Bohr had approximated the motion of the electron as a circling about a stationary nucleus. Actually, both electron and nucleus circle about their common center of mass. The lighter the nucleus, the more pronounced its motion. We can take this motion into account in Bohr's formula by replacing the mass, m, of the electron with a so-called "reduced mass," μ, whose value depends slightly on the mass, M, of the nucleus:

$$\mu = \frac{mM}{m + M} \tag{20}$$

(Note that reduced mass has the dimension of mass.) Thus corrected, Bohr's formula *does* precisely predict the observed values of the Rydberg constant as well as the observed spectra of various one-electron atoms.

Problem 18-11. (a) The mass M of the hydrogen nucleus (a lone proton) is 1840 times the mass m of an electron. Calculate the reduced mass of the hydrogen atom in terms of electron-mass units. ($m = 1$, $M = 1840$; i.e., 1 amu = 1840 electron-mass units.)

(b) Calculate the reduced mass of the He$^+$ ion.

(c) Calculate the reduced mass of a hypothetical extremely light atom whose nucleus has the mass of an electron.

In 1932 Harold Urey applied the reduced-mass effect to look for a new isotope of hydrogen, one with twice the mass of ordinary hydrogen. Lines in the spectrum of this substance should be slightly shifted from those in the spectrum of ordinary hydrogen. Urey was able to observe this shift and became the discoverer of *deuterium,* as the new isotope was called (*see* Chapter 11).

A MODIFICATION: ELLIPTICAL ORBITS

Despite many successes, Bohr's model of circular orbits failed to account for certain fine details in the spectrum of hydrogen. In 1915–1916 this led the German scientist Arnold Sommerfeld (1868–1951) to modify Bohr's original model.

Paschen (of the Paschen series) had measured spectra of hydrogen atoms and helium ions with the highest precision. To his surprise, he found that all "lines" were actually composed of *several* very fine lines close together. The lines were said to have "fine structure." For example, all lines of the Balmer series appeared as two lines, called *doublets.* This meant that the energy-level diagram was more complex than had been thought. Perhaps there were more orbits, with slightly different energy levels, than those shown by Bohr.

Bohr had previously considered elliptical orbits, analogous to Kepler's elliptical orbits for planets (p. 104). But Bohr believed that these all led to the same energy levels, so he did not consider them further. Now, however, Sommerfeld independently reconsidered the idea of elliptical orbits. He demonstrated that different orbit shapes do in fact have slightly different energies, depending on the eccentricity (flatness) of the ellipse. This could account for the observed fine structure of spectral lines.[9]

From the limited number of lines in the hydrogen spectrum Bohr had deduced that only a limited number of circular orbits existed. Similarly, Sommerfeld deduced from the limited number of fine-structure lines that only a limited number of elliptical shapes existed.

[9] The energy-level shift is due to the velocity-dependence of mass in accordance with Einstein's Theory of Relativity (1905). Electrons in elliptical orbits move fast enough to have their masses increased; this lowers their energies slightly (*see* Equation 15).

A short time later Sommerfeld and others introduced a further modification to Bohr's picture. It had been discovered that lines in an emission spectrum were split even more when the exicted atoms were placed between the poles of a magnet. Scientists believed that this splitting was caused by different spatial orientations (tilts) of some of the orbits. Again, the presence of only a limited number of splittings indicated that only a limited number of positions of the orbits were possible. Thus the orbits in Bohr's model of one-electron atoms apparently had quantized sizes, quantized shapes, and quantized tilts.

SUMMARY

Bohr's quantum theory was generally successful in explaining the various features of the hydrogen-atom spectrum: the presence of discrete emission or absorption lines; the grouping of these lines in series; the decreasing separation of the lines in each series, with convergence toward a limit at the low-frequency, long-wavelength, end of each series; and, finally, the precise frequency values of the individual spectral lines.

Even though Bohr had originally designed his model merely to explain the hydrogen spectrum, it accomplished quite a bit more than that:

1. The theory yielded a numerical value of the Rydberg constant, which agreed with the value measured by Balmer, Rydberg, and Ritz.

2. The model explained similar spectra of other one-electron "atoms" such as the He^+, Li^{++}, and Be^{+++} ions. In the process, it verified the values of Moseley's atomic numbers.

3. It accounted quantitatively for slightly different values of the Rydberg constant for the heavier one-electron "atoms." The shift was due to the relative motions of nuclei and electrons—the so-called reduced-mass effect.

4. Theoretical calculations yielded a value for the diameter of the hydrogen atom in accord with diameters measured by collision cross-sections.

5. The model predicted the results of electrical experiments by Franck and Hertz. These experiments involved inelastic collisions of electrons with atoms and measured excitation and ionization energies of atoms.

6. Later, splitting of spectral lines was discovered and could be explained by elliptical in addition to circular orbits.

7. Still later, the observation of line splitting in the presence of magnetic fields could be explained by different orbital tilts.

It seemed unlikely that such good agreement between theory and observation could be merely fortuitous; the agreement suggests, rather, that Bohr was on the right track.

Nevertheless, the model had a number of shortcomings. Although it predicted the frequencies of spectral lines, it gave no information about their intensities. The model indicated which individual transitions an electron *could* make in returning from an upper to a lower level (e.g., $3 \rightarrow 1$, or $3 \rightarrow 2 \rightarrow 1$,), but not which ones it *did* make.

Another major shortcoming of Bohr's theory was the arbitrary nature of its postulates: only certain allowed orbits, only certain shapes, only certain orientations. True, the model worked—it predicted results in accord with experiment. But *why?*

Moreover, Bohr's model dealt only with the simplest of atoms, those having only one electron. (The various orbits we discussed are *possible* orbits of that *one* electron; only one orbit is occupied at a time.) What about atoms with more than one electron? Bohr's theory did not deal with these. His model of discrete orbits did, however, guide the thinking of scientists toward an improvised model for atoms having more than one electron. That is the topic of the next chapter.

ANSWERS TO SELECTED PROBLEMS

18-1.   (a) $33.3 \times 360° = 12{,}000 \text{ deg/min.}$
            (c) $33.3 \times 360/60 = 200 \text{ deg/sec.}$

18-2.   (a) $25{,}000 \text{ mi}/24 \text{ hr} \approx 1050 \text{ mph}$
            or, $v = \omega r = 0.262 \text{ rad/hr} \times 4000 \text{ mi} \approx 1050 \text{ mph.}$
            (b) $\approx 17{,}000 \text{ mph.}$

18-5.   (a) $r_1 = n^2 \dfrac{h^2}{4\pi^2 m Z e^2} = \dfrac{(6.63 \times 10^{-27})^2}{4 \times 3.14^2 \times 9.11 \times 10^{-28} \times (4.80 \times 10^{-10})^2}$
            $= 0.53 \times 10^{-8} \text{ cm.}$
            (c) $r_3 = 9r_1 = 4.8 \text{ A.} \qquad r_n = n^2 r_1.$

18-6.   (a) $v_1 = \dfrac{1}{n} \dfrac{2\pi Z e^2}{h} = \dfrac{2 \times 3.14 \times (4.80 \times 10^{-10})^2}{6.63 \times 10^{-27}}$
            $= 2.18 \times 10^8 \text{ cm/sec.}$
            (b) $\dfrac{v_1}{c_1} = \dfrac{2.18 \times 10^8}{3 \times 10^{10}} \, 100 \approx 0.73\%.$

18-8.   (b) $15{,}234 \text{ cm}^{-1}.$

18-10. (a) $10.20 \; ev.$
            (b) $82{,}258 \text{ cm}^{-1}.$
            (c) $13.60 \; ev.$

18-11. (c) 0.5 m.

Chapter 19

ATOMS WITH MORE THAN ONE ELECTRON

HINTS OF ELECTRON-SHELL STRUCTURE

Evidence from Ionization Energies

Bohr's theory applied only to the hydrogen atom and to ions containing one electron. Naturally, scientists were eager to see whether the theory could somehow be extended to atoms with more than one electron. Could they, for example, devise a Bohr-type model that would predict the ionization energy of helium?

Bohr's model correctly predicted the ionization energy of the helium *ion* (He$^+$)—that is, the energy of the one remaining electron orbiting the helium nucleus. Because the charge, Z, of this nucleus is double that of hydrogen, its electron's energy is four times (Z^2) that of the hydrogen electron: $4 \times (-13.60) = -54.4$ *ev*. Therefore it takes 54.4 *ev* to remove this second and last electron (from $n = 1$ to $n = \infty$), yielding doubly ionized He^{++}. We call this quantity the *second ionization energy* of helium.

What can we say about the *first ionization energy* of helium—that is, the energy required to remove one electron from the *neutral* atom? It should be less than the second ionization energy because the double-positive charge of the nucleus is partly offset by the still-remaining electron. Let us attempt a rough calculation with the following model.

Let us assume that the two electrons are equidistant from the nucleus —i.e., in the same orbit. Because of mutual repulsion, they would be

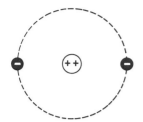

Fig. 19-1. A heuristic model of the helium atom.

opposite each other most of the time (*see* Figure 19-1). The energy of an orbiting electron is given by Equation 15 of Chapter 18:

$$E_n = - \frac{2\pi^2 m Z^2 e^4}{n^2 h^2}$$

In the ground state ($n = 1$) this equals $-13.6Z^2$ ev. For ionized helium we set $Z = 2$. In neutral helium, however, the nuclear charge is partly offset by the other electron. If that other electron were *at* the nucleus, it would reduce the effective charge (Z) to $+1$. But because it is twice as far as the nucleus from the other electron, its effect is reduced by a factor of 4 (Coulomb's Inverse-Square Law of Electrostatic Forces). Therefore it cancels not one charge but only $1/4$ charge, leaving an effective charge of $+1.75$. Accordingly, in our model the energy of the two electrons together is $-2 \times 13.6(1.75)^2$ ev, or -83.2 ev. This would be the energy required to remove *both* helium electrons simultaneously, keeping them opposite each other as we pull them out.

Now we are in a position to figure out how much energy would be required to remove only *one* electron from neutral helium. The energy required to remove both electrons is the same whether we remove them simultaneously or in sequence. Hence the energy required to remove the first electron equals 83.2 ev minus the energy required to remove the second electron. The latter value, the second ionization energy, is 54.4. Therefore, the first ionization energy in our model is $83.2 - 54.4$, or 28.8 ev.

Our predicted value is in reasonable accord with the measured value of 24.6 ev (Table 19-1). Therefore, we tentatively conclude that our model is approximately correct and that the two electrons of helium are in the same orbit.[1]

---

[1] The lower observed value might be explained with our model by the fact that the electrons are not always opposite each other. Greater proximity would result in greater repulsion, lowering the total ionization energy from 83.2 to the required 79.0 ev ($79.0 - 54.4 = 24.6$).

Table 19-1. Experimentally Determined Ionization Energies, in Election Volts, of the first 18 Elements of the Periodic Table

The first ionization energy of an atom is the energy required to remove one electron from the neutral gaseous atom. The second ionization energy is the energy required to remove one more electron from the ion produced by the first ionization. Third and higher ionization energies are defined similarly. Especially note the following:

1. reading downward, the abrupt decrease in the first ionization energies at Li and again at Na—indicating that a new outer shell was started in building up these elements;

2. reading across, the abrupt increase at the dashed lines in the energy required for removal of the next electron—indicating that this electron is tightly held in a completed shell (the innermost shell for elements 3 to 9, the second shell for elements 11 to 17);

3. the ionization energy of the last electron equals $13.6Z^2$, in accord with Bohr's model for one-electron ions.

Z	Element	1st	2nd	3rd	4th	5th	6th	7th	8th
1	H	13.6	(13.6×2^2)						
2	He	24.6	54.4	(13.6×3^2)					
3	Li	5.4	75.6	122	(13.6×4^2)				
4	Be	9.3	18.2	154	218	(13.6×5^2)			
5	B	8.3	25.1	38	259	340	(13.6×6^2)		
6	C	11.3	24.4	48	64	392	490	(13.6×7^2)	
7	N	14.5	29.6	47	77	98	552	667	(13.6×8^2)
8	O	13.6	35.1	55	77	114	138	739	871
9	F	17.4	35.0	63	87	114	157	185	954
10	Ne	21.6	41.1	64	97	126	158	207	238
11	Na	5.1	47.3	72	99	138	172	208	264
12	Mg	7.6	15.0	80	109	141	186	225	266
13	Al	6.0	18.8	28	120	154	190	241	285
14	Si	8.1	16.3	33	45	167	205	246	303
15	P	10.5	19.7	30	51	65	220	263	309
16	S	10.4	23.4	35	47	72	88	281	329
17	Cl	13.0	23.8	40	54	68	97	114	348
18	Ar	15.8	27.6	41	60	75	91	124	143

Let us see if we can extend this picture to lithium. A similar calculation for three electrons in the same orbit yields a first ionization energy of about 35 *ev*. This value is not in accord with observation. The very low measured first ionization energy of lithium, 5.4 ev (Table 19-1), implies that the most easily removable electron is in a much more remote orbit. The high second and third ionization energies of lithium (75.6 and 122 ev), on the other hand, indicate that the remaining two electrons are together in a smaller orbit. The picture accordingly is: two electrons in the same inner orbit, or inner *shell*, and the third electron in an outer shell.

Let us see what additional ideas may be obtained from Table 19-1. We observe that first ionization energies generally increase from lithium (5.4 ev) to neon (21.6 ev), except for a slight drop at boron and oxygen. After neon there is a sharp drop (similar to that after helium), followed by resumed increase. Applying our previous logic, we might deduce the following about electrons added in "building up" elements. The second electron goes into the same shell as the first; the shell then apparently is "filled," and the next electron starts a new shell. This shell in turn is filled when eight electrons have been added (neon, element 10), and a third shell is started with sodium.

Our picture is supported by the second, third, and higher ionization energies in the other columns of Table 19-1. In lithium, for instance, the first electron, alone in the outer shell, is easily removed (5.4 ev); the remaining two, in the inner shell, require more energy for removal (75.6 and 122 ev). In beryllium the *two* electrons in the outer shell are easily removed (9.3 and 18.2 ev); the inner two have high ionization energies (154 and 217 ev). In boron three electrons are in the outer shell (8.3, 25.1, and 38 ev). Skipping to sodium (Na, element 11), only one outer electron is easily removed (5.1 ev); the other electrons are held more strongly (47.3 ev, etc.). The *last* ionization energies listed for the first six elements are in accord with Bohr's model of one-electron atoms: ionization energy $= 13.6 \ ev \times Z^2$.

The first ionization energies of elements 1 through 92 are presented graphically in Figure 19-2. The sawtooth nature of the graph reveals that the overall pattern set by the first dozen elements is continued—ionization energies increase up to a point, a sharp drop occurs, then further increase follows. On the basis of an electron-shell model we could make the following interpretation. The innermost shell accommodates only two electrons and is filled at helium ($Z = 2$). A new shell is started at lithium ($Z = 3$) and is filled at neon ($Z = 10$). A third shell is started at sodium ($Z = 11$) and is filled at argon ($Z = 18$). The fourth shell runs from potassium ($Z = 19$) to krypton ($Z = 36$). And so on. Minor zigzags in the curve might indicate subshells. Thus it appears that the first six *principal shells* accommodate 2, 8, 8, 18, 18, and 32 electrons, respec-

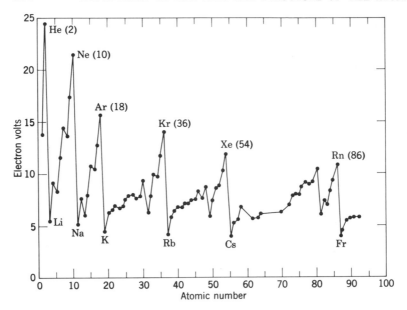

Fig. 19-2. Variation of the first ionization energy with atomic number. Elements with low ionization energies (near the bottom of the graph) have an easily removable electron; elements with high ionization energies hold their electrons relatively tight.

tively. This pattern suggests that the seventh shell is not yet full with only 6 electrons at uranium ($Z = 92$), the last element known in the 1920s.[2]

Our tentative conclusions are borne out by further consideration of the ionization energies in Fig. 19-2. The low ionization energies of the alkali metals—Li, Na, K, etc.—mean easy removability of one electron. In our picture, these atoms have one lone electron in an outer shell; this remote electron "feels" only one of the positive charges of the nucleus, because of the screening effect of the other shells. A general slight decrease of first ionization energies in the Li-Na-etc. series, and in other series, would be due to increased distance from the nucleus; all else being the same, the farther an electron is from the nucleus, the more easily it can be removed.

Noble Gases

The high ionization energies of He, Ne, Ar, etc., indicate that it is difficult to remove an electron from these atoms. This electrical stability is

---

[2] In recent years twelve more elements have been made artificially.

paralleled by very great chemical stability. In fact, these atoms are so stable that they resist combining with other atoms or with each other—the gases are monatomic. Until 1962 they were known as the "inert gases." In that year chemists finally succeeded in preparing some xenon compounds (*see* Chapter 14). Now the name *noble gases* is preferred.

We interpret the chemical inertness of the noble gases as an inherent stability of certain electron configurations—namely, filled electron shells. Other chemical elements react in such a way that their atoms tend to assume the electron configuration of the noble gas nearest in the periodic table. Alkali metals, in the first column of the periodic table, can reach this configuration by losing one electron; alkaline earth metals, in the second column, by losing two electrons. The low first ionization energies of Li, Na, K, etc., and the low first *and* second ionization energies of Be and Mg (*see* Fig. 19-2 and Table 19-1) indicate that these elements do indeed give up one or two electrons easily and thus tend to achieve stable noble-gas configurations.

Positive and Negative Ions in Solutions and Crystals

Even stronger evidence for the idea of stable noble-gas configurations, with filled electron shells, is furnished by the behavior of atoms in solution (*see* Chapter 7). The elements following each noble gas in the periodic table form positively charged ions in solution: The first element after each noble gas loses one electron and forms a singly charged ion (Li^+, Na^+, K^+, etc.); the second element forms a doubly charged ion (Be^{++}, Mg^{++}, Ca^{++}, etc.). The elements preceding the noble gases, on the other hand, gain electrons and form singly or doubly charged negative ions (F^-, Cl^-; and $O^=$, $S^=$). These positive and negative ions have the same numbers of electrons as the nearest noble gases; we say that they have attained noble-gas configurations.

Atoms tend to attain noble-gas electron configurations not only in solution but also in chemical compounds. Solid sodium chloride, NaCl, is a crystal lattice of Na^+ and Cl^- ions; both ions have noble-gas configurations.

Problem 19-1. (a) What charge would you expect on the aluminum ion (Al) in solution? (Refer to a periodic table. As a check, refer to Faraday's electrolysis experiments, discussed in Chapter 7.)

(b) What charge would you expect on the bromine ion (Br)?

(c) On selenium (Se)?

Problem 19-2. (a) Would you expect the hydrogen atom preferentially to lose or to gain an electron? Which ions would be formed in either case?

(b) What experimental evidence can you cite in support of your idea? (Recall the discussion of electrolysis in Chapter 7. Then look up "hydrides" in a freshman chemistry textbook.)

Sharing of Electrons: Covalent Bonds

When atoms in chemical compounds do not form ions, they usually fill their electron shells by *sharing* electrons. We might picture the hydrogen molecule, H_2, thus:

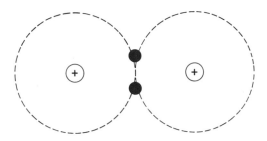

The two electrons are shared so that each hydrogen atom has two electrons, like the nearest noble gas, helium. This type of bond, involving shared electron pairs, is called a *covalent bond*.

It should be apparent that the idea of orbits does not make sense for such molecules. (Should the electrons move in a figure eight?) It is preferable to indicate the sharing of electrons by a simple *electron-dot formula,* thus: H:H. At this stage of our discussion it is not necessary to know just how these electrons move.

A fluorine atom has one electron less than neon. It, too, can fill its shell by sharing an electron pair with another atom—for instance, in the formation of the fluorine molecule, F_2.

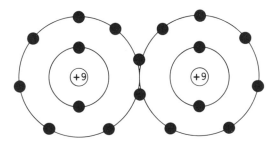

Again, an electron-dot formula showing only the outer electrons is a preferred representation.

$$: \overset{..}{F} : \overset{..}{F} :$$

The filled inner shells are not shown in electron-dot formulas. Note that the number of electrons in the diagram must equal the total outer-shell electrons contributed by the individual atoms—7 + 7 = 14.

Other compounds involving shared electron pairs are hydrogen fluoride and hydrogen chloride, both gases at room temperature and ordinary pressure.

$$H : \ddot{\underset{..}{F}} : \qquad H : \ddot{\underset{..}{Cl}} :$$

Problem 19-3.

Draw electron-dot formulas for:

(a) chlorine molecules, Cl_2;

(b) bromine molecules, Br_2;

(ç) hydrogen bromide, HBr.

An oxygen atom lacks two electrons from the neon configuration. It can complete its shell by sharing *two* electron pairs, as in water, H_2O.

$$H : \ddot{\underset{..}{O}} :$$
$$H$$

A nitrogen atom is three electrons short and can share three electron pairs, as in ammonia, NH_3.

$$H : \ddot{N} : H$$
$$H$$

A nitrogen atom can also share three electron pairs with another nitrogen atom to form an N_2 molecule with a "triple bond."

$$: N ::: N :$$

Experiments reveal that such multiple bonds are much stronger than single bonds. (*See* Example 27-1.)

Problem 19-4. Draw electron-dot formulas for:

(a) hydrogen sulfide, H_2S;

(b) methane, CH_4;

(c) oxygen fluoride, F_2O;

(d) hydrogen peroxide, HOOH.

Problem 19-5. How many fluorine atoms would bond covalently with one carbon atom? Write the electron-dot formula for the compound.

Problem 19-6. Draw electron-dot formulas for these compounds with multiple bonding:

(a) carbon monoxide, CO;

(b) hydrogen cyanide, HCN;

(c) carbon dioxide, OCO.

We see that the observed combining powers of the elements and the occurrence of multiple bonds are on the whole in accord with the electron-shell model. Whether or not this model is entirely correct, it certainly is helpful in accounting for many properties of the elements. In particular, periodicity becomes more understandable if we explain chemical properties as related principally to the outermost shells of atoms.

EXPLANATION OF X-RAY SPECTRA

Recall from Chapter 11 that Moseley caused elements to emit characteristic X-ray frequencies when he bombarded them with high-speed electrons (cathode rays). The X-ray frequencies increased regularly with increasing atomic number. This was revealed through a graph or through the equation

$$\nu = 2.48 \times 10^{15} (Z - 1)^2 \text{ sec}^{-1}$$

where ν is the observed frequency, Z is the atomic number of the element, and the factor 2.48×10^{15} is Moseley's empirical proportionality constant.

X-ray frequencies do not vary periodically with atomic number, as do chemical properties. Nonetheless, an electron-shell model allows us to explain the steady increase of v *with* Z and, what is more, to *derive* the value of Moseley's empirical constant.

Consider the following explanation. A high-energy electron from a cathode-ray beam knocks an electron out of a target atom's innermost shell. Now, when an electron from the next shell ($n = 2$) makes a transition to the vacancy in the lower-energy innermost shell ($n = 1$), the emitted radiation has the following frequency (*see* Chapter 18).

$$\nu = \frac{\Delta E}{h} = \frac{E_2}{h} - \frac{E_1}{h} = \frac{E_1}{h}\left(\frac{1}{2^2} - \frac{1}{1^2}\right)$$

For an element of atomic number Z, E_1 equals $-217.3 \times 10^{-13}Z^2$ erg. Substituting this expression and the value of Planck's constant ($h = 6.63 \times 10^{-27}$ erg-sec) into our frequency equation, we obtain

$$\nu = \frac{-217.3 \times 10^{-13}\text{erg}}{6.63 \times 10^{-27} \text{ erg-sec}} \, Z^2\left(\frac{1}{2^2} - \frac{1}{1}\right)$$

$$= -3.28 \times 10^{15} \, Z^2 \, (-0.75) \text{ sec}^{-1}$$

$$= 2.46 \times 10^{15} \, Z^2 \text{ sec}^{-1}$$

The constant 2.46×10^{15} is almost equal to Moseley's 2.48×10^{15}. Moseley's factor $(Z - 1)^2$ replaces our Z^2 because of partial screening of the nuclear charge by the other inner-shell electron.

With the help of a shell model we thus have explained some features of Moseley's X-ray spectra and even have derived Moseley's equation. This certainly is further strong support for some sort of shell model for multielectron atoms.

"MAGIC NUMBERS"

The concept of electron shells was based on experimentally determined ionization energies. From such data we deduced that the two electrons of helium are at equal distances from the nucleus; they occupy the same "shell." The third electron of lithium, however, goes into a new, farther-out shell, because lithium has a very low first ionization energy and high second and third ionization energies. In succeeding elements, the additional electrons also go into the second shell, which apparently is full when it contains eight electrons; further electrons go into a new shell, until that is filled. And so on.

Cursory examination of ionization energies led us to the tentative conclusion that the first six electron shells contain 2, 8, 8, 18, 18, and 32 electrons, respectively. Closer examination of ionization energies and some additional considerations beyond our present scope (*see* Chapter 24) indicate that the *maximum* population in the first four shells is 2, 8, 18, and 32 electrons.

Like Balmer, we can try to find some arithmetic formula to fit this sequence. Two simple relationships become apparent. First, when we divide these numbers by two, they become 1, 4, 9, 16—which are the squares of 1, 2, 3, 4. In other words, the maximum population of each of the first four shells is $2n^2$, where $n = 1, 2, 3, 4$. Second, each of the numbers 2, 8, 18, 32 can be formed by adding successive numbers as follows.

$$2 = 2$$
$$8 = 2 + 6$$
$$18 = 2 + 6 + 10$$
$$32 = 2 + 6 + 10 + 14$$

Each added number is four greater than the previous one—a definite pattern.

The meaning of this pattern may not be evident at this stage of our discussion. The key is not to be found in the Bohr theory itself. But we shall show later how outgrowths from Bohr's original theory finally led to an explanation of these "magic numbers," as well as of many other observations.

THE SHELL MODEL AND CHEMICAL PROPERTIES

The shell model explains chemical periodicity. The periodic table, originally formulated by Mendeleev in 1869, was based strictly on the chemical properties of elements, especially their combining powers. We now can see a direct correlation between electron structure and chemical families. All members of the first family—Li, Na, K, etc.—have one electron in their outer shell. Elements of the second family—Be, Mg, Ca, etc.—have two electrons in the outer shell. And so on. Members of the unreactive noble-gas family have full outer shells. Apparently chemical behavior is determined primarily by the outer shell.

The shell model also explains the formation of ions in molten salts and solutions. Members of the lithium family, with only a single electron in the outer shell, can easily give up that electron, as indicated by their low first ionization energies. Consequently they can easily form singly charged positive ions: Li^+, Na^+, etc. Faraday's experiments with electrolysis in 1833—long before the periodic table and the discovery of electrons —revealed that one faraday of electricity discharges a mole of these elements at the negative electrode. Atoms with two electrons in the outer shell have low first and second ionization energies and form doubly charged positive ions: Be^{++}, Mg^{++}, etc. It requires *two* faradays to deposit one mole of these atoms. On the other hand, elements with nearly full shells seemingly attract electrons to complete their shells. Fluorine, for example, exists as F^- in the liquid state and is discharged at the positive electrode.

We must emphasize that atoms do not lose electrons spontaneously. Even for first-column elements the ionization process requires several electron volts of energy; however, an electron-attracting atom or molecule can pull the electron away from these elements with relative ease.

We can now also understand combining power. Sodium, for instance, easily gives up an electron; chlorine tends to take on an electron. Hence sodium gives up an electron to chlorine, yielding sodium ions and chlorine ions

$$Na + Cl \rightarrow Na^+ + Cl^-$$

which then form sodium chloride crystals ($NaCl$). For another example, calcium gives up *two* electrons fairly easily, while fluorine tends to take on only one electron. Thus a calcium atom may donate one electron to each of two fluorine atoms,

$$Ca + 2F \rightarrow Ca^{++} + 2F^-$$

in the process of forming calcium fluoride crystals (CaF_2).

The noble gases, with filled electron shells and high ionization energies, show no tendency to lose or gain electrons. Additional electrons would have to go into a new shell. Thus these elements are chemically unreactive.

The elements with half-filled shells, in the middle of the periodic table, can either lose or gain several electrons to achieve noble-gas configurations. Which process occurs depends on whether the other substances participating tend to attract or donate electrons. Actually, these so-called *amphoteric* elements (from the Greek *amphoteros*, meaning "both") tend not to take on electrons entirely, but to complete their shells by sharing electrons.

This sharing of electron pairs, called a covalent bond, takes place between atoms having approximately equal electron-attracting or electron-donating power. It results in such molecules as H_2, N_2, F_2, CO, and NO. Note that covalent bonding occurs especially between elements near each other in the periodic table.

Most atoms must share more than one electron pair to complete their shells. This makes multiple bonds possible. For example, nitrogen must share three electron pairs to complete its valence shell. It can form three bonds with three separate atoms, as in ammonia, NH_3; or it can form a triple bond with another nitrogen atom—N_2.

Back in 1811, scientists were loath to accept Avogadro's suggestion that molecules of some elements are diatomic (*see* Chapter 3, p. 31). Bonding between *unlike* atoms, as in HF, might have been explained as due to electrostatic attraction between positive and negative atoms, but what could hold two identical atoms together, as in H_2 and F_2? Furthermore, if there were some mysterious force bonding these atoms, why would that process stop after the formation of diatomic molecules? Today, on the basis of the shell model, we can understand why there is an H_2 molecule and not an H_3 molecule, and why helium and the other noble gases are monatomic. And we can also understand why there is an H_2O molecule, and not an HO or an H_3O molecule.

A simple electron-shell model enabled scientists to understand many observations. Many other questions, however, remained unanswered. We shall return to the electron-shell model in Chapter 24 *et seq.*, after laying a firmer foundation for this model with the help of quantum mechanics.

We shall then better understand not only the types of bonding between atoms, but also their spatial arrangements.

CHARACTERISTICS OF A GOOD SCIENTIFIC THEORY[3]

In Bohr's theory we recognize many of the attributes of a good scientific theory.

First of all, *many seemingly unconnected facts are correlated in a logical structure.* Correlation alone is not enough. The correlation of spectral frequencies by Balmer, Rydberg, and Ritz can hardly be called a theory—there was no logic involved. Balmer's formula was merely a correlation of numbers, little more than an arithmetic trick. Rydberg and Ritz's expansion of Balmer's formula into a more general formula (the one containing n_1 and n_2) and their Combination Principle still lacked a truly logical structure. We should more properly call their formula a *law*—i.e., a generalization of empirical knowledge. Bohr's theory, on the other hand, definitely involved a logical structure and also correlated many seemingly unconnected facts—namely, spectral frequencies of hydrogen and other atoms, electrically measured excitation and ionization energies, the hydrogen radius, and fine structure of spectral lines.

Second, a theory is particularly valuable when it *suggests new relations and stimulates further experimentation.* Bohr's theory was devised to explain spectra, but it presaged the concepts of excitation and ionization energies and probably guided the thoughts of Franck and Hertz in devising their electron-collision experiments.

Third, a crucial test for a theory is whether it *leads to specific predictions of new phenomena which can be checked by test or experiment.* Bohr's theory adequately met this test. It predicted values of the hydrogen radius, the Rydberg constant in single-electron atoms, excitation and ionization energies, etc. All these were verified experimentally.

A fourth attribute of a good theory is *simplicity.* A theory that involves few basic assumptions and is simple in its structure is more acceptable than one that proposes a different mechanism for each observed fact. Scientists can grow ecstatic over a theory that is economical of their thought processes. The Bohr theory has this kind of "beauty." The entire model was based on only three postulates: definite orbits (quantized angular momentum), no radiation while the electron remains in any one orbit, and quantized emission or absorption when the electron jumps from one orbit to another.

[3] The author acknowledges reference to G. Holton and D. H. D. Roller, *Foundations of Modern Physical Science*, Reading, Mass.: Addison-Wesley, 1958, Chapter 8.

A fifth attribute is *plausibility*. But that is a controversial matter and a rather subjective one. The idea of the earth revolving about the sun seemed preposterous at one time. Avogadro's theory of polyatomic molecules of elements was not plausible to Dalton, who preferred his own simpler ideas. The idea of light waves in empty space was difficult to accept. A photon, which has momentum but no mass and behaves like a particle *and* like a wave, was beyond the credibility of many scientists. And then came this man Bohr, who proposed that circling electrons in atoms did not radiate—in spite of Maxwell's theory and Heinrich Hertz's experiments with radio waves! We see that the criterion of plausibility must be applied with caution. But it *is* nice when a theory fits into a generally accepted frame of reference. Scientists, like others, do not like to discard accepted principles that have withstood the test of time. Before proposing revolutionary ideas, we should see whether we can't after all explain an observation through existing theory.

We can, of course, *make* theories plausible. One approach is to follow the motto Strength Lies Not in Defense But in Attack.[4] If we can prove that existing theory is untenable, people will be more receptive to new theories. That was approximately the situation in 1913; Rutherford's experiments had strongly indicated a nuclear atom with orbiting electrons; but observation revealed that the electrons did not radiate continuously and did not fall into the nucleus. So, almost by the process of elimination, Bohr arrived at his postulate of stable orbits.[5] In addition, of course, he and his supporters bolstered their theory with *postive* evidence.

A final attribute of a good theory is *flexibility*. Is it adaptable to new situations, can it be modified to encompass new data, can it grow? Bohr's theory satisfied this criterion superbly. The theory was able to explain newly discovered variations in the Rydberg constant, and the model could be modified to include atoms with more than one electron. Truly, it was a remarkable achievement!

SUMMARY OF PART THREE: WHERE WE STAND AND HOW WE GOT THERE

With the conclusion of Part Three, let us review the flow of events that started with the recognition of atoms and culminated in a moderately successful model of atomic structure.

---

[4] Adolf Hitler, *Mein Kampf*.

[5] Bohr's approach represents an application of the method of Sherlock Holmes: "When you have eliminated the impossible, whatever remains, *however improbable, must be the truth*." (S. Holmes, quoted by Dr. Watson in A. C. Doyle, "The Sign of the Four".)

Around 1800 Dalton had proposed the idea that matter is composed of little atoms; that all the atoms of any one element are identical; and that different elements are made up of different atoms, distinguishable by their atomic weights. This theory underwent a half-century of trial and further development, finally yielding an acceptable table of atomic weights.

Chemists then realized that arranging the elements in the order of their atomic weights resulted in rhythmic repetition of similar properties. This periodicity of properties was expressed in 1869 in Mendeleev's periodic table, which arranged the elements in rows and columns of chemical families.

Scientists wondered what the cause of this regularity might be. The key presumably was to be found in the atom's structure. In Parts Two and Three, accordingly, we traced two developments that finally led to a useful model of the atom. These developments took place concurrently, sometimes independently but often not.

The first investigational sequence, Part Two, culminated in 1911 in Rutherford's recognition that the atom consists of a tiny, compact, positive nucleus and very light, negative electrons at considerable distances from the nucleus—like a tiny solar system. Because the evidence that eventually led to this model—electrolysis, cathode rays, canal rays, X rays, radioactivity, and alpha-ray scattering—was mostly electrical in nature, we began Part Two with a study of simple electrical phenomena and some basic concepts of physics (motion, Newton's laws, force, momentum, energy). Parts One and Two present an interesting parallel: In Part One we established the discrete nature of matter, its atomicity; in Part Two, while investigating the electrical nature of matter, we discovered the atomicity of electricity as well.

The other chain of events, Part Three, led in 1913 to a picture of the atom's external electrons in well-defined orbits: Bohr's model of the hydrogen atom. Because this model was based mostly on spectroscopic evidence, we began Part Three by exploring the nature of light.

The investigation of light represents a most interesting cross-linked chain of historical developments. Rather arbitrarily, we might start with Galileo, the rebel, who not only espoused the Copernican solar system but in 1610 also found a rather unorthodox use for the newly invented telescope—namely, to look up at the heavens. Galileo's improved measurements of planetary motion enabled Kepler to deduce that the planets move in slightly elliptical orbits, and to develop between 1609 and 1618 a series of laws that accurately describe the motions of all planets in the solar system. Kepler also devised a scale model of the solar system, with the relative distances of all the bodies fixed.

Kepler did not, however, know his scale reduction. He needed to know one absolute distance in the solar system. This was supplied a half century later through a measurement of the Earth-Mars distance, from which all the other distances then could be deduced, including that from the earth to the sun. It was this key distance that enabled Roemer in 1676 to determine the speed of light from the eclipse cycle of Jupiter's moons, which in turn had been discovered by Galileo in 1610.

Newton tried to find an explanation for the orderliness of the solar system, as described by Kepler's laws. He recalled some earlier experiments of Galileo's, plus some of his own on the relation of forces and motions, and in 1687 came up with his Laws of Motion.

In the meantime, while attempting to improve the optics of the telescope, Newton became interested in the nature of light itself and proposed a corpuscular theory of light. At about the same time wave theories of light were proposed by Grimaldi, Hooke, and Huygens. Between 1800 and 1819 Young and Fresnel discovered interference of light and developed a more sophisticated, mathematical wave theory of light. But still, the wave theory was not generally accepted. One of the key differences between the wave and corpuscular theories was the predicted velocity of light in water. The corpuscular theory predicted that light would speed up as it entered water from air, while the wave theory predicted a slowdown. When Foucault in 1850 actually measured the speed of light in water and found it to be considerably slower than in air, the corpuscular theory of light finally had to be discarded.

On the basis of the wave theory Maxwell formulated his electromagnetic theory of radiation in 1864. Maxwell's theory predicted the existence of long-wave radiation, and Heinrich Hertz set out to discover such waves. He succeeded in 1887, thus lending further support to a wave theory of light.

The same series of experiments that supported the wave theory also led Hertz to discover the photoelectric effect—that is, the ejection of electrons from metals by light. Paradoxically, it was not possible to explain this effect on the basis of light waves; waves simply could not concentrate enough energy in one location to eject an electron from a surface. A modified corpuscular theory of light had to be reintroduced. In 1905 Einstein took this drastic step by proposing the quantum theory of light. According to Einstein, radiation comes in packages, called photons or quanta, which—strangely—have wave properties as well as some properties of particles. Einstein's photons added the concept of discrete packages of light to the existing concepts of discrete packages of matter and of electricity.

Meanwhile, Michelson, who had refined Foucault's method of measuring the speed of light, found in 1887 (together with Morley) that the speed of light in a vacuum is invariable and does not depend on the relative motion of the observer and the source. In terms of Newtonian mechanics, this new finding was in direct conflict with the "aberration of light," a phenomenon discovered in 1728 by the English astronomer James Bradley (1692–1762) (Fig. 19-3). Einstein finally solved this dilemma, also in 1905, through his Special Theory of Relativity, which now replaces Newton's Laws of Motion for certain situations.

Let us now return to Bohr's theory. To devise an atom model that would fit spectroscopic data, Bohr made some arbitrary assumptions: He allowed only certain orbits (with quantized angular-momentum and energy values); he simply postulated that these orbits were stable (in spite of classical electromagnetic theory); and he assumed that energy was emitted and absorbed only in quanta, and only when the electron made a transition from one orbit, or energy level, to another.

Bohr could defend his quantum theory for atoms by referring to the successful but equally arbitrary quantum theories of Planck and Einstein. In 1900 Planck had postulated quantized energy levels to explain radia-

Fig. 19-3. Aberration of rain. The source of the rain is apparently displaced in the direction of the observer's motion. The angle of deflection depends on the velocities of the raindrops and of the observer. In 1728 Bradley used the aberration of starlight to determine the speed of light—arriving at a value in substantial agreement with Roemer's (1676).

tion from solids (blackbody radiation): Oscillating charges in solids have energies which are integral multiples of hf, where f is the frequency of oscillation and h is an empirical proportionality constant; radiation is emitted only when an oscillator makes a transition from one allowed energy level to another.

Planck's theory said nothing about the nature of the emitted radiation. That was left to Einstein, who introduced the concept of quantized radiation to explain the photoelectric effect: Radiation comes in packages (quanta, photons) of energy $h\nu$, where ν is the observable frequency of the radiation and h again is an empirical proportionality constant.

It is significant that the same value of h is obtained from seemingly unrelated experiments on the photoelectric effect and on blackbody radiation. In each case h is an empirical constant of proportionality. Then Planck's constant, h, shows up again in Bohr's theory, involving quantized angular momentum of orbiting electrons ($nh/2\pi$). Planck's constant appears repeatedly in many seemingly unrelated contexts and evidently is a fundamental constant of nature, like the speed of light, c, and the unit of elementary charge, e.

The strongest argument for Bohr's theory was that it worked. It explained not only the spectrum of hydrogen—for which it was specifically designed—but also other spectra, the hydrogen radius, and excitation and ionization energies. Later on, Sommerfeld slightly improved Bohr's model by introducing elliptical orbits—partly on the basis of Einstein's Theory of Relativity.

But the Rutherford-Bohr-Sommerfeld model of the hydrogen atom had severe limitations, the main one being that it applied only to one-electron atoms. True, it could be enlarged into a vague shell model of many-electron atoms, but this was far from satisfactory. There was no theoretical foundation for the "magic numbers" of electrons allowed in the various shells. And the model did not really answer the question that had been burning in scientists' minds since Mendeleev had formulated the periodic table in 1869: "Why should periodicity exist?" Nor did it answer the later question: "Why are there stable configurations and magic numbers"?

Another weakness of the model was its arbitrariness: Why should orbits be restricted to certain sizes, shapes, and orientations? What accounts for integral relationships? Why, for example, is angular momentum quantized in *whole*-number multiples of $h/2\pi$? Why not, say, $2.43h/2\pi$?

The model also left some secondary questions unanswered. If the electron can be only in definite orbits, by what physical path does it go from one orbit to another? Why doesn't it fall into the nucleus? What governs the intensities of spectral lines?

Apparently, classical theory—even as modified by Planck, Einstein, Bohr, and others—was not able to give more answers. Classical theory had been pushed as far as it could go, and a completely new approach was needed. This new approach came to be known as "quantum mechanics" and is the subject of Part Four.

ADDITIONAL READING FOR PART THREE

Arons, Arnold., and Alfred M. Bork, *Science and Ideas,* Englewood Cliffs, N.J.: Prentice-Hall, 1964 (paperback). Selected readings on the philosophy and history of science. See especially the article by C. C. Gillispie on Newton.

Darrow, Karl K. "The Quantum Theory." *Scientific American,* March 1952; W. H. Freeman Reprint No. 205.

Encyclopedia Britannica, S. V., "light"; "photoelectricity"; "photography"; "spectroscopy."

Halliday, David, and Robert Resnick, *Fundamentals of Physics,* John Wiley and Sons, New York; 1970.

Hochstrasser, Robin M. *Behavior of Electrons in Atoms: Structure, Spectra and Photochemistry.* New York: W. A. Benjamin, 1964 (paperback). Chapters One and Two. Simple but not superficial.

Oldenberg, O. H., and G. Holladay. *Introduction to Atomic and Nuclear Physics.* New York: McGraw-Hill, 1967.

Popper, Karl, R. *The Logic of Scientific Discovery.* New York: Harper and Row, 1965 (paperback).

Reid, R. W. *The Spectroscope.* New York: The New American Library, 1965 (paperback). A brief, popularly written book.

Richtmyer, Floyd K., et al. *Introduction to Modern Physics,* New York: McGraw-Hill, 1969. A somewhat advanced text, well indexed.

Scientific American, September 1968. A special issue devoted to "Light."

PART FOUR

"Understanding" the Atom: Quantum Mechanics

Quantum mechanics and the Theory of Relativity are perhaps the two most important scientific developments since the time of Newton. Today quantum mechanics is not only a working tool of chemists, but its advent has brought about a change in our outlook on science in general. Gone, at least in some fields, is the deterministic flavor of Newtonian mechanics, to be replaced by the indeterminism of the Uncertainty Principle and the probability calculations of quantum mechanics. De Broglie, one of the founders of the new mechanics, said: "The discovery of quanta . . . seems to require of scientific thought one of the greatest changes in orientation which it has ever had to make in its secular effort to adapt as closely as possible our idea of the physical world to the requirements of our reason."[1] Moreover, the ramifications of quantum mechanics transcend the limits of the physical sciences and extend to the life sciences and even to metaphysics.

Part Four explains in simple terms the basic principles of wave mechanics and applies these principles to the hydrogen atom. The meaning of the quantum numbers n, l, m, and s is made clear, and the Uncertainty Principle" is derived and used to clarify the concept of electron "clouds," which has replaced the idea of classical electron orbits. We shall then explain chemical periodicity and even some of the minor zigs and zags of the ionization-energy plot. This treatment helps us later to understand many features of bonds between atoms—in other words, the formation and behavior of molecules (discussed in Part Five). Finally, we shall gain some insight into the puzzling particle-wave duality of light. In the next chapter this duality is extended to matter itself.

[1] Louis de Broglie, *Matter and Light: The New Physics*, (Dover Publications, original edition, in French, published in 1937), p. 261.

273

Chapter 20

MATTER WAVES

By 1920 the apparent dual nature of light and radiant energy was generally accepted, although not understood. Waves explained diffraction and intereference, but "corpuscles" (quanta, photons) explained the photoelectric effect and blackbody radiation.

In 1923–1924 the French physicist Louis de Broglie (1892–   ) suggested in his doctoral thesis that matter also might possess dual characteristics: Perhaps moving particles, such as electrons, are accompanied by a wave or possess wave characteristics. This seemingly bizarre concept led Erwin Schrödinger, as we shall see, to develop his marvelously successful wave mechanics. Subsequently, de Broglie's strange hypothesis that streams of particles behave like waves was confirmed in the laboratory by diffraction experiments.

DE BROGLIE'S SUGGESTION

Let us follow de Broglie's line of reasoning. Light or electromagnetic radiation in general, whether considered as Maxwellian electromagnetic waves or as Einsteinian photons, has momentum, p,

$$p = E/c \tag{1}$$

where E is energy and c the velocity. According to Einstein's quantum theory, the energy of electromagnetic radiation is

$$E = h\nu \tag{2}$$

where h is Planck's constant and ν is the frequency. That is, a photon, or quantum, of light having energy E is associated with a wave of frequency $\nu = E/h$.

Combining Equations (1) and (2) yields

$$p = \frac{E}{c} = \frac{h\nu}{c} \tag{3}$$

From the relationship of frequency (v), wavelength (λ), and velocity (c),

$$\nu\lambda = c, \quad \text{or} \quad \frac{\nu}{c} = \frac{1}{\lambda} \tag{4}$$

we obtain the following relationship between the wavelength and momentum of a photon:

$$p = \frac{h\nu}{c} = \frac{h}{\lambda} \tag{5}$$

or

$$\lambda = \frac{h}{p}$$

That is, a photon having momentum p is associated with a wave of wavelength λ.

De Broglie suggested that a particle of matter having momentum $p = mv$ similarly may be associated with a wave of wavelength

$$\lambda = \frac{h}{p} = \frac{h}{mv} \tag{6}$$

In some ways, de Broglie's armchair hypothesizing is reminiscent of the Greek philosophers, who were free to make the most fantastic and dogmatic statements about nature. But by de Broglie's time the experimental method reigned supreme. Theories not only were tested by experiment but were actually considered to be little more than speculation if they did not suggest further development, which in turn would be subject to experimental confirmation.

ELECTRON DIFFRACTION

Experimental verification of de Broglie's hypothesis came three years later in 1927, through the exploitation of a laboratory accident. In 1925 Clinton J. Davisson (1881–1958) and Lester H. Germer (1896–   ), at the Bell Telephone Laboratories in New York, were studying the reflection of electrons from a nickel target. After a minor laboratory mishap they heated the nickel to repurify it. But when they continued their experiments and reflected electrons from this target, they obtained

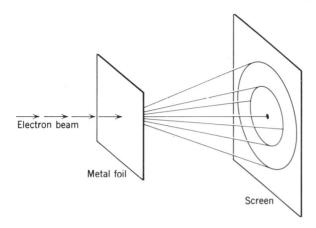

Fig. 20-1. A beam of electrons passing through a metal foil is diffracted into a circular pattern.

strange patterns. Apparently the heat had caused the nickel atoms to re-arrange themselves to form a large polycrystalline mass. At the time, Davisson and Germer did not investigate further. However, one-and-a-half years later, after reading some papers on wave mechanics by Schrödinger (*see* Chapter 22), they performed some systematic follow-up experiments and proved that the strange patterns were diffraction patterns.

At about the same time in Scotland, G. P. Thomson (1892–   ) independently demonstrated diffraction of electrons by passing them through thin sheets of solids (Fig. 20-1).

As we stated earlier, such diffraction patterns can be satisfactorily explained only by assuming that the beam possesses wave character. Davisson and Germer's and Thomson's experiments therefore confirmed de Broglie's suggestion that matter possesses wave character.[1]

To obtain good diffraction patterns, the intervals in diffraction gratings should be of approximately the same order of magnitude as the radiation wavelengths (*see* p. 183). It happens that crystals have just the right spacings between layers of atoms to act as diffraction gratings for moderate-voltage electrons, as well as for X rays (Chapter 13). As a result, electron diffraction, like X-ray diffraction, has become a powerful tool for the determination of crystal and molecular structure (*see* Chapter 27).

[1] It is ironic that the *wave* properties of the electron were confirmed by the son of the man who three decades earlier had experimentally measured the *particle* properties of that same electron (J. J. Thomson's m/e experiment, Chapter 8).

Example 20-1. An electron is accelerated by a potential of 100 volts. Calculate (*a*) its velocity and (*b*) its wavelength.

Solution.

(a) From the definition of the energy unit "electron volt," it can be shown that the kinetic energy, $1/2\ mv^2$, of an electron of mass m and of charge e is related to the voltage v as follows.

$$\frac{1}{2}\ mv^2 = \frac{eV}{300}$$

Then,

$$v = \sqrt{\left(\frac{V}{150}\right)\left(\frac{e}{m}\right)}$$

Substituting numerical values for e and m (Chapter 8),

$$e = 4.80 \times 10^{-10}\ \text{statcoulomb}$$

$$m = 9.11 \times 10^{-28}\ \text{gram}$$

we obtain

$$v = \sqrt{\left(\frac{V}{150}\right)(5.27 \times 10^{17})} = 5.92 \times 10^7\ \sqrt{V}$$

or

$$v \approx 6 \times 10^7 \sqrt{V}$$

Thus, when $V = 100$ volts, the velocity of the electrons is

$$v = 6 \times 10^7 \sqrt{100} = 6 \times 10^8\ \text{cm/sec}$$

(b) The de Broglie wavelength is

$$\lambda = \frac{h}{mv} = \frac{(6.63 \times 10^{-27}\ \text{erg-sec})}{(9.11 \times 10^{-28}\ \text{g} \times 6 \times 10^8\ \text{cm/sec})}$$

$$\lambda \approx 1.2 \times 10^{-8}\ \text{cm} = 1.2\ \text{A}$$

The wavelength of electrons can also be calculated directly, without first calculating the velocity, from the formula

$$\lambda\ (\text{in angstroms}) = \frac{12.3}{\sqrt{V}}$$

Thus, for $V = 100$, $\lambda = 1.23$ angstroms. This wavelength is of the same order of magnitude as the size of an atom and interatomic distances in crystals.

Problem 20-1. Calculate the velocity and the de Broglie wavelength of electrons accelerated by potentials of (a) 10 volts and (b) 1000 volts.

MATTER WAVES IN GENERAL

Nothing in de Broglie's theory limits it to electrons, or even to charged particles. Hence, if de Broglie's theory is correct, neutral molecules should also exhibit wave properties under suitable conditions. Such effects have indeed been observed for atoms.

Example 20-2. Calculate the wavelength of a beam of helium atoms having a velocity of 1.64×10^5 cm/sec. (This is a velocity of about 1 mile per second, or 3600 mph, which is about five times the speed of sound.)

Solution.

The mass M of the helium atom (4 amu) is 4 grams divided by Avogadro's number:

$$M = \frac{4}{6.0 \times 10^{23}} = 6.7 \times 10^{-24} \text{ g}$$

$$\lambda = \frac{h}{Mv} = \frac{(6.63 \times 10^{-27})}{(6.7 \times 10^{-24} \times 1.64 \times 10^5)}$$

$$= 6 \times 10^{-9} \text{ cm} = 0.6 \text{ A}$$

The existence of de Broglie's waves is not even limited to particles of atomic sizes. However, from the formula $\lambda = h/Mv$ and the small value of h ($\approx 10^{-26}$ erg-sec), it is evident that wavelengths of macroscopic bodies are negligibly small.

Example 20-3. Calculate the de Broglie wavelength of a 40-g bullet traveling at 1000 m/sec.

Solution.

$$\lambda = \frac{h}{Mv} = \frac{(6.6 \times 10^{-27} \text{ erg-sec})}{(40 \text{ g} \times 10^5 \text{ cm/sec})}$$

$$= 1.7 \times 10^{-33} \text{ cm.}$$

Compared to the size of the bullet, perhaps 2 cm long, this wave is much too short to be observed.

Problem 20-2. Calculate the de Broglie wavelength of a 5-ounce ball moving at 40 mph. (Hint: Convert to the metric–cgs–system).

RELATION TO BOHR'S ATOM MODEL

The reader is justified in asking what all this has to do with atomic structure. Consider the two following examples.

Example 20-4. In Problems 18-5 and 18-6 you were asked to calculate the radius and velocity of the electron in the $n = 1$ orbit of a Bohr hydrogen atom. The correct answers were 0.53 A and 2.18×10^8 cm/sec, respectively.
(a) What is the wavelength, λ, of this electron?
(b) What is the circumference, C, of the $n = 1$ orbit?
(c) What is the ratio of C to λ?

Solution.

(a) $\lambda = h/mv = (6.63 \times 10^{-27})/(9.11 \times 10^{-28} \times 2.18 \times 10^8)$
$$= 3.33 \times 10^{-8} \text{ cm} = 3.33 \text{ A}$$
(b) $C = 2\pi r = 2 \times 3.14 \times 0.53 = 3.33$ A
(c) $C/\lambda = 1$

Example 20-5. (a) What is the wavelength of the electron when it is in the $n = 4$ orbit of a Bohr hydrogen atom?
(b) What is the circumference of that orbit?
(c) What is the ratio of C to λ?

Solution.

(a) The velocity is proportional to $1/n$ (see p. 241). Hence the velocity in the fourth orbit is 1/4 of the velocity in the first orbit. But the wavelength is inversely proportional to velocity ($\lambda = h/mv$); therefore it is four times the wavelength of the electron in the first orbit:

$$\lambda_4 = 4\lambda_1 = 4 \times 3.33 \text{ A} = 13.3 \text{ A}$$

(b) The radii of orbits in Bohr's hydrogen atom are proportional to n^2 (*see* p. 241). Therefore,

$$r_4 = 16 r_1$$

and
$$C_4 = 16 C_1 = 16 \times 3.33 \text{ A}$$
$$\text{(c) } C_4/\lambda_4 = 4$$

In both examples the ratios of C/λ are whole numbers—1 and 4, respectively. We can easily prove that the ratio of C to λ always is an integer, because $\lambda_n = n\lambda_1$ and $C_n = n^2 C_1$; therefore,

$$\frac{C_n}{\lambda_n} = \frac{n^2 C_1}{n\lambda_1} = n\left(\frac{C_1}{\lambda_1}\right) = n$$

or,

$$n\lambda = C$$

In other words, in Bohr's model an orbit circumference contains an integral number of electron wavelengths.

We can also derive this relationship directly from Bohr's angular momentum quantum condition,

$$mvr = \frac{nh}{2\pi}$$

Combining this with de Broglie's relation

$$\lambda = \frac{h}{mv},$$

and remembering that

$$C = 2\pi r$$

we obtain

$$\frac{C}{\lambda} = \frac{2\pi r}{h/mv} = mvr\,\frac{2\pi}{h} = \frac{nh}{2\pi}\,\frac{2\pi}{h} = n$$

Again, this means that an orbit contains an integral number of wavelengths.

This result is illustrated in Fig. 20-2, which shows the fourth and fifth orbits of a Bohr hydrogen atom drawn to scale, with the electron sketched as a wave rather than as a particle. The fourth orbit contains four complete wavelengths; the fifth contains five wavelengths.

It is evident from this illustration that the condition $C/\lambda = n$ limits the number of possible orbits. The small insert of Fig. 20-2 illustrates what would happen if the circumference did *not* contain an integral number of wavelengths: The waves would interfere destructively and cancel each other, in effect destroying the electron, which "is" the wave.[2]

Thus, by considering the electron as a wave, we automatically allow only certain orbits. Bohr in 1913 had limited the number of orbits by

[2] We shall elaborate on this theme in Chapter 22's discussion of standing waves.

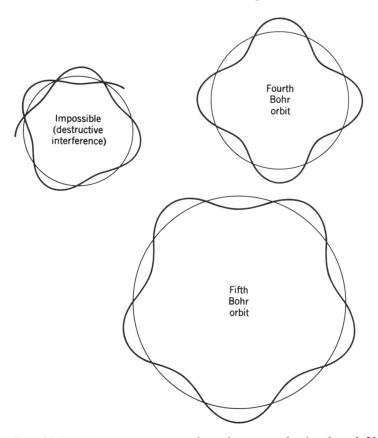

Fig. 20-2. Wave representation of an electron in the fourth and fifth orbits of a hydrogen atom (drawn to scale).

artificially introducing the arbitrary quantum condition $mvr = nh/2\pi$. In fact, we can now derive this quantum condition:

If
$$n\lambda = C \quad \text{and} \quad \lambda = \frac{h}{mv}$$

then
$$n\left(\frac{h}{mv}\right) = 2\pi r$$

or
$$\frac{nh}{2\pi} = mvr$$

In a sense, then, de Broglie's waves make Bohr's quantum condition plausible.

DENSITY AND PROBABILITY DISTRIBUTION PATTERNS

To understand the following, let us take a closer look at an electron diffraction experiment of the type performed by G. P. Thomson (Fig. 20-1). The electrons are diffracted into a circular pattern, which appears as a set of concentric circles on a fluorescent screen or a photographic plate. A cross-sectional view of this experiment is shown in Fig. 20-3a. A, B, B', C, and C' are the bright circles seen edgewise; in between the brightness is low. This information is also presented graphically in the intensity plot of Fig. 20-3b.

We can explain the observed intensity pattern on the basis of constructive and destructive intereference, in a manner identical to the explanation of optical diffraction patterns (Chapter 12). At positions A, B, and C, electron waves arrive in phase and reinforce each other; in between they arrive out of phase and destroy each other. Thus we can easily understand and explain the observed phenomenon if we think of electrons as waves.

In terms of classical particles we cannot really explain the observed phenomenon; we can merely describe what happens: At positions A, B, and C, many electrons arrive; in between no electrons appear. On this basis, the wavy graph of Fig. 20-3b is a plot of *electron density*.

Now let us perform an imaginary experiment, and consider the "diffraction" of one electron at a time. We cannot predict where a single electron will hit the target—it can hit anywhere on circles A, B, or C. We can, however, predict where it will *not* hit—namely, the areas between circles A, B, and C.

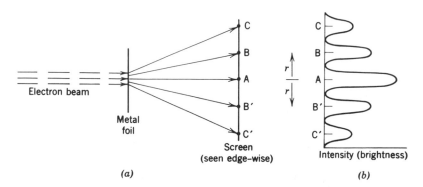

Fig. 20-3. (a) Cross-sectional view of Thomson's electron-diffraction experiments. (b) Plot of the measured intensity, or brightness, as a function of the radial distance, r, from the central dot A.

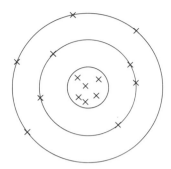

Fig. 20-4. Probable impact sites of the first few electrons in an electron-diffraction experiment.

Let us then say that the first electron hits in area *A*, the second in *B*, the third in *A* again, the fourth in *C*, and so on. Figure 20-4 shows on an enlarged scale the impacts of the first few electrons. The greatest probability of impact is at the center (position *A*), with decreasing probability in circles *B* and *C*.

If we continue this experiment longer and take a time-exposure photograph of the flashes caused by the impact of each electron on the fluorescent screen, we obtain exactly the same pattern as in Figs. 20-1 and 20-3. The result is the same whether we diffract a beam of many electrons or one electron at a time and sum up our observations over a period of time.

The plot of Figure 20-3*b*, then, represents both the *density distribution* of an electron beam after diffraction and the *probability distribution* of the impact of single electrons. In other words, *the two distribution concepts are equivalent.*

THE MEANING OF MATTER WAVES

To understand electron-diffraction patterns we can think of electrons in two ways. Either we think that moving electrons simply *are* waves and that the observed patterns result from constructive and destructive interference of these waves; or else we think of the electrons as *accompanied* by waves that guide the motion of the electron. Physicists still disagree about the proper interpretation of these waves—whether the wave *is* the electron, or whether it is a physically real guiding wave accompanying the particle, or whether it is merely a mathematical probability function. Therefore let us be fully clear what is *real* and what exists only in our minds. The observed intensity pattern of an electron beam, shown in the plot of Fig. 20-3*b*, is quite real and experimentally observable. We can account for this pattern *if* we can represent the electron by a wave. No one has ever seen such an electron wave. However, whether an electron wave actually exists is not important. What is important is that, by assuming (i.e., postulating) the existence of such a wave, we can explain diffraction and also can predict other observable phenomena. Physicists may disagree about the interpretation of matter waves, but they do agree about the usefulness of the concept for describing and predicting experimental results.

Should we, then, discard the concept of the electron as a particle? Where did we, in the first place, get the idea that the electron is a particle? The concept owes its origin to the well-established beliefs in the corpuscular nature of matter. The electron was regarded as one of the building blocks of matter and therefore was visualized as a particle. Let us review briefly the experimental evidence for the corpuscular nature of the electron:

1. Quantitative experiments with electrolysis (Faraday, 1833): Electricity (and matter) comes in packages.

2. The photoelectric effect (H. Hertz, 1887): Electric charge comes in packages (which are identical to Thomson's electrons).

3. Measurement of the mass-to-charge ratio (J. J. Thomson, 1897): Cathode-ray electricity comes in packages of definite mass (actually, m/e); electrons in cathode rays follow definite paths.

4. Determination of the electron's charge (Millikan, 1910): Electricity comes in lumps of definite charge.

5. Tracks in cloud chambers (C. T. R. Wilson, 1911): Definite trajectories.

6. Compton effect (1923, interaction of radiation with electrons—not included in this text): Explained on basis of collisions of corpuscular electrons with "corpuscular" light.

Certainly a substantial body of evidence for a corpuscular electron—but only circumstantial! No one has ever seen an electron particle. Yet, like the electron wave, it has been a most useful concept.[3]

Therefore, in the future we shall let convenience determine which concept of the electron we wish to employ. We shall think of the electron as a wave, whenever that concept is useful to us. When it is not useful—

[3] We tend to forget the steps that led to a conclusion and remember only the conclusion itself. Faraday, whose quantitative work on electrolysis helped to establish the electron's corpuscular nature, wrote in the *Philosophical Magazine* in 1844 about ". . . the constant tendency of the mind to rest on an assumption, and, when it answers every present purpose, to forget that it is an assumption. . . . It, in such cases, becomes a prejudice, and inevitably interferes, more or less, with a clear-sighted judgment."

for example, in the photoelectric effect—we shall remember that the wave concept was a mere creation of de Broglie's imagination and shall consider the electron as a particle.[4]

SUMMARY

In 1923 de Broglie suggested that electrons are accompanied by waves, or behave as if they are waves. This suggestion was experimentally verified in 1927 through electron diffraction experiments by Davisson and Germer in the United States and by G. P. Thomson in Scotland. The Americans used crystals as reflection gratings for electron beams, whereas Thomson used very thin metal foils as transmission gratings.

All moving objects are accompanied by waves whose wavelengths are

$$\lambda = \frac{h}{Mv}$$

But these waves are immeasurably small for any other than atomic and subatomic particles.

On the basis of the electron's wave character, it was possible to derive and explain Bohr's seemingly arbitrary quantum condition (1913) that the electron's angular momentum in the hydrogen atom is quantized in units of $h/2\pi$ and that only certain orbits are possible.

On the one hand, de Broglie's proposal of matter waves deepened the particle-wave conflict that we first encountered with light. On the other hand, it was a progressive step toward the new quantum mechanics. In the next chapter we shall see how a suggestion by Heisenberg helped to resolve the puzzle of particle-wave duality.

ANSWERS TO SELECTED PROBLEMS

20-1. (a) 1.9×10^8 cm/sec; 3.9 A.
(b) 1.9×10^9 cm/sec; 0.39 A.

20-2. 2.6×10^{-32} cm.

[4] Sir William Bragg, the famous X-ray crystallographer, reportedly quipped that on Mondays, Wednesdays, and Fridays we regard the electron as a particle, and on Tuesdays, Thursday, and Saturdays, as a wave. (Apparently this statement was made before the introduction of the five-day work week.) Somebody else has proposed the term "wavicle" for the electron.

Chapter 21

THE UNCERTAINTY PRINCIPLE

According to classical Newtonian mechanics, if we know the exact position and velocity (speed and direction) of a particle, we can, at least theoretically, predict its position at any instant thereafter. Yet, in Chapter 20's imaginary diffraction of single electrons, we could not predict exactly where an individual electron would go. Newtonian mechanics apparently does not apply to small particles like electrons.

In 1927 the young German physicist Werner Heisenberg (1901–1965) showed that we can *never* measure exactly both the position and velocity of a particle, because we change its position and/or velocity by our very act of measuring. The ultimate limitation to any measurement effort is expressed quantitatively in Heisenberg's Uncertainty Principle.

LIMITATIONS OF MEASUREMENT

If we wish to determine the exact thickness of a very thin wire, application of calipers or a micrometer would slightly compress the wire, giving us a low reading. If instead we try to determine the thickness optically under a microscope, the light beam of the microscope will heat the wire, expanding it, and will therefore give us a high reading. Either way, our measurement process affects the measurement.

When we photograph a rapidly moving object at ordinary shutter speeds, it appears blurred because of its movement during the exposure time. Higher shutter speeds, on the other hand require "faster" film with attendant greater graininess, which yields a "fuzzy" picture. There is an inherent conflict between determining the object's position and its shape.

The moon is visible to us only by reflected sunlight. This sunlight exerts momentum, or pressure, on the moon. Hence, by the time the light from the moon has reached us, the moon has been pushed into a different orbit. Indeed, this is a trivial example. The impact of photons on the moon does not jar it so much that we cannot determine its exact position at any instant, because the sun's radiation pressure on the moon amounts only to 7,000 tons, a very small amount compared to the moon's total mass. But what about a body as small as an electron? During the hypothetical observation of an electron by reflected light, the rebounding photons would shove the electron about in an unpredictable manner and impose a definite upper limit on the precision of our measurements.[1]

THE "GAMMA-RAY MICROSCOPE"

In the following imaginary experiment we shall attempt to measure precisely both the position and the velocity of an electron by means of a hypothetical supermicroscope. General principles of optics tell us that we cannot locate the electron much more accurately than $\pm\lambda$, where λ is the wavelength of the light used in the microscope. Therefore, we would try to make λ as small as possible—perhaps using very short-wavelength gamma rays. At least in principle, we could thus locate the electron to any desired degree of accuracy through a "gamma-ray microscope" (which exists only in our imagination).

Here a new predicament develops. We observe the electron's position by means of a reflected photon. After a known time interval we observe a second position by means of another reflected photon, and calculate the electron's average velocity, v, from the formula

$$v = \frac{\Delta s}{\Delta t}$$

where Δs is the change in the electron's position during the time interval Δt. But a photon of short wavelength has high energy and, when colliding with an electron, will cause it to recoil in an indeterminate manner. A photon of wavelength λ has momentum $p = h/\lambda$. In a photon-electron collision some unknown fraction of this momentum is transferred to the electron, changing its momentum by as much as $\Delta p = h/\lambda$, the momentum of the incident photon. Accordingly, our effort to measure very accurately the position of the electron by using short wavelengths of light comes at the expense of uncertainty in momentum, or velocity.

---

[1] The tails of comets are visible evidence of radiation pressure. A comet is a swarm of relatively small bodies held together loosely by mutual attraction. As the comet in its orbit approaches the sun, the increased radiation pressure pushes the small particles away from the comet, forming a long tail. This tail always points away from the sun.

The uncertainty in position, Δs, is approximately equal to the photon's wavelength, λ; the uncertainty in momentum, Δp, is h/λ. The product of the two uncertainties is approximately equal to Planck's constant:

$$\Delta s \Delta p \approx (\lambda)\left(\frac{h}{\lambda}\right) = h$$

Thus we cannot simultaneously measure precise position and precise momentum by this method. A variety of other imaginary experiments points to the same conclusion. To date, nobody has devised a valid experiment, even an imaginary one, that would allow simultaneous measurement of position and momentum of anything with an accuracy greater than that allowed by the *Uncertainty Principle*.

$$\Delta s \Delta p \approx h \qquad (1)$$

We therefore believe that the Uncertainty Principle has general validity as a fundamental principle of nature. Planck's constant, h, represents an absolute limit to the simultaneous measurement of position and momentum, a limit we may approach under favorable conditions but can never exceed.[2]

The expression

$$\Delta s \Delta p \approx h$$

is only one form of Heisenberg's Uncertainty Principle, which is also known as the Principle of Indeterminacy. Since $p = mv$ and m is a constant $\Delta p = m\Delta v$. Therefore, Equation (1) may be written alternatively as

$$\Delta s \Delta v \approx \frac{h}{m} \qquad (2)$$

Example 21-1. The speed of an electron is 500 m/sec $\pm$ 0.01%.

(a) What is the absolute uncertainty of the velocity in m/sec; in cm/sec?

(b) With what limiting uncertainty can we determine this electron's position?

Solution.

(a) $\Delta v = 500$ m/sec $\times$ 0.0001 = 0.05 m/sec = 5 cm/sec.

(b) $\Delta s \Delta v = h/m$

$$\Delta s = \frac{h}{m\Delta v} = \frac{6.6 \times 10^{-27} \text{ erg-sec } (= \text{g-cm}^2/\text{sec})}{(9.1 \times 10^{-28} \text{ g}) \times (5 \text{ cm/sec})} = 1.4 \text{ cm}$$

[2] The uncertainty h represents merely an approximate figure, an order of magnitude. Various authors, depending on the derivation used, will instead quote values of $h/2\pi$ or even $h/4\pi$, etc.

In other words, there is no hope of determining this electron's position with an accuracy better than about 1/2 inch. If we could look at it, this electron would appear as a fuzzy 1/2-inch object, fuzzy like the moving object we photographed with the grainy film. Under these conditions we certainly wouldn't be able to see the electron as a tiny particle.

Example 21-2. A bullet weighing 50 g (about 2 oz) has a speed of 500 m/sec ± 0.01%. What is the limiting uncertainty in its position? (This example differs from the foregoing one only in the mass of the particle involved.)

Solution.

$\Delta s = 2.6 \times 10^{-29}$ cm.

That is a very small uncertainty. (The diameter of the nucleus is on the order of 10^{-13} cm.) We see that for ordinary macroscopic bodies, like bullets, the Uncertainty Principle is of no consequence; hence we are justified in using classical mechanics. This example illustrates the *Correspondence Principle:* For macroscopic bodies quantum mechanics and classical mechanics give the same result.

Example 21-3. (a) Recall that the radius of the Bohr hydrogen atom is 0.53 A. From this value and de Broglie's equation, $\lambda = h/mv$, calculate the velocity and momentum of the electron in the ground state of the hydrogen atom ($n = 1$).
(b) Assume that the velocity of this electron is known with an accuracy of ± 10%. Calculate the uncertainty in the electron's position.
(c) Calculate the uncertainty in the electron's position if the velocity is known to ± 1%.

Solution.

(a) We have shown in the previous chapter that the wavelength of the orbiting electron is equal to the circumference divided by the orbit's quantum number:

$$\lambda = \frac{C}{n}$$

In the first orbit ($n = 1$),

$$\lambda = C = 2\pi r = 2\pi \times (0.53 \times 10^{-8} \text{ cm}) = 3.3 \times 10^{-8} \text{ cm}$$

Also,

$$\lambda = \frac{h}{mv} = \frac{h}{p}$$

Hence the momentum,

$$p = \frac{h}{\lambda} = \frac{6.6 \times 10^{-27}}{3.3 \times 10^{-8}} = 2.0 \times 10^{-19} \text{ g-cm/sec}$$

From $p = mv$, we calculate the velocity,

$$v = \frac{p}{m} = \frac{2.0 \times 10^{-19}}{9.1 \times 10^{-28}}$$

$$= 2.2 \times 10^{8} \text{ cm/sec}$$

(Check this value against the velocity calculated by a different method in Problem 18-6.)

(b) $\Delta v = 10\%$; therefore, because $\Delta p = m\Delta v$, also $\Delta p = 10\% = 0.10 \times (2.0 \times 10^{-19}) = 2.0 \times 10^{-20}$ g-cm/sec

$$\Delta s = \frac{h}{\Delta p} = \frac{6.6 \times 10^{-27}}{2.0 \times 10^{-20}} = 3.3 \times 10^{-7} \text{ cm}$$

$$= 33 \text{ A}$$

(c) $\Delta v = 1\%$; therefore Δs is ten times the previous value, or 330 A.

Solution (b) indicates an uncertainty of 33 A in the electron's position, sixty times as great as the Bohr radius. We see that we have no way of determining whether this electron is anywhere near its "orbit." If we specify the velocity more precisely, the uncertainty becomes even greater—600 Bohr radii in solution (c)! Under these conditions the orbital electron appears as a huge fuzzy cloud and not as a tiny particle in a neat orbit.

The solution of Example 21-1, then, indicates that the Uncertainty Principle does not allow us to picture an electron as a point particle. Rather, it appears as a large, fuzzy object (1/2-inch diameter, in the illustrated case). Example 21-3 exposes the atom itself, not as a compact little planetary system but as a spread-out charge cloud. The interpretation of these charge clouds is taken up in Chapter 23.

THE RELATION OF PARTICLE AND WAVE VIEWPOINTS

It should be clearly understood that the Uncertainty Principle is not a statement of technical incompetence; it does not imply that we are not clever enough to measure things properly. It is not a practical limitation imposed by the crudeness of our apparatus. The limitation is a theoretical one. The position and velocity of an atomic particle can *never* be specified simultaneously. Preciseness of position *and* velocity simply does not exist.

The Uncertainty Principle, once we accept it, may help us to understand particle-wave duality. After all, what *is* a particle? After some consideration you will probably conclude that a particle has definite mass (inertia) and definite size and shape; that is, it has sharp boundaries (unlike, say, a fog with fuzzy edges). In other words, a particle occupies a definite part of space or is moving from a definite part of space to another along a definite path.

But no one has ever determined these characteristics for the electron. No one has ever *seen* an electron or measured its actual size and shape. If we attempt to confine an electron to a small space and stop it, we get a smeared-out pattern. And the Uncertainty Principle tells us that we will never be able to get different results. Does *this* seem like a particle?

In the foregoing chapter we reviewed the electron's history and pointed out that the concept of the electron as a particle owes its origin to the well-established belief in the corpuscular nature of matter. The electron was regarded as one of the building blocks of matter and therefore was visualized as a particle. The evidence supporting this belief was rather indirect, such as Thomson's m/e experiment. He found that the electron behaved *like* a particle—it had a measurable mass (so does fog) and it seemed to follow a trajectory (but how well defined?) Let's face it: The particle is a classical concept, and the Uncertainty Principle tells us that our classical concepts of particles do not apply to "things" having atomic dimensions. We really should not use the term "particle" for such a "thing."

The same sort of thinking can be applied to the term "wave." In some respects electrons and light behave *like* waves. They seem to behave like the waves we can see and feel, like waves in air, like waves on water, like vibration waves along a taut string. But no one has ever *seen* a light wave or an electron wave.

The terms "particle" and "wave" are carry-overs from another realm of experience and strictly speaking should not be applied to the realm of atomic dimensions. Thus we really need not be disturbed about the *so-called* particle-wave duality. We can continue to use the terms strictly for convenience, if we remember the inherent limitations. For some phenomena it still is convenient to think of "things" like electrons and photons *as if* they were particles; for other phenomena, *as if* they were waves.

Even for more mundane substances we use varied descriptions. Our choice depends in part on our background and inclination and in part on convenience of application. For example, the author's small daughter describes water as "what you wash your hands in, and what you drink when you can't have a Coke." Her older sister says water is "wet, tastes good, and is clear—except in Boston, where it's milky." (Note that definitions, like "laws" of nature, are subject to revision as we gain experience!) The engineer describes water as a liquid having such-and-such viscosity, index of refraction, density, etc.[3] The chemist simply says water is "H_2O."[4] Is any of these descriptions less real, less correct than the others? They are all "correct"; their *usefulness*, however, depends on the purpose for which we use them.

To summarize, the Uncertainty Principle helps us to circumvent the quandary of particle-wave duality. It answers the question, "How can an electron be both a particle and a wave?", by saying that the question is asked in the wrong way. The electron *is* not a particle or a wave. It *behaves* sometimes like a macroscopic particle and sometimes like an ordinary wave. But the concepts of particle and wave can be applied only to macroscopic things like bullets and water. An electron is *not* one of those big things; it is a subatomic object, and it can't be described completely with old-fashioned classical terms and classical mechanics.

Moreover, we should abandon the picture of corpuscular electrons circling around nuclei in well-defined orbits. We must use an alternative description of the atom consistent with properties that can be measured experimentally. Such an alternative description is quantum mechanics, taken up in the next chapter.

THE UNCERTAINTY PRINCIPLE VERSUS CAUSALITY

When a physical scientist performs an experiment, he expects that repetition of the experiment under identical conditions will yield identical results: same cause, same effect. This direct relation between cause and effect is known as causality. Does the Uncertainty Principle conflict with the principle of causality? According to the Uncertainty Principle we cannot predict the future of a particle or a system of particles because we cannot determine the present state of the system with absolute precision. In other words, the future of the particle or system is indeterminate!

Isn't the scientist's basic motivation grounded in a belief in the orderliness of nature, in the law of cause and effect? His highest aspiration is

[3] What *is* the density of water?

[4] If pressed further, the chemist would explain that a water molecule is made up of two hydrogen atoms and one oxygen atom—meaning 2 grams of hydrogen combined with 16 grams of oxygen. In his everyday thinking, however, "H_2O" *is* water. This thinking is analogous to a mathematician or a physicist thinking in terms of mathematical equations; to him, the mathematical representation *is* the system.

to discover the universal laws that control nature and to use these to predict future events. If there is no underlying law of causality, what is the scientist looking for? If effects are independent of causes, if the outcome of experiments is unpredictable, then why look for "laws of nature"? Scientists by no means agree on the answers, and volumes have been written on this topic.[5]

However, let the student be assured that science is not a will-o'-the-wisp. For one thing, scientists can adopt the point of view expressed by Max Born, one of the founders of quantum mechanics. Born said that the law of cause and effect may indeed be operating in the realm of atomic and subatomic events, but we cannot *predict* the effects because of the Uncertainty Principle.[6] The author suspects this argument to be a circular argument or an empty one. Most scientists agree that it is meaningless, in the context of science, to talk about things which can never be observed or verified. Such things remain figments of our imagination and, as far as scientists are concerned, do not exist.[7]

Another point of view, reflecting man's reluctance to accept indeterminism in nature, is based on the belief that indeterminism is only an expression of our incomplete understanding of the universe and that at some future date we may be clever enough to discover a deterministic picture of nature. Few hold this view today, although Einstein was one who did (Einstein argued, "God does not play dice").

Most scientists side-step the philosophical implications of the Uncertainty Principle. Whether nature at the atomic level is indeterminate or whether it only seems so is of secondary importance to these scientists. In their opinion the scientific method still is valid and useful. One object of scientific theories is to predict future events. And we can still do that, in spite of the Uncertainty Principle. True, we cannot predict where an individual electron will go in a diffraction experiment; but we *can* predict accurately the *density distribution* of a beam containing many electrons, and we *can* predict precisely the *probability* of finding an individual electron at a given position. In other words, at the atomic level the determinism of classical mechanics has given way to the probability statistics of a new mechanics. The development of this mechanics, called quantum mechanics, is the subject of the next chapter.

[5] The student is urged to look up some of this material. The two works of de Broglie listed on p. 331 and those of Heisenberg and Popper will serve as a starting point.

[6] Max Born, *Atomic Physics*, 5th ed. (New York: Hafner Publishing Company), p. 100.

[7] P. W. Bridgman calls this point of view the *operational viewpoint:* A concept—say, "element" or "velocity"—has meaning only when we prescribe some operation for detecting and perhaps measuring what is being defined.

Chapter 22

WAVE MECHANICS

Quantum mechanics was developed between 1925 and 1927 by two groups, which came up with different formulations. One formulation, called *matrix mechanics*, was begun by Heisenberg in 1925 (at age twenty-four) on the basis of concepts later embodied in his Uncertainty Principle. It was then developed further by Heisenberg and Max Born (1882–1970) in Germany, and by P. A. M. Dirac (1902–   ) in England.

Meanwhile, following up de Broglie's idea about the wave character of electrons, the Austrian Erwin Schrödinger (1887–1961) discovered and developed his *wave mechanics*, which differed from Heisenberg's formulation in using differential equations rather than matrix calculations. Dirac later showed that matrix and wave mechanics are equivalent in content although different in form. We shall concentrate on Schrödinger's wave mechanics because it is more easily visualized.

This chapter lays the groundwork for understanding the quantum-mechanical picture of the atom, which is taken up in the following two chapters. We shall analyze the nature of waves, especially standing waves—the kind of waves in a vibrating violin string. We shall see that the standing waves of a given system may have only certain wavelengths and no others; in other words, wavelengths, and therefore frequencies and energies, are quantized.

We shall illustrate the application of wave mechanics by a simple example, the "particle in a box." That illustration clearly reveals the divergence of the quantum-mechanical and classical descriptions of an electron, and their convergence for macroscopic bodies.

TRAVELING WAVES

Waves may be classified as traveling and standing waves. Ocean waves, for example, are traveling waves. Looking down on them from an

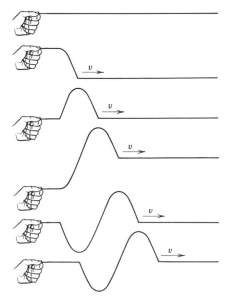

Fig. 22-1.   A wave traveling along a string.

airplane, we see a regular pattern of evenly spaced waves. Let us say that the distance between two crests—the wavelength, λ—is 10 feet and that these crests are moving along at a speed, v, of 30 ft/min. At the beach we may time the arrivals of succeeding crests to determine their frequency, ν, and find that one wave arrives every 20 seconds, or three per minute. The relationship $\nu \lambda = v$ must hold, as it does for light waves. In this case,

$$3/\text{min} \times 10\,\text{ft} = 30\,\text{ft/min}$$

Other examples of traveling waves are waves traveling along a vibrating string and sound waves, perhaps traveling down a long tube (Fig. 22-1).

STANDING WAVES

We obtain standing waves if we restrict waves by imposing physical barriers, such as fastening both ends of a string on which waves are moving or capping both ends of a tube in which sound waves are traveling, as in an organ pipe. As we shall see, only certain wavelengths, or frequencies, can now exist; all other waves are canceled by destructive interference when they are reflected from the barriers.

Consider a vibrating violin string. Both ends are fastened and cannot move. If we pluck the middle of this string, it will move up and down until the vibration dies out. There is no crest traveling along the string— the wave is a standing wave (Fig. 22-2). Plucking the string harder will increase the amplitude but otherwise will alter nothing.

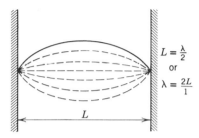

$L = \frac{\lambda}{2}$

or

$\lambda = \frac{2L}{1}$

Fig. 22-2. A vibrating string whose ends are fastened down. The solid line represents the position of the string at the time of plucking. The dotted lines represent various positions during the vibrational cycle. This string is vibrating at the fundamental frequency, also called the first harmonic (*see* text).

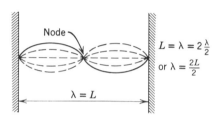

$L = \lambda = 2\frac{\lambda}{2}$

or $\lambda = \frac{2L}{2}$

$\lambda = L$

Fig. 22-3. A string vibrating at the second harmonic frequency.

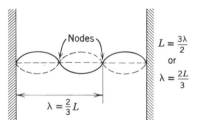

$L = \frac{3\lambda}{2}$

or

$\lambda = \frac{2L}{3}$

$\lambda = \frac{2}{3}L$

Fig. 22-4. A string vibrating at the third harmonic frequency.

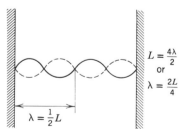

$L = \frac{4\lambda}{2}$

or

$\lambda = \frac{2L}{4}$

$\lambda = \frac{1}{2}L$

Fig. 22-5. A string vibrating at the fourth harmonic frequency.

If we pluck the string not in the middle but a quarter of the way from either end, we obtain the vibration shown in Fig. 22-3. The center of the string does not move up and down during the vibration; the area of the string that does not move during the vibration is called a *node*. This vibration has half the wavelength of the previous one, and therefore has twice the frequency; accordingly it is called the *second harmonic frequency*. The vibration shown in Fig. 22-2 is called the *first harmonic frequency*, also known as the *fundamental frequency*. Other vibrations with more nodes are possible. Figs. 22-4 and 22-5 picture the third and fourth harmonic frequencies, respectively.

Fig. 22-3 shows a *whole* wave whose wavelength, λ, is equal to the length, L, of the string. In Figure 22-2, which shows only half a wave, the wavelength is twice L; this is the maximum wavelength possible. It is evident that allowed vibrations contain an integral number of loops, that is, half waves. Symbolically,

$$L = n\left(\frac{\lambda}{2}\right), \qquad \text{where } n = 1, 2, 3 \ldots$$

Thus the only possible wavelengths are

$$\lambda = \frac{2L}{n}$$

In other words,

> in standing waves the wavelengths are restricted to discrete values and are said to be "quantized."

Frequencies are similarly quantized. The vibration illustrated in Figure 22-2 represents the fundamental frequency, or first harmonic, ν_1; the vibrations in the subsequent figures represent the second, third, and fourth harmonics:

$$\nu_2 = 2\nu_1, \quad \nu_3 = 3\nu_1, \ldots \nu_n = n\nu_1$$

The production of standing waves is responsible for the characteristic sound of each piano string. For example, when we hit a C-string, we produce a standing wave with a frequency of about 260 vibrations per second. In addition to this fundamental, we also excite vibrations with wavelengths equal to one-half, one-third, one-fourth, etc., of the wavelength of the fundamental (*see* Figs. 22-2 to 22-5). These vibrations are harmonics with frequencies $2 \times 260 = 520$, $3 \times 260 = 780$, etc., vibrations per second. It is these harmonics—also called "overtones"—that lend richness to a musical instrument.

Standing sound waves are not due to mechanical holding down of the ends of a string but are the result of constructive and destructive interference. The waves are reflected back and forth in the pipe. Only those meeting the quantum condition

$$\lambda = \frac{2L}{n}$$

are reinforced by reflection; the others destructively interfere after reflection and quickly die out. Thus an organ pipe of, say, 8-foot length can have sound waves of the following wavelengths:

16 ft ($n = 1$, fundamental, or first harmonic)

8 ft ($n = 2$, second harmonic)

16/3 ft ($n = 3$, third harmonic)

4 ft ($n = 4$, fourth harmonic)

16/5 ft (etc.)

Interference and the production of standing waves also explain why only certain wavelengths exist in each orbit of Bohr's model of the hydrogen atom (*see* Chapter 20). Orbiting electrons produce, or we might say "are," standing waves. Figure 22-6 illustrates the standing wave pattern of the electron in the fifth orbit of a Bohr hydrogen atom. The circle contains five whole waves; i.e., the wavelengths is one fifth of the circumference. Were this not so, if the circle contained some fractional waves, then on successive passes these waves would interfere with

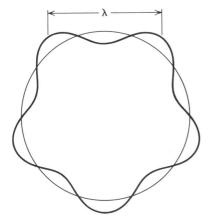

Fig. 22-6. Standing wave in the fifth orbit of a Bohr hydrogen atom.

each other, destroying the electron, which "is" the wave. This situation was shown in the insert of Fig. 20-2. In other words, stable orbits may not contain fractional waves. Only those orbits are stable that meet the quantum condition

$$C = n\lambda$$

where C is the orbit circumference.

Quantum restrictions clearly apply only to electrons or other particles in periodic (i.e., recurring) motion—for example, an orbiting electron or one bouncing back and forth between two walls. *Nonperiodic motion is not quantized;* an electron on a straight path, like a wave on the open sea, is not subject to destructive interference and may have *any* wavelength and therefore *any* velocity and *any* energy.

For violin strings each wavelength or vibrational frequency is associated with a different vibrational pattern—that is, with a different spatial distribution of the string (*see* Figs. 22-2 to 22-5). But, as we pointed out, only certain wavelengths or frequencies are possible; hence only certain wave patterns or spatial distributions of the string are possible. Likewise, in atoms only certain electron wavelengths are possible; each of these is associated with a different spatial distribution, which is called an *orbital*. The orbitals are responsible for many of the chemical properties of atoms. We shall return to these orbitals in the next chapter.

To summarize, we have explained why only certain wavelengths are possible in systems involving standing waves. The introduction of integers for characterizing orbital electrons arises in the same natural way as, for example, it arises when considering a vibrating string. Quantization is a natural consequence of the assumption of wave mechanics; unlike Planck and Bohr, we did not have to introduce quantization as an ad hoc assumption.

THE PARTICLE IN A BOX

We shall now apply the principles of wave mechanics to the "particle in a box." Imagine a submicroscopic particle—say, an electron—traveling along a straight path and being reflected back and forth from the opposite walls of a small box of length L. Newton's classical mechanics would allow us to calculate the exact pathway of the particle if we knew its exact position and velocity at a given time; but this is impossible according to Heisenberg. Quantum mechanics allows us to calculate only the *probability* of finding the particle in a certain place.

According to classical mechanics, the particle may have any velocity and therefore may have any kinetic energy whatsoever. It is equally likely to be found in any segment of its path. Not so according to wave mechanics. Only those velocities and energies are possible for which the particle's associated waves ($\lambda = h/mv = h/p$) exactly fit into the box. Just as for a vibrating violin string, the only waves allowed in the box are those of wavelength

$$\lambda = \frac{2L}{n}$$

where the quantum number n is an integer. Since the particle's wavelengths may be only $\lambda = 2L/n$, its momentum, p, likewise is restricted to certain values:

$$p = \frac{h}{\lambda} = n\,\frac{h}{2L}$$

This in turn means that velocities are restricted to

$$v = \frac{p}{m} = n\,\frac{h}{2mL}$$

Kinetic energy also is restricted to definite values:

$$KE = \frac{1}{2}mv^2 = \frac{1}{2}m\left(\frac{nh}{2mL}\right)^2 = n^2\,\frac{h^2}{8mL^2}$$

Thus, according to wave mechanics, wavelength, momentum, velocity, and kinetic energy all have quantized values for the particle in a box.

When the particle in a box is in its lowest allowed energy state, E_1 ($n = 1$), its accompanying wave (Fig. 22-7) looks just like the first of our pictures of the vibrating string (Fig. 22-2).

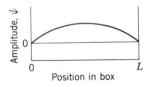

Fig. 22-7. The wave representation of the particle in a box in its lowest energy state, E_1 ($n = 1$). The Greek letter ψ (psi) conventionally designates wave amplitude.

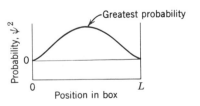

Fig. 22-8. The probability distribution of the particle in a box in its lowest energy state. The particle is most likely to be in the center of the box. The probability of finding the particle at either wall is zero.

In Schrödinger's wave mechanics the probability of finding the particle at a given position is proportional to the *square* (ψ^2) of the wave amplitude (ψ), just as in Maxwell's electromagnetic theory the intensity of a light wave is proportional to the square of its wave amplitude. Thus, to obtain the probability of finding the particle at a given position in the box, we must square the value of the wave amplitude at that position. When we do this for each position in Fig. 22-7, we obtain the *probability distribution* shown in Fig. 22-8. From the hump in the curve we see that, when the particle is in its lowest energy state, it is most likely to be in the center of the box. This is in direct conflict with classical ideas, which would have the particle distribute its presence uniformly along its path.

If we make enough energy available to the particle to promote it to the next higher energy state, E_2 ($n = 2$), the wave representation and probability distribution would be as shown in Fig. 22-9. Note that,

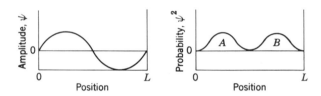

Fig. 22-9. Wave representation and probability distribution of the particle in a box in its second-lowest energy state, E_2 ($n = 2$). The particle is most likely to be either at position A or B. The probability of finding the particle in the center of the box is zero.

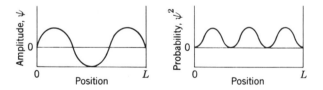

Fig. 22-10. Wave representation and probability distribution of the particle in a box in energy state E_3.

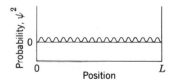

Fig. 22-11. Probability distribution of the particle in a box at some higher energy level. The probability distribution approaches the classical picture of equal probabilities for all positions.

although the wave amplitude is partly negative (at the wave trough), the probability is always positive, because the square of even a negative number is positive. With that much energy, E_2, the particle is most likely to be found *not* in the center of the box but at the 1/4 and 3/4 positions (A and B in Fig. 22-9); the probability of finding the particle in the center actually is *zero*. Again, the quantum-mechanical picture disagrees with the classical picture.

We create a paradox—note that *we* are *creating* the paradox!—when we combine the classical picture of a corpuscular electron with the quantum-mechanically derived probability distribution: How does an electron or other particle get from position A to position B *without* going through the center? In terms of particles, we cannot explain. If, however, we employ the particle's alter ego, the wave, then there is no problem at all. Wave motion can pass from one side of the box to the other right through a node (the place where the string, or wave, is "held down").

The wave representation and probability distribution of the particle in energy state E_3 ($n = 3$) are shown in Fig. 22-10. At still higher energy levels the quantum-mechanical picture begins to correspond more and more closely to the classical picture of equal probabilities for all positions (Fig. 22-11). This is another example of Bohr's Correspondence Principle, which in effect states that qantum mechanics reduces to classical mechanics when the quantum numbers are large. In other words, quantum mechanics is universally applicable, but for higher energy levels and for macroscopic systems the quantum-mechanical picture becomes indistinguishable from the classical picture. (*See* Example 21-2 and Examples 22-2 and 22-3). For these cases, it is often more convenient to substitute classical (Newtonian) calculations.

One important feature of the particle in a box remains to be emphasized.

The lowest allowed kinetic energy of the particle is not zero,
but $E_1 = h^2/8mL^2$. Hence the particle is *never at rest*. This again is in conflict with classical theory, but is in accord with the Uncertainty Principle. If the particle were standing still, both its position and momentum (zero) would be specified exactly—and this is not possible. It is this minimum residual energy that accounts for the atom's stability. The "orbiting" electron in the ground state ($n = 1$) cannot give up its energy. Hence it cannot fall into the nucleus.[1]

---

[1] This point will come up again later in connection with molecular vibrations—molecules never stop oscillating either. This principle also caused Planck to revise his original theory of blackbody oscillators. In the revised theory, zero-energy was not an allowed level; the lowest allowed level was equal to $1/2 \ E_o$ (*see* Fig. 17-2).

So far we have dealt with a one-dimensional "box." To describe a particle in a normal three-dimensional $L_a \times L_b \times L_c$ box, three quantum numbers are needed (say n_a, n_b, and n_c), each representing an independent integral number. The kinetic-energy expression for a particle in a three-dimensional box is:

$$KE = \frac{h^2}{8m}\left(\frac{n_a^2}{L_a^2} + \frac{n_b^2}{L_b} + \frac{n_c^2}{L_c}\right)$$

After considering some numerical examples, we shall be ready to extrapolate our reasoning to the electron in an atom, which is, in a sense, a three-dimensional "box."

Example 22-1. Calculate (a) the minimum kintetic energy and (b) the minimum velocity that an electron may have in a one-dimensional "box" of 1-inch length (approximately 2.5 cm).

Solution.

(a) $KE = n^2h^2/8mL^2$. The lowest energy is in the ground state, where $n = 1$:

$$E_1 = \frac{h^2}{8mL^2} = \frac{(6.63 \times 10^{-27} \text{ erg-sec})^2}{8 \times (9.1 \times 10^{-28} \text{ g}) \times (2.5 \text{ cm})^2} \approx 10^{-27} \text{ erg}$$

or approximately 10^{-35} calories—a very small quantity.
(b) $v = nh/2mL$

$$v_1 = \frac{h}{2mL} = \frac{6.63 \times 10^{-27} \text{ erg-sec}}{2 \times (9.1 \times 10^{-28} \text{ g}) \times 2.5 \text{ cm}} = 1.5 \text{ cm/sec}$$

In other words, an electron in a 1-inch box cannot travel more slowly than about 1/2 inch per second. It cannot stand still. This is in accord with the Uncertainty Principle. The uncertainty in the electron's position is $\Delta s = 2.5$ cm, because we know it is somewhere in the box. The uncertainty in its velocity is 3.0 cm/sec, because we don't know whether it's going forward or backward at 1.5 cm/sec. Then $\Delta s \Delta v = (2.5 \text{ cm}) \times (3.0 \text{ cm/sec}) = 7.5 \text{ cm}^2/\text{sec}$. This value is approximately equal to h/m:

$$\frac{h}{m} = \frac{6.6 \times 10^{-27} \text{ erg-sec}}{9.1 \times 10^{-28} \text{ g}} = 7.3 \text{ cm}^2/\text{sec}$$

Thus

$$\Delta s \Delta v \approx \frac{h}{m}$$

which is one form of Heisenberg's Uncertainty Principle (Equation (2) of Chapter 21).

Example 22-2. Calculate the approximate minimum velocity of a 5-ounce ball ($\sim$ 150 g) in a 3-foot ($\sim$ 100 cm) "box."

Solution.

Approximately 2×10^{-31} cm/sec.

That is *very* slow! According to the Uncertainty Principle and quantum mechanics, an object can never stand still. This applies as well to our 5-ounce ball. Its minimum speed, however, is so small that it *appears* to be standing still. Once more, this example demonstrates the Correspondence Principle: We are quite safe in using classical mechanics for macroscopic bodies.

Example 22-3. (a) Calculate the approximate kinetic energy of a 5-ounce ball bouncing back and forth at 40 miles per hour ($\sim$ 1800 cm/sec) between two walls. (We want the *actual* kinetic energy, not the allowable minimum.)

(b) Calculate the quantum number n for this system if the walls are about 3 feet apart.

Solution.

(a) $KE = mv^2/2 = 150 \times 1800^2/2 \approx 2.5 \times 10^8$ ergs (or about 2 foot-pounds).

(b) $KE = n^2h^2/8mL^2$

$$n = \sqrt{\frac{8mL^2KE}{h^2}} = \sqrt{\frac{8 \times 150 \times 100^2 \times (2.5 \times 10^8)}{(6.6 \times 10^{-27})^2}} \approx 10^{34}$$

The latter calculation can also be performed in a more direct manner by substituting $mv^2/2$ for the energy:

$$n = \sqrt{\frac{8mL^2(mv^2/2)}{h^2}} = \sqrt{\frac{4m^2v^2L^2}{h^2}} = \frac{2mvL}{h} \approx 10^{34}$$

This is a huge quantum number. The probability distribution curve has 10^{34} ripples in the interval of 3 feet. In effect that is a uniform distribution (compare to Fig. 22-11), approaching agreement with classical mechanics.

SUMMARY

From our analysis of standing waves, the quantization of wavelength,

$$\lambda = \frac{2L}{n}$$

followed directly. Recognizing the electron in an atom as a standing wave, and using an analogy to a vibrating string, we obtained quantized wavelengths for the electron in an atom:

$$\lambda = \frac{C}{n}$$

where C is the circumference of a Bohr orbit.

Because frequency, velocity, energy, and momentum are directly related to wavelength, they are also quantized. Depending on the system's parameters (e.g., length of string or path, and mass of particle), only certain "characteristic" frequencies (that is, characteristic of the system), certain characteristic energy values, and certain characteristic momenta are allowed.

We applied wave-mechanical concepts to a small particle bouncing back and forth inside a box. Several interesting features emerged.

1. Quantum mechanics will not tell us where the particle is at any instant, but it will tell us the mathematical probability of finding the particle at a given location. The probability is proportional to the square of the amplitude of the particle's associated de Broglie wave.

2. The probability distribution depends on the energy level, or quantum number, of the system. The probability distribution has one maximum for the lowest energy state, called the ground state. For the second-lowest state ($n = 2$), the probability distribution has two maxima with a minimum of zero between them, coincident with a node in the particle's associated wave. The $n = 3$ state has three maxima in its probability curve, and two nodes in its wave. And so on. These probability distributions are quite different from those predicted by classical mechanics, which says that on the average the particle is equally likely to be any place in the box.

3. The probability distribution of high-energy particles (large n) approaches the uniform distribution predicted by classical mechanics. For these cases the quantum-mechanical description corresponds closely to the classical picture, and it may be not only more convenient but also sufficiently accurate to substitute classical (Newtonian) calculations for the more precise quantum-mechanical ones.

4. The particle can never stand still. It always has a certain minimum velocity. A particle standing still would violate Heinsenberg's Uncertainty Principle, because we could then determine its precise position and velocity.

5. Because the particle can never stand still, it will always have some residual kinetic energy. This minimum kinetic energy, associated with

the $n = 1$ state, is variously called ground-state energy or zero-point energy. Analogously, in an atom an electron cannot give up its ground-state energy; hence, it cannot fall into the nucleus. This principle accounts for the atom's stability.

6. All energy levels are whole-number-squared multiples of the ground-state energy:

$$E_n = n^2 E_1$$

One of the beauties of wave mechanics is that quantization of wavelengths, velocities, frequencies, momenta, and energies is a direct natural consequence of the basic assumptions of wave mechanics. The introduction of integers arises in the same natural way as, for instance, in a vibrating string. Quantum numbers do not have to be introduced artificially by ad hoc assumptions, as previous workers in the field had to do—for instance, Planck for his blackbody oscillators and Bohr and Sommerfeld for their orbiting electrons. In the wave-mechanical view, "orbiting" electrons are simply standing waves inside an atom, like standing waves in an organ pipe.

Now that these principles have been established, we can proceed to a discussion of the hydrogen atom as seen by quantum mechanics.

Chapter 23

THE WAVE-MECHANICAL PICTURE OF THE HYDROGEN ATOM

In principle, classical Newtonian mechanics allows us to calculate the exact future positions, velocities, and energies of all bodies in a system, provided we know their exact positions and velocities at some previous moment and the nature of the interaction forces among these bodies. Thus classical mechanics applies very well to the solar system. Exact astronomical measurements of the relative positions and velocities of the sun and planets, together with the Inverse-Square Law of Gravitational Forces, have led to remarkably accurate predictions of planetary positions. Because of the Uncertainty Principle this approach cannot be applied to atomic systems, because we can never simultaneously determine precise positions and velocities of the subatomic particles involved. In fact, we are not even justified in treating the electron as a conventional "particle."

Even though we cannot determine exact positions and velocities in atomic systems, quantum mechanics does allow us to calculate the *probability* of finding an electron at any position at a given instant, or, what is equivalent, the average *electron density* over a given span of time.

This non-Newtonian approach also enables us to calculate the allowed energy levels of the electron in a hydrogen atom and the spatial distribution of the electronic charge for each of these energy states. These spatial distributions we shall henceforth call *orbitals* rather than orbits.

WAVE EQUATIONS

A basic postulate of wave mechanics is that any particle can be represented by a wave of amplitude, ψ, and that the square of this wave's

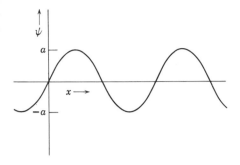

Fig. 23-1. A simple sine wave, whose equation is $\psi = a \sin x$. At a given position, x, the wave's amplitude is ψ, which attains a maximum value of a. Such a wave, according to wave mechanics, represents a particle in uniform rectilinear motion. The probability of finding the particle at a given position x is proportional to ψ^2, the square of the wave amplitude at position x.

amplitude represents the probability of finding the particle at that position. For example, the wave for a particle in rectilinear motion is a "sine wave" (Fig. 23-1), represented by the equation

$$\psi = a \sin x$$

That equation tells us that the amplitude ψ varies sinusoidally with the distance x and has a maximum value of $\psi = a$. For other systems, the wave form and its equation may be quite complex. In the simple case of the particle in a box we obtained wave forms by analogy to vibrating violin strings. For more complex cases we must set up and solve the wave equations. The calculations of wave mechanics usually are quite laborious, although in recent years computers have made them more manageable.

These calculations, like those of classical mechanics, depend on an input of experimental data. For our solar-system example, the data include positions, velocities, and inverse-square gravitational forces. For the particle in a box, the input data are the particle's mass, m, and the length of the box, L. With these data we can solve the wave equation and obtain the allowed energy states, E_n, and a mathematical expression, $\psi_n = \ldots$, for each of the allowed energy states. Figs. 22-6 to 22-9 are graphs of several such ψ expressions. The ψ^2 graphs in these figures indicate the probabilities of finding the particle at various positions in the box. That is, the ψ^2 plot shows how the particle distributes its presence on the average; the ψ^2 plot, in other words, represents the overall spatial distribution of the particle.

For the electron in the hydrogen atom, the input data consist of the mass, m, and charge, e, of the electron and the inverse-square force of electrostatic attraction between the electron and the nucleus ($F = Ze^2/r^2$). With these data the solution of the wave equation is straightforward. As we did for the particle in the box, we obtain the possible energy levels of the hydrogen atom and the spatial distribution of the electronic charge for each of the allowed energy states.

Note that quantum mechanics does not define a path or orbit for either the particle in a box or the electron in the hydrogen atom. That would be a classical description applicable to a classical particle, which the electron is *not*. Instead, quantum mechanics only allows us to determine the probability of finding the electron at a particular location.

THE FOUR QUANTUM NUMBERS: n, l, m, s

Possible Values of n, l, m

For the particle in a one-dimensional box, the allowed energies, E_n, and spatial distributions are a function of an integral quantum number, n: $E_n = n^2 h^2 / 8mL^2$. For a three-dimensional box, as we saw in Chapter 22, the energies are a function of three independent quantum numbers, which we called n_a, n_b, and n_c. "Independent" here means that the three numbers of a set determining the total kinetic energy may each equal *any* integer, such as $n_a = 7$, $n_b = 4$, and $n_c, = 13$.

For the electron in the hydrogen atom, which is also a three-dimensional system, the allowed energy levels and spatial distributions are also a function of three quantum numbers. These are conventionally labeled n, l, and m. In the hydrogen atom, however, the quantum numbers are not quite independent of each other. (This is inherent in the mathematics.) They are limited to the following integral values:

n, known as the *principal quantum number*, may have *any* positive nonzero integral value:

$$n = 1, 2, 3, 4, \ldots .$$

l, known as the *angular momentum quantum number* (or azimuthal quantum number or orbital quantum number), may have any positive integral value from zero up to one less than the value of n for that state (i.e., a total of n possibilities):

$$l = 0, 1, 2, \ldots, (n - 1)$$

m, known as the *magnetic quantum number* (or magnetic momentum quantum number), may have any integral value from $-l$ through zero to $+l$ (i.e., a total of $2l + 1$ possibilities):

$$m = 0, \pm 1, \pm 2, \ldots \pm l$$

For example, when $n = 1$, l and m each must equal 0. When $n = 2$, l may equal 0 or 1; when $l = 0$, m must equal zero, but when $l = 1$, m may equal $+1$, 0, or -1. Thus when the lone electron in the hydrogen atom is in the $n = 2$ state, it may have any one of four sets of quantum numbers (*see* Table 23-1).

Table 23-1. Some Possible Combinations of the Quantum Numbers n, l, and m for the Electron in the Hydrogen Atom

n	l	m	No. of combinations
1	0	0	1
2	0	0	
2	1	-1	
2	1	0	4
2	1	$+1$	

Problem 23-1. Extend Table 23-1 to include the cases for $n = 3$ and for $n = 4$. (Hint: there should be nine sets of quantum numbers with $n = 3$.)

Physical Meaning of the Quantum Numbers

In both the old Bohr theory and new quantum theory the meaning of n is similar. It is an index of orbital size. The higher the n values, the greater is the probability of finding the electron farther from the nucleus. Also, the electron energy in one-electron atoms depends primarily on the value of n:

$$E = -\frac{2\pi^2 m e^4 Z^2}{n^2 h^2}$$

This equation is obtained by solving the wave equation for the hydrogen atom and, astonishingly, is identical with the equation Bohr had obtained earlier from his orbiting-electron model. Energy levels derived from this equation are in accord with experimental values.[1]

The quantum number l is associated with orbital shapes. A value of zero denotes a spherically symmetric shape, like a ball; other values of l are associated with various nonspherical shapes; that is, the smallest orbital of the hydrogen atom ($n = 1$) is spherically symmetric ($l = 0$). When the electron is in the next-higher energy state ($n = 2$), the charge

---

[1] In Chapter 16 we used spectroscopic data to obtain an empirical energy-level diagram. In Chapter 18 we showed how Bohr devised an atom model from which he could *derive* the proper energy levels. We realize today that we must discard that model because of the Uncertainty Principle. So far, the new quantum-mechanical model has withstood the test of time, and we believe it will continue to do so.

distribution is spherical when $l = 0$, but is nonspherical when $l = 1$. Still more orbital shapes are possible for higher-energy states (*see* Problem 23-1).[2]

The energy of *one-electron atoms* depends primarily on the n value and is only very slightly affected by the l value. (Different orbital shapes produce minor effects on the energy levels and thereby are related to the observed fine structure of spectral lines.)[3] In *multielectron atoms,* however, orbital shapes do affect energy values considerably, as discussed in the next chapter.

The quantum number m represents different spatial orientations of the electron orbitals. A spherically symmetric charge distribution ($l = 0$) clearly can have only one orientation ($m = 0$). Less symmetric shapes, however, may have various orientations; so, for example, an orbital shape characterized by $l = 2$ may have five different orientations, associated with m values of 2, 1, 0, -1, -2.

Differences in orientation are related to a slight raising or lowering of energy levels only when the atom is exposed to a magnetic field, as in Zeeman's experiments. In 1896 Pieter Zeeman (1865–1943) discovered that lines in an emission spectrum are split if the light source (i.e., the sample, or flame) is placed between the poles of a magnet. Note that the Zeeman effect was observed before the existence of the Bohr or Rutherford atom models (1913, 1911) and even before Thomson's "discovery" of the electron (1897). It is because of this effect that m is called the "magnetic" quantum number.

"Nicknames" of Orbital Electrons

We commonly use shorthand labels for the n and l values of an electron. These labels are number-letter combinations in which the number is the n value and the letter represents the l value, as follows.

l value: 0, 1, 2, 3, 4, 5, 6, etc.

label: *s p d f g h i*, etc.[4]

Thus an electron with $n = 1$ and $l = 0$ is designated as a 1s electron; $n = 4$, $l = 2$ is designated as 4d; etc.

---

[2] Orbital shapes are discussed in greater detail in a later section of this chapter.

[3] Sommerfeld had tried to explain fine structure in the hydrogen spectrum by introducing elliptical orbits (Chapter 18).

[4] The first four letters—*s, p, d,* and *f*—are carry-overs from spectroscopy, where they stood for *s*harp, *p*rincipal, *d*iffuse, and *f*ine.

Problem 23-2. (a) What are the "nicknames" for electrons with $n = 2$, $l = 0$; and $n = 2$, $l = 1$?

(b) What are the n and l values for $4s$, $4p$, and $4f$ electrons?

Problem 23-3. Can there be a $1p$ electron; a $2d$ electron? (Why not?)

The Fourth Quantum Number, s

In 1925 a fourth quantum number, s, was introduced. Millikan and others in 1924 had discovered certain fine-structure features of spectra (the lines split into two lines); these were interpreted a year later as the result of what might be considered electron spin (rotation of an electron on its own axis). This spin may take only two directions—"clockwise" and "counterclockwise." It is described by the spin quantum number s (or, spin angular momentum quantum number), which may have a value of $+1/2$ or $-1/2$. The spin quantum number s, sometimes denoted as m_s, must not be confused with the shorthand label s for an electron having $l = 0$.

Table 23-2 illustrates the possible combinations of the four quantum numbers that hydrogen's one electron may have. It follows that there are two possible combinations with energy E_1 (i.e., for $n = 1$); eight with energy E_2 ($n = 2$); eighteen with energy E_3; and so on. Although we have discussed only one-electron atoms, we are getting a hint about the "magic numbers" for the "shells" of multielectron atoms. We shall return to this point in the next chapter.

Table 23-2. Some Possible Combinations of the Four Quantum Numbers in One-Electron Atoms

n	l	m	s	No. of combinations
1	0	0	$+\frac{1}{2}, -\frac{1}{2}$	2
2	0	0	$\pm\frac{1}{2}$	
2	1	-1	$\pm\frac{1}{2}$	8
2	1	0	$\pm\frac{1}{2}$	
2	1	$+1$	$\pm\frac{1}{2}$	

Problem 23-4. How many possible combinations of quantum numbers are there with $n = 4$?

SHAPES OF ORBITALS

The application of quantum mechanics to the particle in a box does not yield a specific path, but rather a probability distribution for the location of the particle. The specific spatial distribution depends on which of the allowed energy states the particle happens to have. Similarly, in atoms the Uncertainty Principle makes it meaningless to talk about definite electron orbits; the uncertainty in position is too great. Instead, the solution of the wave equation provides us with a probability distribution of the electron in the atom. The specific spatial distribution again depends on the energy state of the electron—i.e., on the values of the four quantum numbers.

To understand the chemical and physical properties of atoms and molecules, we must know something about these probability distributions for the various energy levels. Let us start by considering the hydrogen atom in the ground state (that is, the $1s$ state, where $n = 1$ and $l = 0$). Quantum mechanics does not provide us with an electron orbit, but it tells us that the electron is most likely to be near the nucleus and progressively less likely to be farther away. There are several ways we can picture this probability distribution.

Dot Diagrams

If we could take a series of instantaneous photographs of the electron, we would obtain a series of dots representing its various positions (just as we obtained the diagram in Fig. 20-4 of the impact sites in the imaginary diffraction of single electrons). The greatest dot density would be near the nucleus, as shown in Figure 23-2.

In three dimensions our picture would be a spherical cloud of dots, with greatest density near the nucleus and fuzzy undefined outer limits, like a fog with decreasing density. Note that *the atom has no definite boundary;* there is a finite, but rapidly decreasing, probability of finding the electron at *any* distance from the nucleus.

By distributing its presence throughout the cloud, the electron in effect is "smeared out" over the entire cloud. Thus the cloud represents not only a probability distribution but, effectively, a charge-density distribution. *The two distribution concepts are equivalent,* as we showed in Chapter 20 (p. 284). The $1s$ orbital of the hydrogen atom, then, is a spherical cloud whose density decreases with distance from the nucleus.

Charge-Density Graphs

We can show the relationship between charge density and orbital radius more precisely by means of a graph. Figure 23-3 shows the $1s$ orbital's charge density—or the equivalent probability, ψ^2—as a function of the radius, r. The plot indicates maximum density *at* the nucleus ($r = 0$) and a vanishing density at large distances ($r \to \infty$).

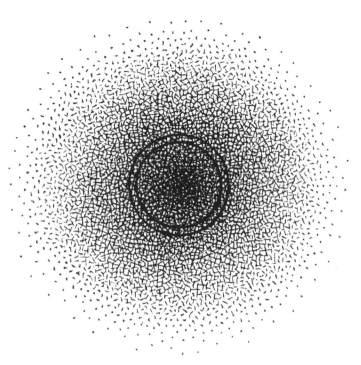

Fig. 23-2. Schematic dot diagram of the electron density of the hydrogen atom in its ground state ($1s$; $n = 1$, $l = 0$). The ring represents a narrow band about the classical Bohr orbit.

We have another, perhaps more useful way of looking at orbitals. We can imagine a series of closely spaced concentric circles around the nucleus and count the number of dots in each ring (Fig. 23-2). This number is a measure of the total probability of finding the electron at that ring's radius. In proceeding out from the high-density center, we find that each ring area increases and its dot density decreases. At first, increasing ring area outweighs decreasing dot density, so that the *total* number of dots at a particular distance from the nucleus increases. The total reaches a maximum and then the rapidly vanishing dot density takes over. For a three-dimensional spherical cloud, this total quantity is the product of ψ^2 and the area of the spherical shell, $4\pi r^2$ (Fig. 23-4).

When appropriate data are used in the calculations, the maximum in the curve occurs when r equals 0.53 angstrom, which is exactly equal to the classical Bohr radius. In other words, we are most likely to find the electron at a distance corresponding to the Bohr radius. But, again, note

Probabilities, or Electron-Density Distributions, in the Hydrogen Atom
in its Lowest Energy State

$(1s; n = 1, l = 0)$

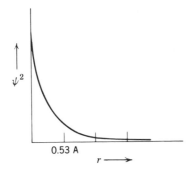

Fig. 23-3. Probability, ψ^2, as a function of radius, r.

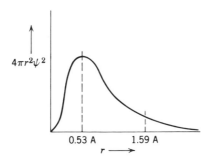

Fig. 23-4. Probability of finding the electron in a spherical shell of radius r.

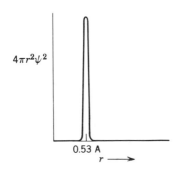

Fig. 23-5. Probability of finding the electron in a spherical shell of radius r in the hypothetical Bohr atom model.

that the atom has no defined outer limits; the charge density still is relatively large at, say, 1.59 A, which is three times the classical Bohr radius. In Fig. 23-5 the probability distribution for the hypothetical Bohr atom is shown for comparison; the probability is 100% at 0.53 A, and zero elsewhere.

Cloud "Photographs"

So far we have portrayed only the spatial distribution of a 1s electron; that is, we have dealt with the ground state of hydrogen. Let us now turn to various excited states of the hydrogen atom.

All s orbitals ($l = 0$) are spherically symmetrical. The 1s orbital is like a ball with fuzzy edges, as discussed before. The 2s orbital is like a ball plus a concentric spherical fuzzy shell—a *spherical* shell, not merely a ring; the ball-and-shell is *one* orbital, not two. The 3s orbital consists of a ball plus two concentric shells.

The other orbitals of the hydrogen atom are not spherically symmetric;

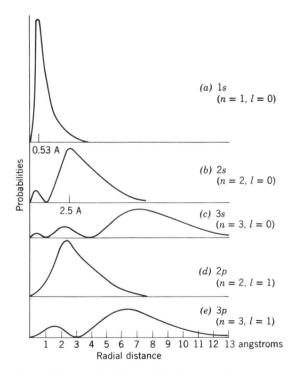

Fig. 23-6. Probabilities of finding the electron at various distances from the nucleus for several orbitals of the hydrogen atom. All figures are drawn to the same scale.

they have directional character. In general, the elongated p orbitals are oriented along either the x, y, or z coordinate axis, depending on whether m equals 0, +1, or −1. We shall see in a later chapter how these orientations account for the directional character of chemical bonds. Orbitals with still larger l values have more complex spatial distributions.

The quantitative aspects of the charge distributions are shown in Fig. 23-6, which (like Fig. 23-4) shows the probabilities of finding the electron at various distances from the nucleus. As stated before, the 1s electron is most often found at a distance of 0.53 A from the nucleus. For the 2s electron, the most probable distance is about 2.5 A, and for 3s it is about 6 A. In both the 2s and 3s orbitals, however, the electron also spends some time quite close to the nucleus, as revealed by the left-most small humps in Figs. 23-6b and c; these humps represent the inner balls in the cloud pictures. Figs. 23-6d and e show the same distributions of the 2p and 3p orbitals. These representations, however, give no hint of the directional character of these nonspherical spatial distributions.

SUMMARY

Solution of the wave equation for the hydrogen atom leads to quantized energy levels. Each energy level is associated with a specific spatial probability distribution of the electron—i.e., with a specific average charge-density distribution, which we call an orbital.

The energy levels and spatial distributions are functions of four quantum numbers, labeled n, l, m, and s. They may have the following values:

$$n = 1, 2, 3, 4, \ldots$$

$$l = 0, 1, 2, 3, \ldots, (n - 1)$$

$$m = 0, \pm 1, \pm 2, \ldots, \pm l$$

$$s = \pm 1/2$$

The principal quantum number, n, determines the energy and orbital size of one-electron atoms. The angular momentum quantum number, l, and the magnetic quantum number, m, are associated with orbital shape and orientation, respectively. The spin quantum number, s, is related to spin of the electron about its own axis.

Orbitals with $l = 0$ (s orbitals) are spherically symmetric. Orbitals with $l = 1, 2, 3, 4$, etc. (called p, d, f, g, h, i, etc., orbitals) are not spherical but have directional character.

The quantum-mechanical atom, unlike the old Bohr atom, has no definite boundary; the probability, or average charge density, is smeared out over the whole of space. We may picture the atom as a cloud with decreasingly dense, fuzzy outer limits. This does not mean that one electron occupies at one time the whole cloud. The cloud density represents the *probability* of finding the electron. However, for many purposes it is useful to think of the atom *as if* its electron were smeared out like a cloud.

Chapter 24

MULTIELECTRON ATOMS AND THE PERIODIC TABLE

In Chapter 23 we discussed the sizes, shapes, and energy levels of the various orbitals that may be occupied by the lone electron of a hydrogen atom. Would it now be reasonable to assume, at least as a start, that similar orbitals are available to the many electrons of other atoms?

Naturally, we expect this heuristic assumption to yield at best a crude facsimile of the actual multielectron atom, because electrons in such atoms repel one another; that repulsion is absent in the hydrogen atom with its single electron. But, if our crude picture corresponds even vaguely to reality, we can refine it by introducing repulsion effects until our predictions are more in line with experimentally observed data.

We shall follow this scheme in the present chapter. But first let us review what we learned about electron shells in Chapter 19.

EVIDENCE OF ELECTRON SHELLS

The most important feature of multielectron atoms is that their electrons are arranged in groups, called *shells*. Each shell apparently may contain not more than a certain number of electrons—two in the innermost shell, eight in the next shell, and so on.

The experimental evidence cited in Chapter 19 for the existence of electron shells came mostly from ionization energies—the energies needed to remove electrons from atoms. The first ionization energy (or energy required to remove one electron from a neutral isolated atom) is plotted in Fig. 24-1, for elements 1 through 92. The graph (which is identical with Fig. 19-2) has a general sawtooth pattern; elements with low ioni-

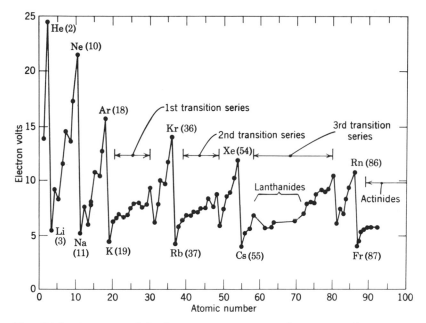

Fig. 24-1. Variation of the first ionization energy with atomic number.

zation energies (near the bottom of the graph) have an easily removable electron; elements with high ionization energies hold their electrons relatively tight.

In Chapter 19 we interpreted these observations as follows. The easily removable electron in the bottom elements is well screened from the positive nucleus by other, closer-in electrons. In the alkali-metal series (Li, Na, K, Rb, Cs, and Fr) this electron is in an outer shell by itself and in effect "feels" only one of the positive nuclear charges. The next-higher elements have two electrons in their outermost shell; these electrons "feel" almost two nuclear charges and are not quite so easily removed. The top of each sawtooth represents elements with seemingly full outer shells: the noble gases (He, Ne, Ar, Kr, Xe, and Rn, with atomic numbers 2, 10, 18, 36, 54, and 86, respectively).

Counting the number of elements between the bottom and top of each sawtooth gives us an idea of the number of atoms in each outer shell—namely, 2, 8, 8, 18, 18, and 32, respectively.

The general idea that electrons are arranged in shells is corroborated by examination of the energies required to remove the first, second, third, and subsequent electrons of any *one* element (*see* Table 19-1). For instance, in carbon (element 6) there is a gradual increase in the energies required to remove successively the first four electrons: 11, 24, 48, and 64 ev, respectively. As the atom becomes increasingly positive as a result of electron removal, the remaining electrons are held tighter. Removal of the fifth and sixth electrons, however, requires *much* more energy—392 and 490 ev, respectively; this fact indicates that these last two electrons are much closer to the nucleus than the previous four. In other words, carbon has four electrons in an outer shell, and two in an inner shell.

The chemical inertness of the noble gases further sustains the idea of shell structure. We ascribe their relative inertness to filled electron shells. It seems that a given shell may contain only so many electrons.

Additional support for the idea of electron shells comes from periodicity of chemical properties, from the formation of ions in solutions and crystals, and from X-ray spectra, as follows.

1. The chemical properties of elements apparently depend primarily on the electrons in the outermost shell of their atoms.

2. The formation of positive and negative ions in solutions and crystals can be explained in terms of an electron-shell model and with the assumption of stable noble-gas configurations.

3. Some features of Moseley's X-ray spectra could be accounted for by an electron-shell model, notably the discreteness of emission lines, the dependence of frequency on atomic number, and the failure of frequencies to vary periodically (as chemical properties do). What is more, we were able to derive Moseley's empirical equation.

We see then that experimental evidence favors some sort of shell model, in which the innermost shell contains only two electrons, and other shells may contain 8, 18, or 32 electrons. We would now like to develop some scheme that will allow us to explain, or derive, these "magic numbers."

THE PAULI EXCLUSION PRINCIPLE

In Table 23-2, we saw that the lone electron in hydrogen has a choice of two different sets of quantum numbers for energy state E_1 ($n = 1$); 8 choices for energy state E_2 ($n = 2$); 18 for energy E_3 ($n = 3$); 32 for energy E_4 ($n = 4$). (Also *see* Problems 23-1 and 23-4.) These numbers of possible combinations of quantum numbers for hydrogen—2, 8, 18, and 32—are precisely the "magic numbers" of electrons allowed in the various shells of multielectron atoms.

This correspondence between the numbers for hydrogen and for multielectron atoms justifies the assumption that multielectron atoms have hydrogen-like orbitals and that each orbital is filled by not more than two electrons.

This idea was proposed formally in 1925 by the Austrian theoretical physicist Wolfgang Pauli (1900–1958). We can state his *Exclusion Principle* in two different ways: (1) No two electrons may have the same set of four quantum numbers; or (2) not more than two electrons may occupy the same orbital (characterized by a set of n, l, and m numbers), and these two electrons must have opposed spins ($s = +1/2$ and $s = -1/2$).

Interestingly, the Exclusion Principle was promulgated shortly before the formulation of quantum mechanics (1925–1926). It is strictly an empirical ad hoc rule, artificially introduced, without theoretical justification. It is compatible with quantum mechanics, but apparently cannot be proved by or derived from quantum mechanics.

We are now ready to consider the electron structures of individual atoms.

ELECTRON CONFIGURATIONS

Beginning with the assumption that multielectron atoms have hydrogen-like orbitals, we shall mentally build up these atoms by filling their orbitals with electrons, one at a time. As we add more electrons we shall consider interactions with the electrons already present. In predicting these interaction effects we shall be guided by the orbital shapes in hydrogen atoms—i.e., by their charge-density distributions. Ultimately, however, we must tailor our multielectron model to fit experimental data, such as ionization energies, spectra, and chemical properties. Remember: *Theory Guides, Experiment Decides.*

We proceed by assigning each electron to the lowest available energy level. We follow this order, rather than starting with higher energy levels, because systems in nature tend to assume their lowest energy state; we say, "Energy runs downhill." Accordingly, the most stable electron configuration of an atom, its ground state, has all electrons in the lowest available energy levels, subject to the limitations of Pauli's Exclusion Principle (no two electrons may have the same set of four quantum numbers).

The ground state of hydrogen is 1s: $n = 1$, $l = 0$, $m = 0$, $s = +1/2$ or $-1/2$. In helium both electrons are in the same orbital with opposed, or "paired," spins: $n = 1$, $l = 0$, $m = 0$, $s = +\frac{1}{2}$ and $-\frac{1}{2}$. We use superscripts in the "nickname" designations to indicate the number of electrons in an orbital; thus helium's ground-state configuration is $1s^2$. Since there are no other orbitals with $n = 1$, helium has a full shell.

The third electron of lithium (Li, Z = 3) then has to go into a new shell. It will occupy the 2s orbital ($n = 2, l = 0, m = 0, s = +\frac{1}{2}$ or $-\frac{1}{2}$); the complete configuration of lithium then is $1s^2 2s$. The additional electron in beryllium (Be, Z = 4) will also be in the 2s orbital and pair its spin with that of the other electron; beryllium has a $1s^2 2s^2$ configuration. With boron (B, 5) we start filling three 2p orbitals ($l = 1; m = -1$, 0, +1; $s = \pm\frac{1}{2}$). We call this group a *subshell*. This subshell is filled with six electrons in neon (Ne, 10). With neon, then, we have completed the second *principal shell* ($n = 2$). The electron configurations of the first ten elements are as follows.

H	–	$1s$	**C**	–	$1s^2 2s^2 2p^2$
He	–	$1s^2$	**N**	–	$1s^2 2s^2 2p^3$
Li	–	$1s^2 2s$	**O**	–	$1s^2 2s^2 2p^4$
Be	–	$1s^2 2s^2$	**F**	–	$1s^2 2s^2 2p^5$
B	–	$1s^2 2s^2 2p$	**Ne**	–	$1s^2 2s^2 2p^6$

The third shell is started with sodium (Na, 11), and the first eight electrons occupy the 3s and 3p subshells, as they did in the lithium series. Argon (Ar, 18) then has this configuration: $1s2s^2 2p^6 3s^2 3p^6$.

Problem 24-1. Write the chemical symbols and electron configurations for the eight members of the sodium series (*see* periodic table).

The third shell that is begun with sodium also has a *d* subshell ($l = 2$), which may contain ten electrons ($m = 0, \pm1, \pm2; s = \pm\frac{1}{2}$). We might expect this subshell to be filled next. Such, however, is not the case. The low ionization energies of the next two elements (potassium, K, and calcium, Ca) indicate that their two electrons have entered a new, farther-out principal shell, not the 3d subshell. Moreover, the chemical similarity of K to Na and Li indicates that K has an analogous configuration in its outer shell—namely, 4s. Likewise, the chemical similarity of Ca to Mg and Be indicates that its outer shell is $4s^2$.

Why is the 4s orbital filled before the 3d orbitals, contrary to the previous pattern? Evidently the 4s orbital in K and Ca has lower energy than the 3d orbitals. But *why* does the 4s orbital have lower energy in these elements but not in hydrogen and other one-electron atoms? The difference is due to interaction of electrons in multielectron atoms, which interaction of course is absent in one-electron atoms.

By way of explanation, consider Fig. 24-2, showing the probabilities of finding the electron at various distances from the nucleus in the 2s and 2p orbitals of hydrogen. The big hump of the 2s orbital is slightly farther from the nucleus than the single hump of the 2p orbital; but the little

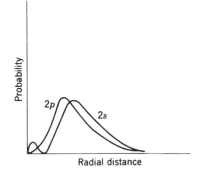

Fig. 24-2. Probabilities of finding the electron at various distances from the nucleus in the 2s and 2p orbitals of the hydrogen atom.

hump of the 2s orbital is much closer. Thus the electron in either orbital is at the same *average* distance from the nucleus and "feels" the same nuclear attraction on the average. Accordingly, the 2s and 2p orbitals have nearly the same energy in one-electron atoms.

However, in an atom like lithium, where the 1s shell is filled, the 2p electron is screened by the inner shell from the full effect of the nuclear charge. The 2s electron, on the other hand, spends part of its time inside the 1s cloud (while it is in the small close-in hump). By penetrating the screening 1s cloud in this way, a 2s electron "feels" more of the nuclear charge, is held more tightly, and thus has lower energy than a 2p electron.

In comparing 4s and 3d orbitals (Fig. 24-3) we see that a 4s electron is on the average much farther from the nucleus; but penetration in multielectron atoms again lowers its energy (note the two small humps close to the nucleus). Evidently the 4s energy thus is actually decreased below the 3d level, since 4s is filled before 3d.

The relative positions of elements in the periodic table tell us in what order orbitals are filled. From this order we can in turn deduce the relative positions of the various energy levels of multielectron atoms (Fig. 24-4).

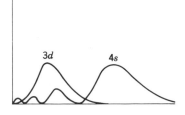

Fig. 24-3. Probabilities of finding the electron at various distances from the nucleus in the 3d and 4s orbitals of the hydrogen atom.

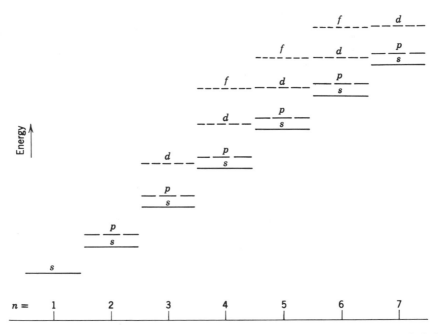

Fig. 24-4. Approximate energy levels of subshells in multielectron atoms. Each dash represents an orbital that may be occupied by two electrons with opposed spins.

Let us return now to the building-up of electrons in atoms. After the 4s orbital is filled by two electrons in Ca, the 3d orbitals are filled. These hold a total of ten electrons. The filling begins with element 21 (scandium) and is completed with element 30 (zinc). This series of ten metals, known as the first "transition series," produces the first leveling of the ionization-energy curve shown in Fig. 24-1.

When the 3d subshell is filled, we resume the normal sequence by filling six electrons into the 4p subshell. These elements, gallium (Ga, 31) through krypton (Kr, 36), have chemical properties similar to other elements with outer p electrons, namely, the series boron to neon, and aluminum to argon (*see* periodic table).

Problem 24-2. Write the electron configuration of the noble gas krypton.

In the next row of the periodic table, starting with rubidium (Rb, 37), an analogous fill-in sequence occurs: first $5s^2$, then $4d^{10}$, and last $5p^6$, which brings us to xenon (Xe, 54).

The fourth shell may have an f subshell ($l = 3$). This subshell is filled after the $6s$ subshell is filled with two electrons (*see* Fig. 24-4). Judging from chemical properties and ionization energies, the sequence, starting with cesium (Cs, 55), is as follows: $6s^2$, $5d$, $4f^{14}$, $5d^9$, $6p^6$—ending with radon (Rn, 86). The 14 elements filling their $4f$ shell (elements 58-71) are shown as a separate row at the bottom of the periodic table, to keep the table from getting too wide. Together with lanthanum (La, 57), which they resemble chemically, these elements are known as the *lanthanides* or *rare earths*.

The next row, starting with francium (Fr, 87) follows a similar filling sequence: $7s^2$, $6d$, $5f^{14}$, $6d$. The fourteen elements analogous to the lanthanide series are called *actinides*. Note that this row is not complete; the last known element is number 104. Elements 93 to 104 are artificially created, short-lived radioactive elements, known only since 1940. More of these artificial elements will probably be "discovered" in the next few years. Even now we can predict their probable electron configurations.

Problem 24-3. Write the probable electron configuration of element 105. Where would it be placed in the periodic table?

Problem 24-4. Write the electron configuration of radon, the heaviest of the noble gases.

Problem 24-5. If a new, heavier noble gas were to be artificially made, what would be its atomic number?

Problem 24-6. Complete the following table of principal-shell electron configurations for four selected families of the periodic table.

Alkali Metals	Alkaline Earth Metals	Halogens	Noble Gases
			He 2
Li 2,1	Be 2,2	F 2,7	Ne 2,8
Na	Mg	Cl	Ar
K	Ca	Br	Kr
Rb	Ba	I	Xe 2,8,18,18,8
Cs	Sr	At	Rn
Fr 2,8,18,32,18,8,1	Ra		

In concluding this discussion of electron configurations we wish to point out the historical sequence. First came our knowledge of the chemical properties of elements, systematized in the periodic table; then we obtained data on ionization energies; then, by extension of Bohr's atom model, Bohr and others managed to "build up" atoms by putting electrons into shells and subshells—always keeping in mind each element's ionization energy and its position in the periodic table. Last of all came quantum mechanics, which "explained" the structure of the atom. We shall now use the electron configurations to help us understand variations of properties within periods and families of the periodic table.

PROPERTIES OF ELEMENTS

The connecting link between atomic structure and observed chemical behavior (*see* Chapter 5) is the fact that elements with similar electron configurations have similar chemical properties. As Problem 24-6 revealed, members of the alkali-metal family all have one electron in the outer shell; the alkaline earths have two; the halogens have seven; the noble gases have eight (except helium). The number of outer-shell electrons appears to be the only structural feature that the members of any chemical family have in common and that distinguishes one family from another. Hence it is reasonable to conclude, as we tentatively did in Chapter 19, that the chemical behavior of elements is determined primarily by the number of electrons in the outermost shell of the atoms. The other shells generally make a difference only in degree, not in kind.

Differences in chemical properties are most pronounced between elements on the far-left and far-right sides of the periodic table, i.e., between atoms with few electrons in their outer shell and those with nearly full shells. The former are metals, the latter are nonmetals. Metals are good conductors of electric current because their outer electrons are only loosely held, so that they are highly mobile. In most nonmetals, on the other hand, the electrons are tightly held, making these substances poor conductors. Metallic atoms have low ionization energies, giving up their electrons easily to form positive ions. Nonmetallic atoms have high electron affinity and consequently tend to form negative ions.

Metallic character of elements decreases as we proceed to the right in the periodic table, toward the nonmetals with their higher ionization energies; but metallic character increases as we go down the periodic table, encountering larger atoms with more easily removable distant electrons. Thus francium, at the lower left of the periodic table, is the most metallic element (Fig. 24-5a). Nonmetallic character changes in the reverse direction. The most nonmetallic element—fluorine—is at the upper right (Fig. 24-5b).

To sum up, electronic structure is directly related to the properties of the elements. This intertwining relationship may be outlined as follows.

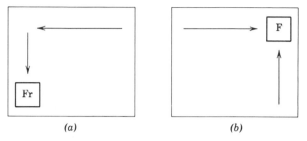

Fig. 24-5. (*a*) Increasing metallic character in the periodic table. Francium, at lower left, is the most metallic element. (*b*) Increasing nonmetallic character. Fluorine, at upper right, is the most nonmetallic element.

1. Atoms of various elements differ from each other in their nuclear charges and numbers of electrons.

2. Electrons in atoms are arranged in orbitals and shells.

3. Each orbital, or electron cloud, is defined by three quantum numbers: n, l, and m.

(a) The principal quantum number, n, determines the electron's energy and average distance from the nucleus.

(b) The angular momentum quantum number, l, determines the orbital's shape (together with n), and also to some extent the electron's energy.

(c) The magnetic quantum number, m, determines the orbital's orientation in space. It affects the energy only in the presence of a magnetic field.

4. Orbitals having identical n values, and therefore similar distances and energies, make up an electron shell. Within shells we recognize subshells having the same l values.

5. According to the Pauli Exclusion Principle, each orbital may be occupied by not more than two electrons, which then have opposite spins (designated by a fourth quantum number, s). Because of the Exclusion Principle, outer electrons do not all "fall" into the inner shell; thus the atom remains stable.

6. Electrons in the outermost shell of an atom are the most important in determining chemical properties. Chemical reactions involve only outer electrons; inner electron shells are only slightly disturbed.

7. Elements in a given vertical column of the periodic table have similar outer-shell electron configurations and similar chemical properties, and are said to belong to the same family.

8. There is usually a regular gradation of properties within a given family.

9. There is also, in most cases, a regular trend in properties within any given horizontal row, called a period, of the periodic table.

These trends enable us to understand much of the chemical behavior of elements and compounds. Further aspects of chemical behavior, however, are best discussed after we have learned something about the nature of the chemical bond, which is the subject of the next chapter.

SUMMARY OF PART FOUR

In Part Four we have seen how the determinism of classical Newtonian mechanics has given way to the indeterminism of quantum mechanics. The classical view of nature held that if we know the exact positions and velocities of all particles in a system (i.e., if we know where they are and whereto and how fast they are going), then we can predict exactly the future of the system. Werner Heisenberg with his Uncertainty Principle, however, showed that we can never measure the exact position *and* velocity of a particle such as an electron.

What is more, because of the Uncertainty Principle, we are not even justified in calling the electron a "particle." We think of a "particle" as occupying a definite part of space and, if moving, as having a definite velocity. But defintie position and definite velocity are two quantities we can never determine simultaneously, and only those quantities that can be measured have any real meaning in science. "Unobservables" must not be postulated. An electron *particle* is such an "unobservable" and therefore does not exist as far as the literal-minded scientist is concerned.

This indeterminism, far from creating a scientific vacuum, has been quite productive. For one thing, it has shown us a way out of the particle-wave dilemma, for electrons as well as for photons. For another, indeterminism coupled with quantum mechanics has produced a picture of the atom that is more satisfactory than the Bohr model. The new quantum-mechanical model, with its charge clouds rather than orbiting electrons, enables us to explain the periodicity of chemical properties and also many specifics of chemical behavior. True, the Uncertainty Principle prevents us from predicting the complete behavior of an individual atom. But we can (ideally) come up with a mathematical *probability* describing how any one atom will behave; i.e., we can predict how *many* atoms will behave on the *average*.[1]

---

[1] It should be noted that the equations of quantum mechanics are very difficult to solve. Exact solutions have been obtained only for one-electron systems; for anything more complex, only approximate solutions have been obtained. However, any desired degree of exactness is possible and depends only on the amount of computational effort one is willing to invest.

Probability statistics applied to large numbers of atoms or molecules leave practically no uncertainty. For example, when we flip a coin, we cannot predict whether it will land as "heads" or "tails"; but when we flip a million coins, we can predict with a high degree of certainty that 500,000 of them will land as "tails." Because the chemist is generally interested in the bulk behavior of very large numbers of atoms or molecules, he is, in practice, not greatly affected by the indeterminism of quantum mechanics.

ADDITIONAL READING FOR PART FOUR

Bridgman, Percy W. "Suggestions from Physics," in A. B. Arons and A. M. Bork (Ed.), *Science and Ideas.* Englewood Cliffs, N. J.: Prentice-Hall, 1964 (paperback). Pp. 10–36 (reprinted from P. W. Bridgman, *The Intelligent Individual and Society.* New York: Macmillan, 1938, pp. 10–47). Contains a good discussion of operational definitions, using various concepts of velocity as illustration. Bridgman discusses the danger of extrapolating ideas to other realms and stresses the need for experimental checks. Also: "Quo Vadis," pp. 270–278, reprinted from *Daedalus,* 87: 85–93.

de Broglie, Louis. *Matter and Light: The New Physics.* Translated by W. H. Johnston. New York: Dover Publications (paperback). Original French edition first published in 1937. A historical account of quantum mechanics by one of its founders. In addition, there are some philosophical chapters at the end.

———. *The Revolution in Physics.* Translated from the French by R. W. Niemeyer. New York: Noonday Press, 1953 (paperback). A historical account of quantum mechanics by one of its founders. The book was originally written about 1936.

Feynman, R. P., et al. *The Feynman Lectures on Physics.* Reading, Mass.: Addison-Wesley, 1963. Written in a conversational style, but not easy reading. See especially Sections 37-4 to 37-8, which analyze imaginary experiments with electrons and slits; and Chapter 38, "The Relation of Wave and Particle Viewpoints." Section 38-6 contains an unusual discussion of unmeasurable things.

Furth, R. "The Limits of Measurement." *Scientific American,* July 1950; Freeman Reprint No. 255.

Gamow, George. *Biography of Physics.* New York: Harper and Row, 1961 (paperback).

———. "The Exclusion Principle." *Scientific American,* July 1959; Freeman Reprint No. 264.

———. "The Principle of Uncertainty." *Scientific American,* January 1958; Freeman Reprint No. 212. A good, simple explanation, containing two good imaginary experiments. The last page contains an interesting account of Bohr's debate with Einstein.

Heisenberg, Werner. *Physics and Philosophy, The Revolution in Modern Science.* New York: Harper and Row, 1962. (Originally published in 1958, paperback.) Stimulating reading. Contains an interesting Introduction by F. S. C. Northrop (Professor of Philosophy and Law at Yale University).

Popper, Karl R. *The Logic of Scientific Discovery,* New York: Harper and Row, 1965 (paperback). *See* especially Section 69 on "Law and Chance"; Sections 73-78, "Some Observations on Quantum Theory"; and New Appendix XI, "On the Use and Misuse of Imaginary Experiments, Especially in Quantum Theory." Popper disagrees with the more commonly accepted interpretation of the Uncertainty Principle. His exposition is nonmathematical but difficult reading.

Schrödinger, Erwin, "What Is Matter?" *Scientific American,* September 1953; Freeman Reprint No. 241. A somewhat rambling dissertation on particle-wave duality by one of its originators.

Sebera, Donald K. *Electronic Structure and Chemical Bonding.* New York: Blaisdell Publishing Company, 1964 (paperback). Chapter 6.

Townes, Charles H. "The Convergence of Science and Religion." *The (MIT) Technology Review,* vol. 63, May 1966. A thought-provoking article.

Young, Louise B., ed. *Exploring the Universe.* New York: McGraw Hill, 1963.

——— ed. *The Mystery of Matter.* New York: Oxford University Press, 1965. Pages 103–131 deal with particle-wave duality and related philosophical aspects, in excerpts by Weisskopf, Schrödinger, Heisenberg, Einstein and Infeld, Jeans, and Rabinowitch.

PART FIVE

Structure of Molecules

Parts One to Four were devoted to the structure of the *atom*. In Part One we discussed the evidence that convinced scientists that atoms are the "ultimate" building blocks of matter. In Parts Two, Three, and Four we traced the chain of investigations and discoveries that revealed that atoms themselves are made up of smaller particles; we showed how scientists eventually were able to deduce our present-day picture of the atom.

We shall now examine the combination of these atoms into molecules. A *molecule* is any group of atoms held together strongly enough to be considered a unit. The attraction between two or more atoms within a molecule is called a *chemical bond*. Examples of molecules range from H_2, O_2, and H_2O, through C_2H_5OH (ethyl alcohol) and $C_{12}H_{22}O_{11}$ (sucrose), to giant protein molecules like hemoglobin (molecular weight about 65,000) and DNA (MW 1 to 6 million). There also are "macromolecules" containing a large but arbitrary number of repeating units. Included in this latter category are ionic crystals, such as sodium chloride (the entire crystal is one cross-linked $(NaCl)_x$ molecule); metals; diamond; and polymers, such as polyethylene—$(C_2H_4)_x$.

First we shall examine the nature of the chemical bond itself. What makes a bond? What kinds of bonds are there? Under what conditions can we expect the various types of bonds to be formed? What is the relation between bond type and properties of molecules?

Next, we shall take a closer look at covalently bonded molecules. We shall examine the shapes of these molecules and the effect of various geometric arrangements on the properties of substances. Our attention will focus on the versatile carbon atom: how carbon atoms form single, double, and triple bonds; how carbon atoms combine with other atoms to form isomers—that is, compounds having identical empirical formulas

333

but different structural or geometric arrangements; for instance, C_2H_6O could be dimethyl ether—$(CH_3)_2O$—or ethyl alcohol—C_2H_5OH.

With this basis we shall be able to understand how relatively few different atoms can account for such a multitude of compounds—literally millions of them. We shall also see how this huge field can be systematized by a few basic principles involving molecular structure.

After so much discussion of molecular structures, we should answer the questions, how do scientists go about finding out what the structures of molecules are? How do they experimentally determine structures, and what tools do structural chemists use in their laboratories? In answer, we shall survey some classical methods of "wet chemistry," as well as more recent instrumental methods of diffraction, mass spectrometry, and molecular spectroscopy.

Chapter 25

BONDS BETWEEN ATOMS: THE FORMATION OF MOLECULES

EARLY THEORIES

The introduction of Dalton's atomic theory shortly after 1800 raised the question, "What forces hold atoms together?" Almost from the beginning, attention was directed to electrical forces. Recall from Chapter 7 that in 1800 Nicholson and Carlisle had found that an electric current decomposed water into hydrogen and oxygen; in the same year Volta discovered the electric battery. Shortly thereafter Humphry Davy used electrolysis to isolate the elements sodium and potassium from their molten hydroxides. If electric currents can tear molecules apart, isn't it reasonable to assume that these molecules are being held together by some sort of electric forces?

In 1811 the Swedish chemist Jöns Berzelius, the inventor of our present-day symbols for the elements, advanced his electrostatic theory of chemical bonding (Chapter 7, p. 118). Berzelius proposed that bonds are the result of electrostatic attractions between positive and negative atoms. He could not, however, explain how normally neutral atoms would acquire positive or negative charges. Perhaps one atom somehow pulled electric charge away from the other. Such a theory, though, could not explain the combination of *identical* atoms to form molecules like H_2. (In 1811 Avogadro had proposed the existence of diatomic molecules of *elements,* in order to save the equal-volumes-equal-numbers (EVEN) theory of gases (*see* Chapter 3.)

This difficulty of the electrostatic-bond theory was compounded with the evolution of organic chemistry in the middle of the last century (*see* Chapter 26), when scientists began to recognize *many* compounds made up of atoms having nearly equal charge-attracting power. There are, for

335

example, an enormously large number of carbon-hydrogen-oxygen compounds. It was difficult to explain the formation of these compounds on the basis of electrostatic bonding. In addition chemists realized that most carbon bonds form 110° angles with each other; such a fixed direction of bonds again was not easily explained through an electrostatic model.

Today we realize that Berzelius was essentially right. But his theory of electrostatic bonding applies only to what we call *ionic bonds*. In addition we recognize two other categories of bonds—*covalent* and *metallic*. The essential features of these three bond types are taken up in the following sections.

THE IONIC BOND

In Chapter 19 we pointed out the great stability of inert-gas configurations. J. J. Thomson, the discoverer of the electron, suggested that elements other than inert gases may achieve noble-gas configurations by gaining or losing electrons. In 1916 the German chemist Walther Kossel (1888–　) developed this idea into a theory of ionic bonding, in essence reviving Berzelius' electrostatic theory.

We can best explain various features of ionic bonding by using an example such as sodium chloride, NaCl. The chlorine atom exhibits a tendency to *add* an electron, thus achieving a more stable inert-gas configuration (Cl^-); this process is accompanied by a *release* of energy. But *removal* of an electron from the sodium atom (Na^+) requires an *input* of energy. Since "energy runs downhill," the reaction proceeds spontaneously only if energy is released during the overall process. Let us therefore look at the energies involved in the formation of sodium chloride.

Energies of chemical reactions are commonly expressed in kilocalories per mole—i.e., per Avogadro's number of molecules or atoms. Heretofore, because we were dealing with single atoms, we needed a much smaller energy unit, the electron volt. 1 *ev* equals only 3.8×10^{-23} kcal. If a reaction requires 1 *ev* per atom, it would require 6×10^{23} *ev* per mole (Avogadro's number of atoms). Converting this number to kcal, we obtain:

$$(6 \times 10^{23} \text{ } ev \text{ per mole}) (3.8 \times 10^{-23} \text{ kcal}/ev) = 23 \text{ kcal per mole}$$

In other words, *1* ev *per atom is equivalent to 23 kcal per mole*. To ease the transition in our thinking, we shall use both units in the following example.

We emphasized earlier that energy is required to remove electrons from atoms. For sodium atoms the ionization energy is 5.1 electron volts (*see* Table 19-1), or 118 kcal per mole. On the other hand, 3.7 *ev* of energy, or 85 kcal per mole, is released when a chlorine atom captures an electron. The production of isolated Na^+ and Cl^- ions thus requires a net energy input and is *not* spontaneous.

Reaction	Energy Change of the System
$Na \rightarrow Na^+ + e^-$	$+5.1$ *ev*, or 118 kcal/mole
$Cl + e^- \rightarrow Cl^-$	-3.7 *ev*, or -85 kcal/mole
$Na + Cl \rightarrow Na^+ + Cl^-$	$+1.4$ *ev*, or 33 kcal/mole

In accordance with convention, we use a plus sign to indicate energy input—raising the system's energy; and we use a minus sign to indicate energy release—lowering the system's energy.

The two oppositely charged ions, Na^+ and Cl^-, attract each other and may form an Na^+Cl^- ion pair—that is, an NaCl gas molecule. This union is a spontaneous process and releases energy—6.1 *ev* per molecule, or 140 kcal/mole. That amount is more than enough energy to make the overall process spontaneous—the formation of gaseous NaCl molecules from gaseous Na and Cl atoms.

$Na(g)$	$\rightarrow Na^+ + e^-$	$+5.1$ *ev*, or	118 kcal/mole
$Cl(g) + e^-$	$\rightarrow Cl^-$	-3.7	-85
$Na^+ + Cl^-$	$\rightarrow NaCl(g)$	-6.1	-140
$Na(g) + Cl(g) \rightarrow NaCl(g)$		-4.7 *ev*, or -107 kcal/mole	

According to our calculation, then, the formation of ionically bonded gaseous sodium chloride from gaseous sodium and chlorine atoms is a spontaneous process, releasing 107 kcal of energy per mole. This calculated value agrees well with the experimentally measured value of 98 kcal/mole.

Normally the chemical reaction between sodium and chlorine does not stop at this point but continues with the formation of sodium chloride *crystals*. We can understand that process when we realize that the "bonding power" of the ions is not saturated with the formation of an Na^+Cl^- ion pair. The positive and negative charges can still attract other opposite charges, for instance to form

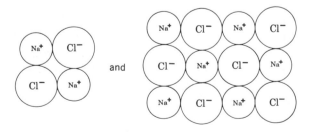

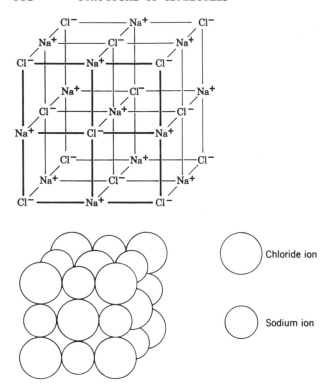

Fig. 25-1. The sodium chloride crystal lattice. Each ion has six oppositely charged ions as nearest neighbors.

This clustering process continues, yielding a three-dimensional cubic lattice in which Na^+ and Cl^- ions alternate regularly (Fig. 25-1).

The formation of these additional bonds releases a considerable amount of energy—an additional 44 kcal per mole—which accounts for the great stability of ionic crystals: To break up the sodium chloride lattice into separate sodium chloride molecules requires 44 kcal per mole.

Lowest energy and therefore maximum stability are achieved when an ion is surrounded as closely as possible by as many oppositely charged ions as possible. The number of ions that can crowd around in this manner depends on the relative sizes of the ions. For sodium chloride maximum stability is achieved when each ion is surrounded by six other ions. In cesium chloride, on the other hand, each ion is surrounded by eight oppositely charged ions (Fig. 25-2).

Note that in such a crystal lattice we cannot recognize any particular pair of ions as a molecule. In sodium chloride each sodium ion is bonded

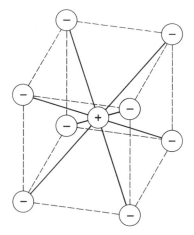

Fig. 25-2. The cesium chloride crystal lattice. Each ion has eight oppositely charged ions as nearest neighbors.

not merely to one chloride ion but to six of them; and each chloride ion, in turn, is electrostatically bonded to six sodium ions. Thus the entire crystal is one huge molecule.

We can now understand some of the properties of ionic crystals. Consider melting points. The melting point of a solid is the temperature at which its molecules have attained enough energy to break away from each other—the rigid solid then becomes a flowing liquid. Ionic crystals have such high melting points (NaCl, 800°; CsCl, 600°) because much energy is required to overcome the cross-linked attractive forces in the crystal lattice.

Ionic crystals are brittle—that is, they cannot be deformed appreciably without breaking. Fig. 25-3a represents a cross section through an NaCl lattice. When we deform the crystal by sliding one layer of ions past

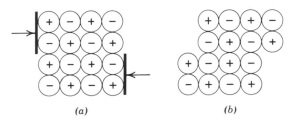

(a)   (b)

Fig. 25-3. Deformation if ionic crystals. In the deformed crystal (b) like charges are next to each other, causing repulsive forces that may be strong enough to cleave the crystal. This accounts for the brittleness of ionic crystals.

another (indicated by arrows), at one point ions of like charge are facing each other (Fig. 25-3b). The resulting repulsive forces may be sufficient to cause the layers to separate, thus cleaving the crystal.

Ionic crystals are nonconductors of electricty. Because ions and electrons are tightly held in place, there are no mobile charges capable of carrying a current. However, when ionic compounds are melted, the resulting mobile ions are good conductors of electricity. During the electrolysis of molten ionic compounds such as sodium chloride, the positive metal ions are attracted to the negative plate (called the cathode), become discharged there, and are deposited as metals; while the negative ions move to the positive plate (the anode). In this way, through the electrolysis of molten sodium and potassium hydroxides, Davy in 1807 for the first time prepared pure sodium and potassium metals.

Example 25-1. We have shown that the formation of solid sodium chloride from sodium *atoms* and chlorine *atoms* releases 151 kcal/mole (107 from the formation of NaCl ion pairs and 44 kcal more from the formation of the crystal lattice.) Normally the reaction would be between *solid* sodium metal and gaseous chlorine *molecules*. It requires 26 kcal/mole to vaporize solid sodium (its *heat of vaporization*) and about 29 kcal to produce a mole of Cl *atoms* (the *bond energy* or *heat of dissociation* of Cl_2 molecules is 57 kcal/mole). From these figures calculate the amount of energy evolved in the production of one mole of solid sodium chloride from solid sodium and gaseous chlorine, according to the reaction

$$Na(s) + 1/2\,Cl_2(g) \rightarrow NaCl(s)$$

Solution.

$Na(s)$	$\rightarrow Na(g)$	-26 kcal/mole
$\frac{1}{2}Cl_2(g)$	$\rightarrow Cl(g)$	-29
$Na(s) + Cl(g)$	$\rightarrow NaCl(s)$	$+151$
$Na(s) + \frac{1}{2}Cl_2(g) \rightarrow NaCl(s)$		$+96$ kcal/mole

THE COVALENT BOND

Ionic substances are formed by the reaction of elements that have great attraction for electrons with elements whose atoms have only small attraction for the electrons in their outer shells. Clearly, such a process cannot explain bonding between identical atoms (for example in H_2, O_2, F_2) or between atoms having similar attraction for electrons (such as C, N, and O).

In 1916, the year of Kossel's theory of ionic bonding, G. N. Lewis (1875–1946) at the University of California proposed a theory of covalent bonding. Like Kossel, Lewis built his theory around the stability of noble-gas configurations. Only, instead of a transfer of electrons, Lewis's theory involved the *sharing* of electrons to achieve noble-gas configurations. Lewis' most important contribution was the recognition of the importance of electron *pairs* in the formation of most covalent bonds.

The Octet Rule

Lewis recognized the stability of configurations in which eight outer-shell electrons surround most atoms. This is the configuration of the noble gases, except helium. According to the *octet rule* an atom tends to form bonds until it is surrounded by eight outer-shell electrons.

For example, two fluorine atoms, each with seven outer electrons, can gain complete octets by sharing an electron pair to form a fluorine molecule, as shown by the following electron-dot formula (introduced in Chapter 19).

$$: \overset{..}{\underset{..}{F}} \cdot \ + \ \cdot \overset{..}{\underset{..}{F}} : \ \longrightarrow \ : \overset{..}{\underset{..}{F}} : \overset{..}{\underset{..}{F}} :$$

Or a fluorine atom can complete its octet by sharing an electron pair with a hydrogen atom, to form hydrogen fluoride.

$$H \cdot \ + \ \cdot \overset{..}{\underset{..}{F}} : \ \longrightarrow \ H : \overset{..}{\underset{..}{F}} :$$

The octet rule of course does not apply to the hydrogen atom, which can have only two electrons in its shell.

Oxygen, with six outer electrons, can complete its octet by sharing an electron pair with each of two hydrogen atoms, to form water.

$$2H \cdot \ + \ \cdot \overset{..}{\underset{.}{O}} : \ \longrightarrow \ \begin{matrix} H : \overset{..}{O} : \\ H \end{matrix}$$

Likewise, nitrogen can share electron pairs with three hydrogen atoms, to form ammonia.

$$3H \cdot \ + \ \cdot \overset{..}{\underset{.}{N}} \cdot \ \longrightarrow \ \begin{matrix} H : \overset{..}{N} : H \\ H \end{matrix}$$

Nitrogen can also share three electron pairs with another nitrogen atom, to form a triply bonded nitrogen molecule.

$$\cdot \overset{..}{\underset{.}{N}} \cdot \ + \ \cdot \overset{..}{\underset{.}{N}} \cdot \quad : N :::N :$$

Experiments confirm the existence of multiple bonds by showing them to be stronger than single bonds (*see* Example 27-1).

The octet rule applies also to larger molecules. Carbon, for instance, has four outer-shell electrons. It can form compounds such as the following.

$$\cdot \overset{\cdot}{\underset{\cdot}{C}} \cdot \quad + \quad 4H \cdot \quad \longrightarrow \quad H : \overset{..}{\underset{H}{C}} : H \qquad \text{(methane)}$$

$$2 \cdot \overset{\cdot}{\underset{\cdot}{C}} \cdot \quad + \quad 6H \cdot \quad \longrightarrow \quad H : \overset{H \ H}{\underset{H \ H}{C : C}} : H \qquad \text{(ethane)}$$

$$2 \cdot \overset{\cdot}{\underset{\cdot}{C}} \cdot \quad + \quad 4H \cdot \quad \longrightarrow \quad \overset{H \ \ H}{\underset{H \ \ H}{C : : C}} \qquad \text{(ethylene)}$$

$$2 \cdot \overset{\cdot}{\underset{\cdot}{C}} \cdot \quad + \quad 2H \cdot \quad \longrightarrow \quad H : C : : : C : H \qquad \text{(acetylene)}$$

Following are some compounds containing more than two elements.

$$2H \cdot \quad + \quad \cdot \overset{\cdot}{\underset{\cdot}{C}} \cdot \quad + \quad \overset{..}{\underset{..}{O}} : \quad \longrightarrow \quad H : \overset{H}{\underset{}{C}} : : \overset{..}{O} : \qquad \text{(formaldehyde)}$$

$$6H \cdot \quad + \quad 2 \cdot \overset{\cdot}{\underset{\cdot}{C}} \cdot \quad + \quad \overset{..}{\underset{}{O}} : \quad \longrightarrow \quad H : \overset{H}{\underset{H}{C}} : \overset{..}{\underset{..}{O}} : \overset{H}{\underset{H}{C}} : H \qquad \text{(dimethyl ether)}$$

$$\text{or} \quad \longrightarrow \quad H : \overset{H \ H}{\underset{H \ H}{C : C}} : \overset{..}{\underset{..}{O}} : H \qquad \text{(ethyl alcohol)}$$

We see that the octet rule helps to systematize information about chemical formulas and about the combining power of elements. In the foregoing formulas hydrogen has two electrons and the other atoms have complete octets. The octet rule allowed us to "predict" that F combines with 1 H, O combines with 2 H, N combines with 3 H, that N_2 has a triple bond, C_2H_6 has a carbon-carbon single bond, C_2H_4 has a carbon-carbon double bond, C_2H_2 has a carbon-carbon triple bond, and that there are two compounds with the formula C_2H_6O—namely, dimethyl ether and ethyl alcohol.

Problem 25-1. Draw simple dot formulas for CH_4, CH_3F, CH_2F_2, CHF_3, CF_4; NF_3, NF_2Cl; NH_3, NH_4^+, BH_4^-.

Problem 25-2. Draw dot formulas for N_2; N_2F_2; N_2F_4.

Problem 25-3. On p. 342 dot formulas appeared for C_2H_6, C_2H_4, and C_2H_2. These compounds have, respectively, carbon-carbon single, double, and triple bonds. Draw a dot formula for a compound containing a carbon-carbon quadruple bond. Is your formula compatible with the octet rule? No compound with a carbon-carbon quadruple bond has even been found. The reason why such a compound does not exist is discussed in the next chapter.

Although the octet rule is a convenient empirical rule, there are many exceptions. Some compounds like

$$Li : H \qquad H : Be : H \qquad : \overset{..}{\underset{..}{F}} : B : \overset{..}{\underset{..}{F}} :$$
$$: \overset{}{\underset{..}{F}} :$$

have incomplete octets. Other compounds may have more than eight electrons surrounding an atom. It takes eight electrons to fill the outer shell of second-row elements ($2s^2p^6$); but third-row elements may have electrons in their $3d$ orbitals, which have only slightly higher energy levels than the $3p$ orbitals (*see* Fig. 24-4). Thus third-row elements sometimes reach stable configurations with eight outer electrons—for example, in the noble gas argon ($3s^2p^6$), or in phosphorous trichloride:

$$: \overset{..}{\underset{..}{Cl}} : \overset{..}{\underset{}{P}} : \overset{..}{\underset{..}{Cl}} :$$
$$: \overset{}{\underset{..}{Cl}} :$$

Other times, third-row elements form stable compounds—such as phosphorous pentachloride, PCl_5—with more than eight electrons. Similarly, SF_2 obeys the octet rule, but SF_4 and SF_6 do not.

Note that the mere existence of variable combining power does not in itself indicate violation of the octet rule. For example, two common compounds of hydrogen and oxygen, H_2O and H_2O_2, both obey the octet rule.

$$H : \overset{..}{\underset{..}{O}} : \quad \text{water} \qquad H : \overset{..}{\underset{..}{O}} : \overset{..}{\underset{..}{O}} : H \quad \text{hydrogen peroxide}$$
$$H$$

Odd-electron compounds represent another exception to the octet rule. Clearly, the rule cannot apply to compounds like nitric oxide, NO, with an odd number of outer electrons ($5 + 6 = 11$). Interestingly, most odd-electron compounds are not too stable but tend to react further to complete their outer shells.

Another shortcoming of the octet rule appears with certain compounds.

Application of the octet rule to sulfur dioxide, SO_2, for instance, would lead us to predict the following possible structures.

:Ö:S̈::Ö   Ö::S̈:Ö: :S̈:Ö::Ö   S̈::Ö:Ö:

Each hypothetical structure has one double and one single bond. Experimental evidence, however, reveals that SO_2 has only *one* type of bond and that this bond is stronger than a single bond. Accordingly, we might picture this molecule as having bonding clouds that consist of an electron pair-and-a-half; or as having two single bonds plus an additional "delocalized" bond distributed between the two bonds, making each what might be called a bond-and-a-half:

$$O\text{---}S\text{---}O$$

The dotted lines here indicate half bonds.[1]

Exactly the same type of bonding is found in the nitrite ion. NO_2^- has the same number of outer electrons and the same geometry as SO_2—namely, five electrons from the nitrogen atom, six each from the two oxygen atoms, and one representing the ion's negative charge—a total of eighteen electrons. The dot formula

$$\left[:\ddot{O}:\ddot{N}::\ddot{O}\right]^-$$

does *not* represent the bonding in NO_2^- because again only one type of bond is found. A better representation would be a picture with one-and-one-half bonds:

$$\left[O\text{---}N\text{---}O\right]^-$$

A similar situation is found in sulfur trioxide, SO_3. The simple dot formula

$$:\underset{:\ddot{O}:S:\ddot{O}:}{\overset{:O:}{}}$$

is misleading because actually the three SO bonds are equivalent. A more appropriate representation would be

---

[1] If you don't like this explanation, ask your instructor to attempt an alternative explanation in terms of "hybridized" orbitals.

where each dotted line now represents a *third* of a bond. In other words, the cloud of the extra electron pair is distributed among three bonds, making each a one-and-one-third bond.

We see from these examples that electron-dot pictures sometimes may be misleading, even though they are useful representations in many other cases. The octet rule is helpful in predicting chemical formulas of compounds, especially of second-row elements. The rule also recognizes that most chemical bonds involve *pairs* of electrons. It is, however, merely an empirical rule and does not tell us *why* electrons are usually paired, nor does it tell us anything about the nature of the bonding force. We shall take that up next.

Problem 25-4. The NO_3^- (nitrate) ion has the same number of outer electrons as the SO_3 molecule. The three NO bonds are equivalent. Draw a bonding diagram for the nitrate ion.

The Nature of the Bonding Force

One way of looking at covalent bonds is in terms of forces. An electron outside two nuclei

would tend to pull them apart, by pulling on the nearby nucleus more strongly than on the far one. On the other hand, an electron between two nuclei tends to draw them together.

The nuclei cannot "fall into the electron" for the same reason that an electron in an atom may not fall into the nucleus: Wave mechanics and the Uncertainty Principle do not allow a particle to give up all its energy (*see* p. 303). We see then that shared electrons act almost like glue in bonding atoms together.

Another way of looking at covalent bonds is in terms of energy. Electrons—or, more precisely, a concentration of charge density—in the space between two atoms decreases the system's energy and consequently makes the bonded configuration more stable than nonbonded atoms. Let us calculate just where we would have to put an electron between two nuclei to obtain lowest potential energy. Coulomb's Inverse-Square Law of Electrostatic Forces (p. 114) tells us that the force, F, between two charges, q_1 and q_2, is inversely proportional to the square of the distance, r, between them.

$$F \propto \frac{q_1 q_2}{r^2}$$

This force formula can be converted into a potential-energy formula.[2] The potential energy of an electron of charge $-e$ and a nucleus of charge $+e$ is $-e^2/r$. For an electron, and *two* nuclei, A and B, separated by a certain distance, the potential energy is

$$-e^2\left(\frac{1}{r_A} + \frac{1}{r_B}\right)$$

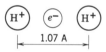

where r_A and r_B are the electron's distances from the two nuclei. The potential energy is lowest (most negative) when both r_A and r_B are smallest, that is, when the electron is closest to *both* nuclei—in other words, when it is in the region *between* the two nuclei. The sharing of an electron by two atoms thus results in lower potential energy, meaning a stable bond.

Such a one-electron bond between two nuclei occurs in the hydrogen-molecule *ion*, H_2^+, found in hydrogen-filled cathode-ray tubes. The bond energy is 64 kcal per mole, the bond distance is 1.07 A (about two Bohr radii).

A more common bond is the two-electron bond. Here, too, the most stable configuration has the electrons between the nuclei, the two electrons being somewhat separated by their mutual repulsion.

One electron makes a good bond, but two electrons make a better one.[3] The two-electron bond is both stronger and shorter: In H_2 the bond energy is 104 kcal per mole (versus 64 for H_2^+), and the bond distance is 0.74 A (versus 1.07 A).

[2] *See* footnote 6, p. 243.

[3] Caution: Do not extrapolate this reasoning to woodworking. Don't say, "If a little glue is good, more is better." Or, worse yet, "If a little solder is good, lots of solder is better."

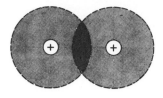

Fig. 25-4. The overlap of two hydrogen 1s orbitals results in greater charge density in the overlap region. This increased charge density acts like glue bonding the atoms together to form the H_2 molecule.

There is no such thing as a three-electron bond. It is ruled out by the Pauli Exclusion Principle, which states that no more than two electrons can occupy the same region of space—i.e., the same orbital—and then only if they have opposite spins (Chapter 24). Lewis in 1916 had recognized the importance of electron pairs, but of course no explanation could be offered until the development of quantum mechanics in 1925–1927.

In summary, we have shown that sharing of electron pairs by two atoms results in lower energy and in attractive forces between the atoms, in other words, in a bond. In quantum-mechanical language we speak of the *overlap* of atomic orbitals, or electron clouds, which we might picture as in Fig. 25-4.[4]

Properties of Covalent Compounds

In most covalent compounds a limited number of atoms are bonded together to form distinct molecules, because the covalent bonding power of atoms is limited and becomes *saturated* (unlike electrostatic bonding in ionic crystals). For example, after two atoms have formed a hydrogen molecule, H:H, no further stable bonds can be formed. To form H_3, H:H·H, the third electron, because of the Exclusion Principle, would have to be in a new higher-energy orbital around the central hydrogen atom; such a high-energy orbital would not result in a stable, low-energy molecule. Similarly, when two fluorines have shared an electron pair and completed their octets,

$$: \overset{..}{\underset{..}{F}} : \overset{..}{\underset{..}{F}} :$$

they show no tendency to form additional bonds; F_3 does not exist. Because of bonding saturation, distinct molecules are recognizable in crys-

---

[4] In addition to this picture of the covalent bond in terms of *overlapping atomic orbitals*, there is an alternative description in terms of *molecular orbitals*. The molecular-orbital theory, developed at about the same time, assumes that the principles of atomic orbitals can be applied equally well to molecules. It treats the molecule as a whole unit and endows it with molecular orbitals and shells, described by sets of quantum numbers—analogous to the orbitals, shells, and quantum numbers of atoms.

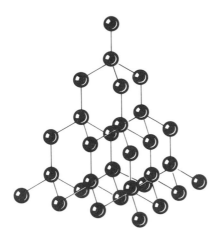

Fig. 25-5. The diamond crystal. Each carbon atom is linked tetrahedrally to four other carbon atoms.

tals of covalent compounds, which we therefore call *molecular crystals*. Such crystals are quite different from ionic crystals. In NaCl, for instance, the whole crystal is one huge molecule, but in a crystal of cane sugar we find discrete molecules of $C_{12}H_{22}O_{11}$.

In such molecular crystals, the separate molecules generally are held to each other by only weak intermolecular forces. As a result molecular crystals, unlike ionic crystals, have low melting points. Examples: fluorine, $-223°C$; iodine, $114°C$; naphthalene ($C_{10}H_8$, mothballs), $80°C$.

Covalently linked solids, like ionic crystals, do not conduct electric current; nor do liquid or molten covalent compounds (but molten ionic compounds do conduct).

Not all crystals containing covalent bonds are molecular crystals. A diamond crystal, for example, consists of covalently bonded carbon atoms, cross-linked to form one macromolecule (Fig. 25-5). There also are covalently linked macromolecules known as *polymers*, such as polyethylene, which is a tough, virtually unbreakable solid.

$$
\begin{array}{cccc}
\text{H} & \text{H} & \text{H} & \text{H} \\
| & | & | & | \\
-\text{C}- & \text{C}- & \text{C}- & \text{C}- \\
| & | & | & | \\
\text{H} & \text{H} & \text{H} & \text{H}
\end{array}
$$

On the other hand, some ionic crystals may have covalent bonds *within* their constituent ions. For example, ammonium chloride, $NH_4^+ Cl^-$, has covalent N-H bonds within the NH_4^+ ion.

Finally, we should point out that the electron pair in a covalent bond need not be shared *equally* by the two elements. True, in H_2 or F_2 the electron pair is shared equally, resulting in a *nonpolar* bond. But in hydrogen fluoride, HF, the fluorine atom has greater electron-attracting power than hydrogen does. Consequently, the electron pair is closer to the fluorine than to the hydrogen atom. Thus the fluorine end of the bond is more negative and the hydrogen end is more positive; we speak of a *polar* bond and say that the molecule has a dipole moment. By measuring the dipole moment of a substance, chemists often gain valuable information about its molecular structure. The effect of dipole moment on the solubility of substances is discussed in the next chapter.

THE METALLIC BOND

The metallic elements, on the left side of the periodic table, have low ionization energies and exhibit little attraction for electrons. Consequently, when two such atoms share an electron pair, only a weak covalent bond is formed. In Table 25-1 contrast the low dissociation energies of gaseous diatomic metal molecules with the much higher energies of covalent and ionic bonds.

Table 25-1. Bond Energies of Diatomic Molecules (kcal/mole)

Metals		Covalent Bonds		Ionic Bonds	
Li_2	25	H_2	104	LiF	137
Na_2	17	F_2	37	LiI	81
K_2	12	Cl_2	57	NaF	107
Rb_2	11	Br_2	46	NaCl	98
Cs_2	10	I_2	36	NaI	71
NaK	14	BrCl	52	CsF	121
NaRb	13	ICl	50	CsCl	101
Pb_2	16	HCl	103	CsI	75
Zn_2	6	HI	71		
Hg_2	1.4	O_2	120[a]		
		N_2	225[a]		
		NO	150[a]		
		CO	255[a]		

[a] Multiple bonds.

When several metal atoms share several electrons, stronger bonds are formed. Solid metals, like ionic crystals, have their atoms arranged in regular geometric lattices forming large crystalline molecules. The bonding electrons are not associated with individual atoms but form a cloud of negative charge distributed throughout the crystal. This cloud of

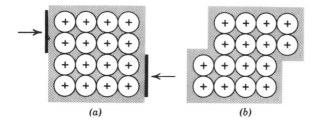

Fig. 25-6. Ductility and malleability of metals. As seen from an atom in the interior or the crystal, there is not much difference between the original state (*a*) and the "deformed" state (*b*). (Contrast this situation with that in deformed ionic crystals, Fig. 25-3*b*.)

mobile electrons accounts directly for the high electrical conductivity of metals.

Our model of positive metal ions surrounded and held together by a sea of electrons also explains the characteristic ductility and malleability of most metals. When we deform a metallic crystal (Fig. 25-6*a*), its atoms slide past each other without great changes in internuclear distances, and only weak forces need be overcome; the resulting structure is not greatly different from the original one (Fig. 25-6*b*). Contrast this picture, explaining ductility and malleability of metals, with Fig. 25-3, explaining the hardness and brittleness of ionic crystals.

THE PERIODIC TABLE AND BOND TYPES

We have already shown how the periodic table, with assistance from the octet rule, helps predict chemical formulas of simple compounds, for example the fluorides of the second-row elements: LiF, BeF_2, BF_3, CF_4, NF_3, OF_2, and F_2. The periodic table also helps correlate information about the nature of these bonds.

Let us look at the bond types of the above series of compounds. We certainly would expect the first one or two, each involving a bond between a very metallic and a very nonmetallic element, to have ionic bonds and therefore to form ionic crystal lattices. On the other hand, the last few compounds involve only nonmetals, and we would consequently expect covalent bonds.

But what about BF_3? The first three ionization energies of boron are 8, 25, and 38 electron volts (*see* Table 19-1). This means it takes a total of 71 *ev* (1600 kcal/mole) to form the B^{+++} ion from an isolated B atom. Such high energy is more than would be "paid back" by the formation of three F^- ions from F atoms (only 3.5 ev for each F^-) and the subsequent formation of ionic bonds. Therefore, BF_3 does not form ionic crystals.

Similar arguments suggest that CF_4, NF_3, and OF_2 also do not form ionic crystals; the total ionization energies of C, N, or O are too high:

carbon—11, 24, 48, and 64 ev, or a total of 147 ev, to form C^{+4}.

nitrogen—14, 30, and 47 ev, or a total of 91 ev, to form N^{+++}.

oxygen—14 and 35 ev, or a total of 49 ev, to form O^{++}.

Beryllium, on the other hand, has ionization energies of 9 and 18 ev; it requires only 27 ev (600 kcal/mole) to form Be^{++}. Apparently the energy *released* by formation of two F^- ions and the subsequent formation of the BeF_2 ionic crystal lattice is greater than 27 ev, so that BeF_2 does form ionic crystals.

Observed melting points give a clue to bond types. The second-row fluorides have the following melting points (°C).

LiF	BeF$_2$	BF$_3$	CF$_4$	NF$_3$	OF$_2$	F$_2$
845	800	−129	−184	−208	−224	−223

The high melting points of LiF and BeF_2 are in accord with our prediction of ionic lattices. The other compounds have very low melting points (all below zero), in accord with our prediction of covalent bonding. In fact, these compounds exist only as individual gaseous molecules at room temperature.

We would expect similar trends to be observed in other rows of the periodic table. We have to allow, however, for increased metallic character as we proceed downward in the periodic table (see Fig. 24-5). Consider the melting points of third-row fluorides:

NaF	MgF$_2$	AlF$_3$	SiF$_4$	PF$_3$	SF$_6$	ClF
995	1263	1290	−90	−160	−50	−155

The somewhat higher melting points of the sodium and magnesium fluorides, as compared with the lithium and beryllium fluorides, probably reflect stronger ionic bonds. Note the high melting point of aluminum trifluoride. It obviously forms an ionic lattice (MP 1290), whereas boron trifluoride does not (MP −129). This is so because of the greater metallic character of aluminum compared with boron, as reflected by its lower first three ionization energies: 6, 19, and 28 ev, or a total of 53 ev (1200 kcal/

mole), to form Al^{+++}; contrasted with 71 ev (1600 kcal/mole) for B^{+++}.

Through these examples we have shown that the periodic table helps to correlate information and make predictions about the nature of bonds in various compounds. With the help of experimentally obtained ionization energies we can understand some of these empirical correlations and refine them. Verification of our predictions of course must rest on experimental evidence, such as observed melting points.

Problem 25-5.   Which of the following compounds contain essentially ionic bonds and which are essentially covalent?

(a) CsBr.   (d) BaSe.

(b) SCl_2.   (e) IBr.

(c) CBr_4.   (f) Na_2O.

Problem 25-6.   Which covalent bond in each pair is the more polar?

(a) C-O, N-O.   (c) P-Cl, P-I.

(b) S-F, S-Cl.   (d) P-O, S-O.

SUMMARY

In general, elements adjacent to each other in the periodic table have similar electron-releasing and electron-attracting qualities and consequently form either covalent or metallic bonds. The farther apart the elements are in the periodic table, the more polar those bonds will be—that is, the more ionic character the bonds will possess. If two elements are sufficiently metallic and nonmetallic, respectively, they will be able to form ionic lattices.

In ionic crystals each ion is surrounded and held in place by a number of oppositely charged ions. The entire crystal lattice thus is one large molecule. This cross-linkage of electrostatic bonds accounts for the high melting and boiling points of ionic crystals and their hardness and brittleness. Because their ions are rigidly held in place, ionic crystals are nonconductors of electricty. When they are melted, however, the now-mobile ions do conduct current.

Molecular crystals differ from ionic crystals in that they contain individual molecules as recognizable entities. These molecules are joined to each other by relatively weak intermolecular forces but have strong internal covalent bonds. Therefore, most such molecular crystals have low melting points, the molecules usually preserve their identity in the molten state, and the liquids are poor conductors of electricity.

Covalent bonds generally involve the sharing of electron pairs by two elements of similar electron-attracting power. We sometimes speak of the overlap of atomic orbitals, resulting in greater charge density in the region of overlap. This increased charge density is like a glue bonding the atoms together.

The metallic bond differs from ionic and covalent bonds in that it involves completely delocalized bonding electrons, not associated with any individual atom or molecule. But, like ions, metallic atoms form regular geometric crystalline lattices. Metallic solids are characterized by electrical conductivity (due to the mobile sea of electrons) and malleability and ductility.

We wish to emphasize that these groupings represent extreme cases. Most covalent bonds, have partial ionic character; certain kinds of covalent bonds involve delocalization of electrons within the molecule (e.g., SO_2); not all covalent bonds result in molecular crystals (e.g., diamond and polymers); not all covalently bonded solids are nonconductors of electricity (graphite). In spite of these and other exceptions it is useful to classify bonds as mainly metallic, ionic, or covalent, and to describe most solids as metallic, ionic, or molecular.

Chapter 26

THE GEOMETRY OF MOLECULES

In the previous chapter we explored the nature of chemical bonds, and we distinguished covalent from ionic bonds. Purely ionic bonding is due to electrostatic attraction among positive and negative ions. These electrostatic forces extend in all directions and cannot be saturated. Hence in a sodium chloride crystal, for example, as many positive sodium ions (Na^+) surround a negative chloride ion (Cl^-) as can find room, and vice versa. We are not justified in speaking of NaCl *molecules* in a crystal, since there is no close relationship between any particular pair of Na^+ and Cl^- ions.

In covalent bonding, on the other hand, there is saturation of bonding power, and molecules usually exist as definite entities. When a hydrogen atom unites with another hydrogen atom to form an H_2 molecule, its bonding power is saturated; no H_3 is observed. Or, when a carbon atom unites with four hydrogen atoms to form a CH_4 molecule, its bonding power is saturated; no CH_5 is found.

Covalent bonds have another feature. Whereas the electrostatic force field of an ion extends in all directions, the covalent bond is characterized by strong directional character, which in turn is principally responsible for the definite geometry of molecules. In this chapter we examine more closely this directional character of covalent bonds and some of its consequences. Experimental aspects of structure determinations are treated briefly in the following chapter.

BOND ANGLES

Bond angles of covalent compounds follow certain patterns. Members of the beryllium family form two covalent bonds at a 180° angle; the molecule is said to be linear. Gaseous $BeCl_2$ is an example.

Cl—Be—Cl
180°

On the other hand, when members of the oxygen family form two bonds, the compounds are nonlinear. In water, for instance, the O——H bonds form a 104° angle.

O
H $\overset{}{\underset{104°}{}}$ H

Why is BeCl₂, linear, but H₂O angular?

Members of the boron family form three covalent bonds lying in a plane with 120° angles between adjacent bonds, as in BF₃.

F
|
B
F   F
120°

But members of the nitrogen family form nonplanar compounds. In NH₃ (ammonia), for example, the nitrogen atom is at the apex of a pyramid; the HNH angles are 107 degrees. (The dotted lines indicate that the third H is not in the plane of the paper.)

N
H   H   H
107°

Why is BF₃ planar, but NH₃ nonplanar?

Finally, in CH₄ (methane) and similar compounds of the carbon family, the four hydrogen atoms are at the apexes of a regular tetrahedron with the carbon in the middle. All bond angles are 109.5 degrees (Fig. 26-1a). Another way of visualizing tetrahedral angles is to put the carbon atom into the center of a cube and then draw bonds to four nonadjacent corners of the cube (Fig. 26-1b).

Chemists have come up with a number of schemes that correlate these observations and allow us to predict structures of other compounds. None of these schemes is entirely adequate. Bonding theory is still in the development stage, although much progress has been made lately through the use of computers.

In Chapter 24 we saw that, as a first approximation, we can think of the electron cloud around a multielectron atom as a superposition of hydrogenlike orbitals. This model in effect says that each electron is unaware of the presence of other electrons. We know, however, that this is not quite true; there *are* interactions among the electrons of a multi-

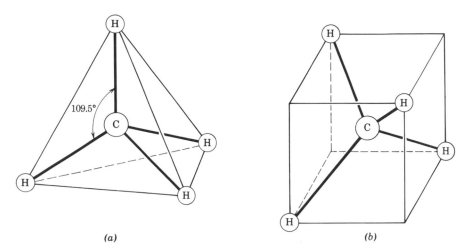

Fig. 26-1. The geometry of CH₄. (*a*) Ball-and-stick model; the hydrogen atoms lie at the apexes of a regular tetrahedron; all HCH bond angles are 109.5 degrees. (*b*) Cube representation.

electron atom. In addition, there are interactions with the electrons of other atoms in a molecule. These electrostatic repulsions among orbital electrons—especially among the electrons in the outer shell—can be used to explain observed bond angles. Naturally such an unsophisticated treatment does not suffice to explain all that is known about the structure and shapes of molecules, but it does allow us to understand many of the observed features of molecular geometry.

When an atom with only two electrons in its outer shell (e.g., beryllium) uses up its bonding power and forms two covalent bonds, electrostatic repulsion causes the electron clouds around the central atom (Be) to move as far away from each other as possible, that is, at a 180° angle to each other (e.g., in BeCl₂–Fig. 26-2).

When a central atom having three outer electrons (e.g., boron) forms three bonds, they also lie at the maximum angular separation possible, which is in a plane with 120° angles between adjacent bonds (e.g., in BF₃–Fig. 26-3). Any deformation out of this plane results in a pyramid-shaped molecule having smaller bond angles.

For an atom forming four bonds (e.g., carbon), the maximum separation is through the nonplanar tetrahedral angle of 109.5 degrees (e.g., in CH₄).

In the previous examples, all outer-shell electrons of each central atom were used in bonding: two electrons of beryllium (there are two 2s elec-

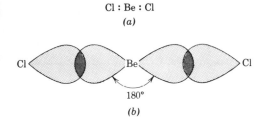

Cl : Be : Cl

(a)

180°

(b)

Fig. 26-2. The linear BeCl₂ molecule. (*a*) Electron-dot diagram. (*b*) Overlapping atomic orbitals.

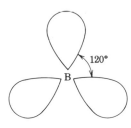

120°

B

Fig. 26-3. The atomic orbitals of boron lie in a plane and are separated by a 120° angle.

trons in the beryllium atom) to form $BeCl_2$; three of boron (two $2s$ and one $2p$ electron; $2s^2p$, in the commonly used shorthand notation) to from BF_3; four of carbon ($2s^2p^2$) to form CH_4. On the other hand, although nitrogen has five outer-shell electrons ($2s^2p^3$), it usually bonds to only three atoms, completing an octet in its outer shell. Consequently only three electron pairs in the molecule are used in bonding, the fourth is an unshared electron pair, as in ammonia, NH_3 (Fig. 26-4). NH_3, unlike BF_3, is *not* a planar compound; the nitrogen atom is not in the plane of the three hydrogens. The HNH angles are 107 degrees.

There is a simple way of explaining this 107° angle. We start with a picture of the four electron pairs arranged at the tetrahedral angle of

H : N̈ : H

H

(a)

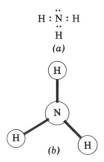

Fig. 26-4. The pyramidal NH₃ molecule. (*a*) Electron-dot diagram. (*b*) Ball-and-stick model.

109.5 degrees, as in CH_4. (As a matter of fact, in the NH_4^+ ion, the angles *are* 109.5 degrees.) The unshared electron pair in NH_3, being closer to the central atom (not pulled out by the attraction of any other nucleus), repels the other pairs more strongly than they repel each other and causes their angles to close from 109.5 to 107 degrees.

A similar explanation can account for the observed 104.5° HOH angle in H_2O—a closing down of the tetrahedral angle bv repulsion of the *two* unshared electron pairs (Fig. 26-5).

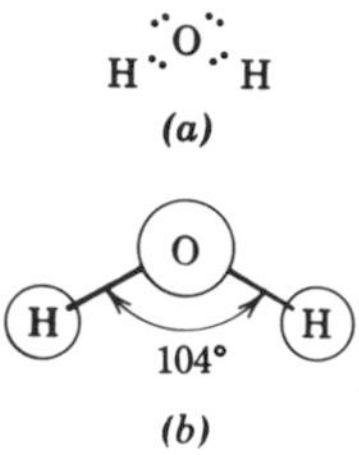

Fig. 26-5. The angular H_2O molecule. (*a*) Electron-dot diagram. (*b*) Ball-and-stick model.

In summary, we can explain the directions of bonds reasonably well through an electrostatic model in which the electron pairs surrounding an atom get as far away from each other as possible: 180 degrees, linear, if there are two pairs of electrons; 120 degrees, planar, for three; 109.5 degrees, nonplanar, tetrahedral, if there is a full octet of four electron pairs. If any of these electron pairs are unshared—not used in bonding— or if the atoms surrounding the central atom are not identical (e.g., CH_3F), then minor changes from the previous angles can usually be explained on the basis of electrostatic repulsions. With the help of this picture of the directed covalent bond, we can now explain some of the features of multiple bonds between atoms.

MULTIPLE BONDS

In Chapters 19 and 25 we indicated the role of octets in the formation of multiple bonds, that is, the sharing of two or three electron pairs in order to reach stable configurations. These configurations frequently, but not always, have eight electrons around one or both atoms. Carbon, for example, can form four single bonds (a *single* bond involves *one pair* of electrons) as in methane (CH_4) and ethane (C_2H_6—Fig. 26-6). These compounds are nonplanar; the bonds around the carbon atoms are tetrahedrally arranged. For convenience we frequently draw the bonds as fol-

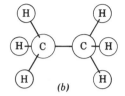
(a)

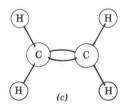
(b)

Fig. 26-6. The C_2H_6 molecule. (a) Electron-dot diagram. (b) Ball-and-stick model.

lows, but we should *not* let such a diagram mislead us into thinking of planar 90° bonds.

$$H—\overset{\overset{\displaystyle H}{|}}{\underset{\underset{\displaystyle H}{|}}{C}}—H \quad \text{and} \quad H—\overset{\overset{\displaystyle H}{|}}{\underset{\underset{\displaystyle H}{|}}{C}}—\overset{\overset{\displaystyle H}{|}}{\underset{\underset{\displaystyle H}{|}}{C}}—H$$

In ethylene, C_2H_4, the two carbons are linked by a double bond (Fig. 26-7). Experiments reveal that the carbon-carbon double bond in ethylene is stronger than the single bond in ethane; it is also shorter (*see* Table 26-1); and ethylene is a planar compound.

$$\overset{H}{}\overset{}{\underset{H}{C}}::\overset{}{\underset{}{C}}\overset{H}{} \qquad \overset{H}{\underset{H}{\diagdown}}C=C\overset{H}{\underset{H}{\diagup}}$$

(a)   (b)

In acetylene, C_2H_2, the carbons are linked by a triple bond, which is stronger and shorter than single or double bonds; and acetylene is a linear compound.

$$H:C:::C:H \quad \text{or} \quad H—C{\equiv}C—H$$

Fig. 26-7. The C_2H_4 molecule. (a) Electron-dot diagram. (b) Bond diagram. (c) Ball-and-stick model.

Table 26-1. Experimentally Determined Parameters of Carbon-Carbon Bonds[a]

Bond	Bond Length (A)	Bond Energy (ev)	Bond Energy (kcal/mol)	Force Constant (mdyn/A)
C—C	1.54	3.6	83	4.5
C=C	1.34	6.3	146	10.8
C≡C	1.20	8.7	200	16

[a] Values were determined for single, double, and triple bonds, respectively, in ethane (C_2H_6), ethylene (C_2H_4), and acetylene (C_2H_2). The first column of numbers lists carbon-carbon bond distances in angstroms. The second and third columns contain *bond energies* in electron volts per molecule and kilocalories per mole—the energy necessary to *rupture* the carbon-carbon bond. The last column presents *force constants* in millidynes per angstrom—the force needed to *stretch* the bond. (Yes, bonds can be stretched. *See* Example 27-1.)

So, we have single, double, and triple bonds, respectively, in C_2H_6, C_2H_4, and C_2H_2. A numerologist might be tempted to eliminate another pair of hydrogens and use the available electron pair to form a carbon-carbon quadruple bond: C≣C (*see* Problem 25-4). Such a bond, however, is not found. Our model of tetrahedral carbon bonds will reveal the reason for the nonexistence of quadruple bonds and explain the observed features of double and triple bonds.

We pictured the formation of single bonds as due to overlapping electron clouds of two atoms (Fig. 26-8). It is not difficult to imagine the carbons rotated slightly so that two pairs of orbits can overlap simultaneously, forming a double bond (Fig. 26-9). Such a bond can be expected to be stronger, and examination of the diagram also indicates a shorter carbon-carbon distance. It is easily shown with the help of a physical model, such as the ball-and-stick type, that the plane of the

Fig. 26-8. Overlapping atomic orbitals form the carbon-carbon bond in C_2H_6.

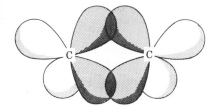

Fig. 26-9. Overlapping atomic orbitals form the carbon-carbon double bond in C_2H_4. All six atoms lie in a plane.

Fig. 26-10. Overlapping atomic orbital form the carbon-carbon triple bond in C_2H_2.

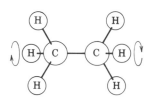

Fig. 26-11. The CH₃ groups in C_2H_6 may rotate in opposite directions about the C—C bond.

overlapping orbitals (shaded) is at right angles to the plane of the other four orbitals (unshaded). In other words, if the shaded orbitals are perpendicular to the paper, the four remaining bonds will be in the plane of the paper. Thus we can account for C_2H_4 being a planar compound. Again, it is easy to show with physical models that three bonds can similarly overlap, forming the triple bond of acetylene with all four atoms in one straight line (Fig. 26-10). Models also reveal that the four tetrahedral bonds of carbon cannot simultaneously overlap; this explains the nonexistence of quadruple bonds.

A final feature of double bonds is restricted rotation about bonds. Groups linked by single bonds generally may rotate, more or less freely, relative to each other. For instance, the two CH₃ groups of ethane may rotate in opposite directions about the C——C bond (Fig. 26-11).

Internal rotation about *double* bonds, however, it restricted. This hindered rotation accounts for the existence of geometric (cis-trans) isomers, as discussed in the next section.

ISOMERS

Chemists and biologists recognize over a million compounds of carbon—the so-called *organic compounds*. This enormous number of compounds exists because chemical bonds can combine in many different structural and geometric arrangements. In this section we discuss some of the possible arrangements.

Formulas

Chemical compounds are characterized by "formulas," but there are several kinds of formulas of varying complexity. The simplest formula of a compound, sometimes called its *empirical formula*, indicates the *relative* number of atoms of each element. It implies nothing about the *actual* number of atoms in the molecule. The empirical formula of hydrogen peroxide (H_2O_2) is HO; of ethane (C_2H_6), it is CH_3; of both nitrogen dioxide (NO_2) and dinitrogen tetroxide (N_2O_4), it is NO_2.

Originally chemists thought that the empirical formula was enough to characterize any compound. However, discovery and analysis of certain compounds suggested revision of this idea. For instance, in 1825 Michael Faraday discovered butene and found that it had the same empirical formula as ethylene, namely, CH_2. Subsequent determinations revealed that butene has twice the molecular weight of ethylene. Today we write their formulas, respectively, as C_4H_8 and C_2H_4. These *molecular formulas* express the *actual* number of atoms of each element in a molecule. Table 26-2 compares the molecular and empirical formulas of some substances.

Table 26-2. Molecular and Empirical Formulas of Selected Substances

Substance	Molecular Formula	Empirical Formula
Phosphorus	P_4	P
Acetylene	C_2H_2	CH
Benzene	C_6H_6	CH
Formaldehyde	CH_2O	CH_2O
Glucose	$C_6H_{12}O_6$	CH_2O
Sucrose	$C_{12}H_{22}O_{11}$	$C_{12}H_{22}O_{11}$
Ethylene	C_2H_4	CH_2
Butene	C_4H_8	CH_2
Cyclobutane	C_4H_8	CH_2
Ammonium cyanate	CH_4N_2O	CH_4N_2O
Urea	CH_4N_2O	CH_4N_2O

Sometimes quite different compounds have identical molecular formulas (*see* the last four compounds in Table 26-2). Two or more substances that have the same molecular formula but different chemical or physical properties are called *isomers* (from the Greek *isos*, "equal," and *meros*, "parts").

Molecular formulas show only the total number of atoms of each element in a molecule, but say nothing about the arrangement of these atoms. We use various types of *structural formulas* to show which atoms are attached to each other. The isomers cis-2-butene and cyclobutane, for example, have identical molecular formulas (C_4H_8) but different structures and different melting and boiling points.

	cis–2–butane	cyclobutane
M.P.	$-139°C$	$-50°C$
B.P.	$-4°C$	$+13°C$

Isomers can be defined in two ways. We can define them operationally as two or more compounds having identical chemical compositions and molecular weights but different chemical or physical properties. (This definition excludes isotopic derivatives, which are not considered isomers.) We can also define isomers theoretically as two or more molecules having the same numbers and kinds of atoms but different arrangements.

Not every compound has isomers. For instance, there is only one CH_4; we can combine a carbon atom and four hydrogen atoms in only one way. The bonds are arranged tetrahedrally around the carbon atom. We showed near the beginning of this chapter how a model of tetrahedral bonds could be constructed: Put the carbon atom into the center of a cube and draw bonds to four nonadjacent corners (Fig. 26-1*b*).

Not only are there no isomers of methane (CH_4), but also there are no isomers of the immediate derivatives of methane—CH_2X, CH_2X_2, CHX_3, and CX_4, where X can be any other atom. This lack of isomers supports the concept of tetrahedral bond angles around carbon atoms. Consider the following argument. It is evident from the cube picture that all four hydrogen atoms are equivalent; i.e., they occupy equivalent positions. It follows from this equivalence that there can also be only one CH_3Cl compound, formed by replacing *any* one of the four hydrogen atoms with a chlorine atom (Fig. 26-12). Now in CH_3Cl the three remaining

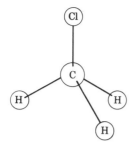

Fig. 26-12. The CH₃Cl molecule. Note that the three H's are equivalent; consequently only one CH₂Cl₂ molecules can be formed by replacing any one of the three H's with a Cl.

hydrogen atoms still are equivalent (as our drawing indicates). None has any special relationship to the chlorine atom; all HCCl angles are equal (109.5 degrees in our model). The compound CH_2Cl_2 can be formed by replacing another of the three hydrogens in CH_3Cl. It does not matter which one is replaced; the *same* CH_2Cl_2 always results. The fact that only one CH_2Cl_2 compound has been found again is good indication that the bonds around carbon are tetrahedral. *If* the bonds were, for example, planar, then *two* CH_2Cl_2 isomers could exist, one with the H's adjacent and the other with the H's opposite each other:

$$\underset{H}{\overset{H}{>}}C\underset{Cl}{\overset{Cl}{<}} \quad \text{and} \quad \underset{Cl}{\overset{H}{>}}C\underset{H}{\overset{Cl}{<}}$$

Problem 26-1. (a) Does (tetrahedral) CH_2FCl have any isomers?

(b) Would a hypothetical planar CH_2FCl have isomers? How many?

Chain Isomers

The number of possible isomers depends in part on the number of carbons in a compound. As a start, let us consider compounds containing only carbon and hydrogen and only single bonds. These compounds are known generically as *alkanes* or *saturated hydrocarbons* ("saturated" because they have only single bonds).

The formulas CH_4, C_2H_6, and C_3H_8 represent only one compound each—no isomers:

$$\begin{array}{ccc}
\overset{\displaystyle H}{\underset{\displaystyle H}{H-C-H}} & \overset{\displaystyle H\;\;H}{\underset{\displaystyle H\;\;H}{H-C-C-H}} & \overset{\displaystyle H\;\;H\;\;H}{\underset{\displaystyle H\;\;H\;\;H}{H-C-C-C-H}} \\[2mm]
\text{methane} & \text{ethane} & \text{propane}
\end{array}$$

But there are two isomers having the formula C_4H_{10}:

$$H-\underset{\underset{\displaystyle H}{|}}{\overset{\overset{\displaystyle H}{|}}{C}}-\underset{\underset{\displaystyle H}{|}}{\overset{\overset{\displaystyle H}{|}}{C}}-\underset{\underset{\displaystyle H}{|}}{\overset{\overset{\displaystyle H}{|}}{C}}-\underset{\underset{\displaystyle H}{|}}{\overset{\overset{\displaystyle H}{|}}{C}}-H \qquad \text{and} \qquad H-\underset{\underset{\displaystyle H}{|}}{\overset{\overset{\displaystyle H}{|}}{C}}-\underset{\underset{\displaystyle C}{|}}{\overset{\overset{\displaystyle H}{|}}{C}}-\underset{\underset{\displaystyle H}{|}}{\overset{\overset{\displaystyle H}{|}}{C}}-H$$

butane

M.P.   −138°C

B.P.   0°C

2-methylpropane

M.P.   −145°C

B.P.   −12°C

Sometimes we indicate such structures through *skeletal formulas* and omit the hydrogen atoms.

$$C-C-C-C \qquad\qquad C-\underset{\underset{\displaystyle C}{|}}{C}-C$$

butane

2-methylpropane

From such skeletal formulas we can deduce the numbers of hydrogen atoms by remembering that each carbon atom has four bonds. We call butane a *straight-chain* compound, in contrast to *branched-chain* 2-methylpropane, even though the "straight" chain really has zigzag shape (Fig. 26-13).[1]

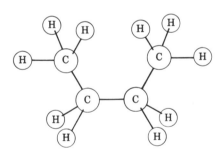

Fig. 26-13. The butane molecule really has zigzag shape, even though we call it a "straight-chain" molecule and represent it as C–C–C–C.

The type of isomerism illustrated by the two C_4H_{10} compounds is called *chain isomerism* because the carbon chains differ.

[1] Equivalent diagrams for butane:

$$C-C-C-C \qquad \underset{\underset{\displaystyle C-C}{|}}{\overset{\overset{\displaystyle C-C}{}}{}} \qquad \underset{\underset{\displaystyle C-C}{|}}{\overset{\overset{\displaystyle C-C}{}}{}}$$

Note that 2-methylpropane cannot be converted into butane by mere rotation of bonds, without breaking a bond.

The number of possible alkane isomers increases rapidly with chain length. C_5H_{12} has three isomers

$$C-C-C-C-C \qquad \underset{\underset{C}{|}}{C-C-C-C} \qquad \overset{\overset{C}{|}}{\underset{\underset{C}{|}}{C-C-C}}$$

and higher alkanes have still more (*see* Table 26-3).

Table 26-3. Isomer Numbers of Some Alkanes

Molecular Formula	Possible Isomers
C_5H_{12}	3
C_6H_{14}	5
C_7H_{16}	9
$C_{10}H_{22}$	75
$C_{20}H_{42}$	366, 319
$C_{30}H_{62}$	4×10^9
$C_{40}H_{82}$	6×10^{13}

Problem 26-2. Draw skeletal formulas for the five isomers of C_6H_{14}.

Functional-Group Isomers

A *functional group* is a recognizable group of atoms that occurs in many molecules and that usually confers upon them a characteristic chemical activity. Examples, are $-OH$ (the hydroxy, or alcohol, group), $-\overset{|}{C}-O-\overset{|}{C}-$ (the ether group), $-NO_2$ (the nitro group), $-C\underset{OH}{\overset{O}{\diagup\!\!\!\diagdown}}$ (the carboxyl group), $-\overset{|}{C}=\overset{|}{C}-$, and $-C\equiv C-$. Hydrocarbons containing multiple bonds are known collectively as *unsaturated hydrocarbons.* Those with double bonds are called *alkenes* (in contrast to single-bonded alkanes); those with triple bonds are called *alkynes.*

Hydrocarbons with two double bonds and those with one triple bond may be isomeric, for example:

$$\underset{}{\overset{\overset{H}{|}}{H-C}}=\overset{\overset{H}{|}}{C}=\overset{}{C}-H \qquad \text{and} \qquad H-C\equiv C-\overset{\overset{H}{|}}{\underset{\underset{H}{|}}{C}}-H$$

And alcohols may be isomeric with ethers:

$$
\begin{array}{ccc}
\text{H} & \text{H} & \\
| & | & \\
\text{H}-\text{C}-\text{C}-\text{OH} & & \text{and} \\
| & | & \\
\text{H} & \text{H} &
\end{array}
\qquad
\begin{array}{ccc}
\text{H} & & \text{H} \\
| & & | \\
\text{H}-\text{C}-\text{O}-\text{C}-\text{H} \\
| & & | \\
\text{H} & & \text{H}
\end{array}
$$

ethyl alcohol   dimethyl ether

Like all isomers, ethyl alcohol and dimethyl ether have the same molecular formula, C_2H_6O, but their properties, including their effects on humans, are decidedly different.

The type of isomerism illustrated here is called *functional-group isomerism*, because the compounds have different functional groups.

Position Isomers

In alkenes and alkynes the double or triple bond can be in various positions along a chain, resulting in what is sometimes called *position isomers*, as in

$$
\begin{array}{cccc}
\text{C}=\text{C}-\text{C}-\text{C} & \quad & \text{C}-\text{C}=\text{C}-\text{C} & \quad \text{C}-\text{C}-\text{C}=\text{C} \\
| & & | & | \\
\text{C} & & \text{C} & \text{C}
\end{array}
$$

When hydrogen atoms are replaced by other atoms or functional groups, the possibilities of position isomerism greatly increase. This is illustrated by the following examples.[2]

$$
\begin{array}{ccc}
| & | & \\
\text{Cl}-\text{C}-\text{C}- & \text{and} & \text{Cl}-\text{C}-\text{C}-\text{Cl} \\
| & | & \\
\text{Cl} & &
\end{array}
$$

$$
\begin{array}{ccc}
| & | & | \\
-\text{C}-\text{C}-\text{C}- & \text{and} & -\text{C}-\text{C}-\text{C}- \\
| & | & | \\
\text{OH} & & \text{OH}
\end{array}
$$

[2] That only one form of $\text{Cl}-\overset{\displaystyle H}{\underset{\displaystyle H}{\text{C}}}-\overset{\displaystyle H}{\underset{\displaystyle H}{\text{C}}}-\text{Cl}$ has ever been isolated (i.e., that $\text{Cl}-\overset{\displaystyle H}{\underset{\displaystyle H}{\text{C}}}-\overset{\displaystyle Cl}{\underset{\displaystyle H}{\text{C}}}-\text{H}$ is not a separate isomer) indicates relatively free rotation about the C-C bond.

$$F-\overset{\underset{|}{|}}{C}-\overset{\underset{}{|}}{C}-\overset{\underset{}{|}}{C}-$$
F

$$-\overset{\underset{}{|}}{C}-\overset{\underset{}{|}}{C}-\overset{\underset{}{|}}{C}-$$
F F

$$-\overset{\underset{}{|}}{C}-\overset{\underset{}{|}}{C}-\overset{\underset{}{|}}{C}-$$
F   F

$$-\overset{\underset{}{|}}{C}-\overset{\overset{F}{|}}{C}-\overset{\underset{}{|}}{C}-$$
F

$$\underset{Br}{\overset{Br}{>}}C=C\underset{}{\overset{}{<}} \quad \text{and} \quad \underset{Br}{\overset{}{>}}C=C\underset{Br}{\overset{}{<}}$$

Problem 26-3. Write structures for the pair of position isomers represented by each formula:
(a) $C_2H_4F_2$.
(b) $C_2H_3F_3$.

Problem 26-4. The molecular formula C_3H_8O includes two types of functional-group isomers; namely, alcohols (containing the $-C-OH$ group) and ethers (containing the $-C-O-C-$ group).
(a) Give condensed structural formulas of the two alcohol position isomers.
(b) How many C_3H_8O ethers are there?

Geometric Isomers

Double-bonded hydrocarbon derivatives frequently exist as *geometric isomers*, which have identical structural linkages in identical sequences but in different spatial arrangements. For example, two attachments on opposite sides of a double bond can be adjacent to each other ("cis") or across from each other ("trans"):

$$\underset{Cl}{\overset{H}{>}}C=C\underset{Cl}{\overset{H}{<}} \qquad \underset{H}{\overset{Cl}{>}}C=C\underset{Cl}{\overset{H}{<}}$$

cis–dichloroethylene   trans–dichloroethylene
M.P.   −80°C   M.P.   −50°C
B.P.   +60°C   B.P.   +48°C

The two structures represent two distinct compounds. We can separate them by distillation and show that they have different properties. Hindered rotation about the double bond (unlike the single bond) prevents the simple conversion of one compound into the other (at room temperature).

Cis-trans isomerism is not restricted to carbon compounds. Dinitrogen difluoride, for example, exists in these two nonlinear forms.

trans
(relatively inactive)

cis
(much more reactive)

Problem 26-5. Which one of these C_4H_8 compounds has cis-trans isomers?
(a) C=C—C—C
(b) C—C=C—C
(c) C—C=C
     |
     C

Problem 26-6. Some compounds of nickel are square planar, instead of tetrahedral like carbon. If A, B, C, and D represent four different atoms, which of the following have geometric isomers? Draw all possible isomers. Contrast these isomer numbers with those for corresponding tetrahedral compounds.
(a) NiA_4 (b) NiA_3B (c) NiA_2B_2
(d) NiA_2BC (e) $NiABCD$

Problem 26-7. Recall that boron trihalides are planar compounds with 120° bond angles. On this basis, which of the following has geometric isomers? (How many?)
(a) BF_2Cl (b) BFClBr

Problem 26-8. Recall that ammonia and its derivatives are not planar but pyramidal. On this basis, which of the following would you expect to have geometric isomers? (How many?)
(a) NF_2Cl (b) NFClBr

Optical Isomers

There is another, more subtle, form of isomerism. Certain compounds (pure or in solution) rotate plane-polarized light and are called *optically active*. They come in pairs that have identical properties in all respects but one: They will rotate a beam of plane-polarized light to an equal degree but in opposite directions. We therefore call them *optical isomers*. Compounds that rotate the plane of polarized light to the right (clockwise) are called dextrorotatory compounds; those that rotate to the left are called levorotatory. Practically all optically active compounds found in plants and animals are levorotatory.

A partial answer to the puzzle of optical activity was found in 1848 by the French chemist Louis Pasteur (1822–1895). He noticed that some crystals of tartaric acid were not identical but were mirror images of each other; they had the same symmetry relationship to each other as our left and right hands. Pasteur separated the crystals carefully by hand, dissolved them in water, and found that, when dissolved, one type was dextrorotatory and the other levorotatory to an equal degree.

Further explanation of optical activity had to wait until the Dutch chemist J. H. van't Hoff (1852–1911) and, independently, the French chemist J. A. LeBel (1847–1930) developed the concept of the three-dimensional tetrahedral carbon atom. Today we realize that optical activity requires molecular asymmetry—the molecules must have neither a plane of symmetry nor a center of symmetry.

A carbon atom with four different attached atoms or groups is a good example of a simple asymmetric molecule. Such a molecule can exist in two different mirror-image forms—i.e., optical isomers, like a right and a left hand (Fig. 26-14).

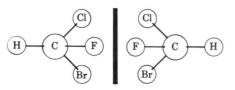

Fig. 26-14. Optical isomers of CHF-ClBr. These molecules are mirror images of each other—like a right and a left hand.

The C—H bond is in the plane of the paper; the C—F bond goes into the plane; and the C—Cl and C—Br bonds come out of the plane. The two molecules cannot be superimposed on each other. If, for instance, the molecule on the right were flipped over so that its C—H bond coincides with the other molecule's C—H bond, then its C—Cl and C—Br bonds would go *into* the plane of the paper.

$$\underset{\text{Br}}{\overset{\text{Cl}}{\text{H}-\text{C}-\text{F}}}$$

Problem 26-9. Which of the compounds in Problems 26-7 and 26-8 would exhibit optical activity and may therefore exist as optical isomers?

Problem 26-10. Of the following compounds, which can have no isomers of any kind?
(a) C_3H_8 (b) $C_3H_6Cl_2$ (c) $C_2H_2F_2$
(d) C_2HF_3 (e) C_2H_5Cl

Problem 26-11. Which of the above may have optical isomers?

Classification of Isomers

Chain isomers, e.g.,

C—C—C—C (straight-chain) and C—C—C (branched-chain)
$\qquad\qquad\qquad\qquad\qquad\qquad\qquad\quad$ |
$\qquad\qquad\qquad\qquad\qquad\qquad\qquad\quad$ C

Functional-group isomers, e.g.,

$\quad$ CH≡C—CH$_3$ and CH$_2$=C=CH$_2$
$\quad$ CH$_3$CH$_2$OH and CH$_3$OCH$_3$
$\quad$ NH$_4$NCO and (NH$_2$)$_2$CO

Position isomers, e.g.,

$\quad$ C=C—C—C and C—C=C—C
$\quad$ F—C—C—F and F$_2$C—C

Geometric isomers, e.g.,

Optical isomers

EFFECT OF MOLECULAR STRUCTURE ON THE BEHAVIOR OF SUBSTANCES

The foregoing discussion of isomers suggests that an almost infinite number of molecular arrangements is possible, especially for carbon-containing compounds. More than one million different compounds of carbon have been identified so far, and the list is growing rapidly.

How can we ever hope to cope with the chemistry of a million compounds? Fortunately, these compounds and their reactions fall into classes. Once we recognize the "common factors" and their effects, we have come a long way in systematizing and thereby simplifying the huge field of "organic chemistry." In the following pages we illustrate, using a few select examples, the relation between molecular structure and behavior of compounds.

Solubility Behavior

The effect of structure on behavior is particularly well illustrated by solubility behavior. A *solution* is defined operationally as a homogeneous mixture of two or more substances, or defined theoretically as a random distribution of the molecules, atoms, or ions of two or more substances.

The dissolved substance is called the *solute,* and the dissolving medium is called the *solvent.* For example, in a saline solution, water is the solvent and sodium chloride is the solute. At times, the designations become somewhat arbitrary. Is, for example, a 50-50 mixture of alcohol and water, a solution of water in alcohol or of alcohol in water? Usually, however, the substance present in larger amount is considered the solvent.

The dissolving of solids in liquids is akin to the melting of solids. Both processes involve breaking of bonds between the units of a solid. These bonds may vary from weak intermolecular forces among the molecules in a molecular crystal to strong electrostatic attractions in an ionic crystal lattice.

For these bonds to be broken the system must absorb energy. In the melting process this energy is furnished in the form of heat, and the magnitudes of melting points are an indication of the relative interparticle forces. Most molecular solids have relatively low melting points, whereas ionic crystals have quite high melting points (*see* p. 351).

Since the solution process need not involve elevated temperatures, where does the energy originate that is needed to break the interparticle bonds of solids? In the following paragraphs we answer this question as well as some questions pertaining to the selectivity of solubility phenomena: Why does sodium chloride dissolve in water but not in carbon tetrachloride? And why is the converse true for substances like gasoline or oil, which dissolve in carbon tetrachloride but not in water?

When water dissolves sodium chloride, the energy required to break the interionic bonds is mainly supplied by the formation of *new* "bonds" between the water molecules and the ions. Water molecules have considerable polar character, called dipole moment—the hydrogen end of the molecule is positive and the oxygen end is negative. As a result, there is electrostatic attraction between an ion and the oppositely charged end of a water molecule, and each ion is surrounded by a cluster of water molecules (Fig. 26-15). Each *ion-dipole attraction* is relatively weak, but collectively they supply a considerable amount of energy—enough to overcome the attractive forces within the NaCl crystal lattice.

From this model of the solution process it is easily seen why carbon tetrachloride does *not* dissolve sodium chloride. The CCl_4 molecule is nonpolar. Consequently ion-dipole forces between CCl_4 and Na^+ or Cl^- ions are *not* strong enough to overcome the attractive forces within the NaCl crystal lattice.

But why is CCl_4 a good solvent for alkanes, and water is *not*? Alkanes have low dipole moments; as a result, their molecules are only weakly attracted to each other. Wouldn't these weak forces be easily overcome

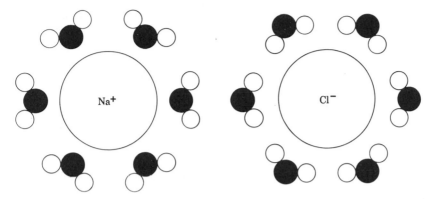

Fig. 26-15. Na⁺ and Cl⁻ ions in water solution. Each ion is caged by a cluster of water molecules attracted by ion-dipole forces. In water molecules the hydrogen end is positively charged, and the oxygen end is negatively charged. Consequently the surrounding water molecules turn their oxygen ends toward Na⁺ and their hydrogen ends toward Cl⁻. Atoms are not drawn to scale.

even by a weak attraction between water molecules and alkane molecules? Maybe so, but a different effect plays the dominant role here. For hydrocarbons to dissolve in water, they would have to "dig a hole" in the water, overcoming the fairly strong *dipole-dipole attractions* among the water molecules. But hydrocarbon-water attractions are not strong enough to do this. Hence, water and hydrocarbons are mutually insoluble.[3]

This dipole-dipole association energy is not present in nonpolar liquids like CCl_4, which accordingly are mutually soluble with other nonpolar liquids like hydrocarbons.

In general, then, a polar solvent may be expected to dissolve only polar solutes or ionic substances; a nonpolar solvent, only nonpolar solutes. *Like dissolves like.*

Problem 26-12. Common lubricating oil is an alkane. To remove an oil spot from your clothing, would you use water or gasoline? Why?

(b) To flush away (quickly!) some hydrochloric acid (HCl) from your clothes, would you use water or carbon tetrachloride?

(c) Polyethylene is a long-chain hydrocarbon (*see* p. 348) and is

---

[3] Even though the dipole-dipole attractions in water are weak compared to normal covalent bonds (4.5 kcal/mole, versus 110 kcal/mole for the O-H bond in H_2O), they are still strong enough to preclude the solution of most nonpolar substances in water.

soluble in water. As a matter of fact, garden hoses are made of poly-ethylene. But would you use a polyethylene hose for *gasoline*?

Problem 26-13. Silver is insoluble in water or gasoline, but it will dis-solve in mercury. (You probably have heard of amalgams, used for filling teeth.) Explain.

With this rule of thumb, *like dissolves like,* we can generally predict what effect various functional groups would have on solubility behavior. Double and triple bonds do not affect polarity greatly; hence alkenes and alkynes would not differ greatly from alkanes in solubility behavior. On the other hand, the introduction of polar groups like $-OH$ (to form alco-hols) or $-NH_2$ (to form amines) does increase water solubility con-siderably. These groups can have dipole-dipole attractions to each other and to water molecules, similar to those in water itself:

$$
\begin{array}{cc}
& \text{H} \\
& | \\
\text{H} - \text{O} \cdots \text{H} - \text{O} \cdots \text{H} - \text{O} & \\
| & | \\
\text{H} & \text{H} \\
\text{water–water} &
\end{array}
\qquad
\begin{array}{c}
\text{C} \\
| \\
\text{C} \\
| \\
-\text{C} - \text{C} - \text{O} \cdots \text{H} - \text{O} \\
| \\
\text{H} \\
\text{alcohol–alcohol}
\end{array}
$$

$$
\begin{array}{c}
\text{H} \\
| \\
-\text{C} - \text{C} - \text{O} \cdots \text{H} - \text{O} \\
| \\
\text{H} \\
\vdots \\
\text{O} - \text{H} \\
| \\
\text{H} \\
\text{alcohol–water}
\end{array}
\qquad
\begin{array}{c}
\text{H} \qquad\qquad \text{H} \\
\backslash \qquad\qquad | \\
-\text{C} - \text{C} - \text{N} : \cdots\cdots \text{H} - \text{O} \\
/ \\
\text{H} \\
\text{amine–alcohol}
\end{array}
$$

Finally, we should point out the effect of carbon-chain length. Hydro-carbon derivatives, such as alcohols, owe their water solubility almost entirely to the polar groups they may contain, such as $-OH$. With in-creasing chain length, the nonpolar part of the molecule increases, while the polar part remains unchanged. The relative decrease in importance of the polar group is accompanied by a decrease in solubility in polar solvents, for example in water (*see* Table 26-4).

The use of soap represents an interesting application of the like-dis-solves-like principle. Soap helps "dissolve" substances like oils, which are ordinarily insoluble in water. The main ingredients of many soaps is sodium stearate, $CH_3(CH_2)_{16}COONa$.

$$CH_3CH_2CH_2CH_2CH_2CH_2CH_2CH_2CH_2CH_2CH_2CH_2CH_2CH_2CH_2CH_2CH_2C\overset{\textstyle O}{\underset{\textstyle ONa}{\diagdown}}$$

The hydrocarbon part of this molecule manages to dissolve in an oil drop, leaving the polar —COONa end protruding. This polar part then dissolves in water, taking the encapsulated oil drop with it.

Table 26-4. Decreasing Water Solubility of Alcohols with Increasing Chain Length

Formula	Solubility[a] (g/100 g H_2O)
CH_3OH	∞ [b]
CH_3CH_2OH	∞
$CH_3CH_2CH_2OH$	∞
$CH_3CH_2CH_2CH_2OH$	7.9
$CH_3CH_2CH_2CH_2CH_2OH$	2.3
$CH_3CH_2CH_2CH_2CH_2CH_2OH$	0.6
$CH_3CH_2CH_2CH_2CH_2CH_2CH_2OH$	0.2
$CH_3CH_2CH_2CH_2CH_2CH_2CH_2CH_2OH$	0.05

[a] The numerical values represent *saturation* solubility; i.e., the maximum amount of solute that can be dissolved in a given amount of solvent at a specified temperature (usually room temperature).

[b] ∞ means "soluble in all proportions."

Chemical Reactivity

The relation of molecular structure and chemical reactivity is well illustrated by the contrast in behavior between alkanes and alkenes. Chemically, the alkanes are quite similar. If we know the chemistry of one we largely know the chemistry of the whole group. This chemistry is relatively simple—alkanes are not very reactive. Although their most important reaction is combustion, for example,

$$C_8H_{20} + 13O_2 \rightarrow 8CO_2 + 10H_2O$$

alkanes do not react with oxygen at room temperature; a high temperature is required to initiate the reaction, indicating the relative inertness of these compounds. (This is fortunate; otherwise we would have considerable problems in handling and storing our common fuels—oil, kerosene, and gasoline.)

Although alkenes are similar to alkanes in physical properties and flammability, they are much more reactive with various *other* reagents because of the "unsaturated" double bond. Characteristic reactions are "addition reactions," to yield "saturated" single bonds. For example, hydrogen halides readily attack the double bond and form an addition product.

$$\mathrm{\underset{/}{\overset{\backslash}{C}}=\underset{\backslash}{\overset{/}{C}}} \quad + \quad HI \quad \longrightarrow \quad -\underset{\underset{H}{|}}{\overset{|}{C}}-\underset{\underset{I}{|}}{\overset{|}{C}}-$$

Hydrogen also adds to the double bond under special conditions—for example, in the presence of catalysts.

$$\mathrm{\underset{/}{\overset{\backslash}{C}}=\underset{\backslash}{\overset{/}{C}}} \quad + \quad H_2 \quad \longrightarrow \quad -\underset{\underset{H}{|}}{\overset{|}{C}}-\underset{\underset{H}{|}}{\overset{|}{C}}-$$

The addition of bromine serves as a convenient test for a carbon-carbon double or triple bond: The reddish color of bromine disappears as this reagent is used up.

$$\mathrm{\underset{/}{\overset{\backslash}{C}}=\underset{\backslash}{\overset{/}{C}}} \quad + \quad \underset{\text{red-brown}}{Br_2} \quad \longrightarrow \quad \underset{\underset{\text{colorless}}{Br \quad Br}}{-\overset{|}{C}-\overset{|}{C}-}$$

Double-bonded compounds can also undergo addition reactions with themselves—for instance, in the polymerization of ethylene to form long-chain, saturated polyethylene.

$$n\ CH_2=CH_2 \quad \longrightarrow \quad \begin{array}{l} -CH_2-CH_2 \\ \quad\quad\quad | \\ \quad\quad\quad CH_2-CH_2 \\ \quad\quad\quad\quad\quad\quad | \\ \quad\quad\quad\quad\quad\quad CH_2-CH_2- \end{array}$$

$$\text{or } (-CH_2-CH_2-)_n$$

Polyethylene is a flexible plastic, rather waxy in feel, tough, and quite unreactive. It is used for "plastic" bottles, tumblers, films and sheets, etc. Teflon,

$$(CF_2{-}CF_2{-})_n$$

a polymer of tetrafluorethylene, is similar in appearance, but chemically even more unreactive, so that it can be used as an inert coating in frying pans.

These few examples illustrate the dependence of chemical reactivity on the presence of "functional groups" such as a double bond. And, on the whole, it makes relatively little difference where in the molecule this functional group is located and what other functional groups are present; usually, even a large molecule containing, for example, a double bond, reacts in a manner characteristic of all alkenes.

SUMMARY

Bond angles, and therefore the geometry of molecules, can be explained on the basis of electrostatic repulsions among the outer orbital electrons in molecules—orbitals surrounding an atom get as far away from each other as possible. This results in 180° (linear), 120° (planar), or 109.5° (tetrahedral) bond angles, respectively, if two, three, or four electron pairs are involved. If any of these electron pairs are not shared, minor changes in bond angles result.

Each bond normally involves a shared electron pair. Double and triple bonds involve overlapping orbitals of two and three electron pairs, respectively. These multiple bonds are stronger and shorter than single bonds.

Isomers are compounds having identical molecular formulas but different structural formulas or geometric arrangements. Various classes of isomers are summarized on p. 371. The possibility of so many different arrangements of atoms is responsible for the large number of compounds found in nature.

Yet the reactions and behavior of these many compounds fall into relatively few classes, which depend to a great extent on the functional groups present in the molecules. So, for example, solubility behavior depends mostly on the presence or absence of polar groups in a molecule. Similarly, chemical reactivity is largely determined by the presence of such functional groups as double bonds, −OH groups, C—O—C groups, etc.

We see, then, that molecular structure can be used as a unifying concept in systematizing our knowledge of the behavior of chemical compounds.

BRIEF HISTORY OF STRUCTURAL CHEMISTRY

The German chemist Friedrich Wöhler (1800–1882) might be considered the father of organic chemistry because he showed, through his laboratory preparation of urea in 1828, that "organic" compounds could be synthesized in the laboratory from "inorganic" chemicals.

$$\text{NH}_4\text{NCO} \xrightarrow{\text{heat}} \begin{array}{c} \text{H}_2\text{N} \\ \diagdown \\ \diagup \\ \text{H}_2\text{N} \end{array} \text{C}{=}\text{O}$$

ammonium   urea
cyanate

Other syntheses of "organic" compounds during the next few years led to abandonment of the "vitalistic theory," which maintained that some of the seemingly unique properties of organic compounds were due to a mysterious "vital force" derived only from living things.

Because most "organic" compounds contained the element carbon, we nowadays define *organic chemistry* as the *chemistry of carbon-containing compounds*. Inclusion or exclusion of compounds like carbon dioxide, carbonates, and metal cyanides is arbitrary and varies with individuals.

Molecular formulas came into general use in the period from 1828 to 1858. They were made possible by improved methods of elemental analysis and the advent of reliable atomic-weight tables, coupled with better methods for determining molecular weights.[4]

In that same period, however, chemists realized that molecular formulas alone are not sufficient to characterize compounds, because of the existence of isomers—compounds having identical molecular formulas but different properties. Wöhler had been one of the first (in the 1820s) to hint at their existence. Other early contributors to that development were Faraday and Berzelius. Evidently there are different ways the same set of atoms can be linked together.

The Scottish chemist A. S. Couper (1831–1892) and the Russian A. M. Butlerov (1828–1886) made the first attempts to write structural formulas using carbon chains, after Couper and the German chemist F. A. Kekulé (1829–1896) had recognized in 1857 that carbon has a combining power of four. The nearly simultaneous proposal of such similar ideas indicates that there was a good communications network and that "the time was ripe."

About the same time, in 1861, J. Loschmidt (the Austrian who first determined Avogadro's number in 1865) introduced the concept of multiple bonds.

In 1874 a third dimension was added to structural formulas when the Dutch chemist J. H. van't Hoff (1852–1911) and the French chemist J. A. LeBel (1847–1930) independently proposed tetrahedral bonds for the carbon atom. This idea actually was a logical interpretation of Louis Pasteur's (French, 1822–1895) discovery of optical isomerism in 1848.

---

[4] Recall, for example, Cannizzaro's method, described in Chapter 4.

In the 1890s this three-dimensional thinking was extended to inorganic compounds, chiefly by Alfred Werner (1866–1919) in Switzerland.

We have already discussed many of the present century's developments pertaining to structural chemistry, albeit in a different context. These developments include:

1. Bohr's atom model (1913).

2. Lewis's theory of covalent bonding, the octet rule, and electron-dot formulas (1916).

3. Quantum mechanics—Heisenberg, Born, Dirac, Schrödinger (1925).

4. Techniques and instrumentation for the experimental investigation of molecular structure, such as dipole-moment measurement, X-ray and electron diffraction, mass spectrometry, and molecular spectroscopy.

In the next, and last, chapter we discuss some of the tools and techniques available to the structural chemist.

ANSWERS TO SELECTED PROBLEMS

26-4. (b) One

26-6. (e) Three geometric isomers:

$$
\begin{array}{ccc}
\text{A} & \text{A} & \text{A} \\
\text{D} - \!\!\mid\!\! - \text{B} & \text{C} - \!\!\mid\!\! - \text{D} & \text{B} - \!\!\mid\!\! - \text{C} \\
\text{C} & \text{B} & \text{D}
\end{array}
$$

If tetrahedral: only a pair of optical isomers.

26-7. Neither.

26-8. Neither.

26-9. Of all compounds, only NFClBr can exist as a pair of optical isomers.

26-10. (a), (d), and (e) have no isomers.

26-11.
$$
\begin{array}{c}
\text{H} \\
\mid \\
CH_3 - C - CH_2Cl \\
\mid \\
Cl
\end{array}
$$
(four different groups attached to central carbon)

Chapter 27

THE MODERN STRUCTURAL
CHEMIST'S TOOLS

In the foregoing chapter we discussed the structure of molecules from a descriptive and semitheoretical point of view. We assumed knowledge of the structure of representative molecules and tried to explain these structures in terms of atomic orbitals and quantum-mechanical principles. On the whole, we arrived at a fairly satisfactory and plausible model that allowed us to correlate and even predict some properties of substances.

In this last chapter we answer the question, *how* do scientists find out what the structures of molecules are. Early evidence indicated that the *composition* of water is H_2O and that of nitrous oxide is N_2O (*see* Chapter 4). But how do we know that the *structure* of water is HOH (O between the H atoms) and not HHO, and that nitrous oxide, on the other hand, is NNO and not NON? Further, how do we know that HOH is a bent molecule ($_H \diagup ^O \diagdown _H$) and that NNO is a linear molecule (N—N—O)? And what experimental confirmation is there of our bonding theory, which tells us that the bonds in HOH are single bonds, whereas the atoms in NNO are joined by multiple bonds?

With the exception of some very big molecules, such as viruses, we have never seen a molecule. Our evidence then is in a sense indirect, but nevertheless generally reliable and often quite precise.[1] Many experi-

---

[1] "Seeing" also is indirect, in a sense. When we "look" at an object under a microscope, our eyes actually are reacting via electrical nerve impulses to *photons* that have been bounced off the object and refracted through a series of lenses. These electrical stimuli then are passed on and interpreted by our brains in order for us to "see." Not such a direct process after all! And it is often subject to error and misinterpretation, such as optical illusion. Perhaps the recording of a diffraction pattern and its subsequent interpretation is no more indirect.

mental methods and tools are available. In the following sections we out-line only a few of these experimental methods: some classical methods of "wet chemistry," diffraction methods (X-ray, electron, and neutron dif-fraction), mass spectrometry, and molecular spectroscopy. These methods may be used separately or in combination. We choose whichever tech-nique best suits the problem at hand—provided our laboratories have been supplied with the necessary equipment.

CLASSICAL METHODS OF "WET CHEMISTRY"

Early attempts to relate physical properties (such as melting points, boiling points, and densities) to structure of molecules met with little success. Before the development of instrumental techniques, the elucida-tion of molecular structure depended almost entirely on the systematic investigation of chemical reactions. And an amazing amount of structural knowledge was actually obtained by the early practitioners of classical wet chemistry. We speak of "wet" chemistry because most reactions are carried out in solution. Today most structural chemists still use classical methods of wet chemistry alongside modern instrumental methods. In the following illustration we isolate an "unknown" compound, obtained by a fermentation process, and determine its structure.

Let us mash some grapes, with our bare feet or otherwise, and let the mash stand for several days. Slow evolution of carbon dioxide indicates that a chemical reaction is taking place. We call it *fermentation*. Even-tually the grape juice develops a new odor and a pleasurable taste. At this stage, we boil the mixture and recondense the vapors, collecting a colorless liquid. We call this process *distillation*. We collect only the first fraction of the liquid and observe that its boiling point (78°C) is less than that of water. What is this liquid?

Since it came from a water solution, does it still contain water? Yes: White anhydrous copper sulfate when added to our liquid is converted into blue hydrated copper sulfate, according to the reaction,

$$CuSO_4 + 5H_2O \rightarrow CuSO_4 \cdot 5H_2O$$

white   blue

We dry the liquid—extract the water—by adding calcium chloride, $CaCl_2$, which combines with water to form hydrated calcium chloride.

$$CaCl_2 + 6H_2O \rightarrow CaCl_2 \cdot 6H_2O$$

We decant or filter the liquid and retest with copper sulfate. The test is negative: Copper sulfate does not turn blue anymore. We shall assume

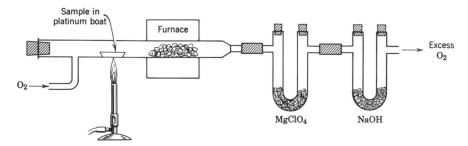

Fig. 27-1. A combustion train for the determination of carbon and hydrogen in combustible substances. The liquid sample is evaporated by the flame and then oxidized by the stream of oxygen in the furnace. (The furnace contains glass wool to cause turbulence and assure intimate mixing of the sample and oxygen.) The product gases contain water vapor, which is absorbed by the magnesium perchlorate ($MgClO_4$); and carbon dioxide, which is absorbed by the sodium hydroxide (NaOH). Weight increases of the U-tubes reveal the amount of H_2O and CO_2 formed in the combustion process.

that we now have a pure chemical substance. (This assumption could be verified by further testing.)

Next, we perform a quantitative elemental analysis of the substance; that is, we find its empirical formula. To determine its carbon and hydrogen content, we burn the liquid in oxygen. The burning process converts the carbon in the sample into carbon dioxide, and the hydrogen into water. We pass the resulting combustion gases first through a tube containing a drying agent, such as anhydrous magnesium perchlorate, which absorbs the water vapor; then through another tube containing sodium hydroxide to absorb the carbon dioxide (Fig. 27-1). The reaction in the second tube is

$$2NaOH + CO_2 \rightarrow Na_2CO_3 + H_2O$$

We can calculate the percentages of hydrogen and carbon in the original sample from the weight changes of the tubes. The sample's oxygen content can also be determined through a similar but more complicated procedure. These percentages add up to 100%, eliminating, the need for further elemental analyses.

carbon	52%
hydrogen	13%
oxygen	35%
total	100%

By dividing these percentages by the corresponding atomic weights, we can calculate an *empirical formula:*

$$C_{52/12}H_{13/1}O_{35/16} = C_{4.53}H_{13}O_{2.18}$$

We then divide these numbers by 2.18 to obtain integral subscripts: C_2H_6O.

The *molecular formula* is some integral multiple of the above empirical formula. A molecular-weight determination reveals how many C_2H_6O units make up our molecule. A molecular weight of 46—determined by methods described in Chapter 4 indicates that the molecular formula of our grape-juice product is simply C_2H_6O.

For such a compound we can draw only two structural formulas:

ethyl alcohol   dimethyl ether

Several chemical tests are available to help us decide whether our fermentation product is an alcohol or an ether. For example, when the compound is treated with metallic sodium, hydrogen gas is produced. In that respect the compound is similar to water, leading us to suspect that, like water, it contains an −OH group; in other words, that it is ethyl alcohol, C_2H_5OH.

$$2HOH + 2Na \rightarrow 2HONa + H_2$$

$$2C_2H_5OH + 2Na \rightarrow 2C_2H_5ONa + H_2$$

An elemental analysis of C_2H_5ONa would confirm that a sodium atom has replaced one hydrogen atom. This fact would indicate that one hydrogen atom in C_2H_6O is different from the other five—true in C_2H_5OH but not in CH_3OCH_3.

Another test: When C_2H_6O is passed over hot aluminum oxide, C_2H_4 is produced. Only one structural formula can be drawn for C_2H_4:

ethylene

Ethylene contains a carbon-carbon bond. Of the two C_2H_6O choices, only ethyl alcohol contains a carbon-carbon bond. Thus once more ethyl alcohol is the more likely choice.

To prove our structure deduction, we could synthesize the substance from simpler, familiar molecules by one or more well-understood steps. For example, ethyl alcohol can be prepared by the addition of H_2O to ethylene in the presence of sulfuric acid catalyst.

$$HOH + H_2C{=}CH_2 \xrightarrow{\text{H}_2\text{SO}_4} CH_3CH_2OH$$

The identity of this synthesized product with our "unknown" can be proved through the identity of their properties, such as melting points, boiling points, density, or spectra.

Our structure determination of ethyl alcohol typifies an approach frequently used by chemists for covalently bonded compounds. There are six basic steps to this approach.

1. *Purification.* We used fractional distillation, followed by drying.

2. *Elemental analysis.* We analyzed for carbon, hydrogen, and oxygen, and arrived at the empirical formula $(C_2H_6O)_n$.

3. *Determination of molecular weight and molecular formula.* Our molecular weight of 46 indicated the molecular formula C_2H_6O.

4. *Postulation of structural formulas.* We considered all structural formulas that were compatible with our general concepts about structures and eliminated any structures that were incompatible with the known properties of the sample (none immediately). We were left with two possible structures.

5. *Selection of the proper structural formula.* In general we first find criteria—chemical reactions, usually—that help us distinguish among the structures under consideration; then we design and perform the appropriate experiments. Usually these experiments are designed to detect the presence or absence of reactive functional groups. We performed two such tests, both designed to detect the presence of an $-OH$ group: reaction of our sample with metallic sodium, resulting in replacement of one hydrogen atom in the molecule; and reaction with aluminum oxide, yielding ethylene by abstracting HOH from the original molecule. Occasionally we can deduce a structural formula by the elimination of the other formulas under consideration. This process has been dubbed the "Method of Holmes."[2]

6. *Synthesis.* It is always elegant and more convincing if the deduced structure can be synthesized by a series of unambiguous and well-understood steps. We synthesized C_2H_5OH by the addition of HOH to C_2H_4. Today's chemist, synthesizing a complex compound, does not need to begin with the simplest compounds; he can start with complex but famil-

---

[2] *See* footnote, p. 267.

iar compounds. Naturally, he should base his work on available knowledge of structures and reactions; there is no need to work in a scientific and intellectual vacuum. The identity of the "unknown" and the synthesized "known" compound can be proved by comparing their properties.

DIFFRACTION METHODS

X-ray Diffraction

X rays were discovered accidentally in 1895 by Wilhelm Konrad Röntgen (1845–1923) while he was experimenting with cathode-ray tubes (*see* Chapter 9, p. 134). In 1912 Max von Laue (1879–1960) established the wave nature of X rays by showing that they can be diffracted just like light (*see* Chapter 13, p. 197). His measurements revealed that X rays are electromagnetic radiation of very short wavelengths, ranging from about 0.01 A to 100 A. (Visible light has wavelengths of 4000–8000 A.)

Laue passed a narrow beam of X rays through a thin crystal and produced a pattern of spots—a diffraction pattern—on a photographic plate (*see* Fig. 27-2). Such diffraction of X rays by crystals is akin to the diffraction of ordinary light by a diffraction grating (*see* Chapter 12, p.

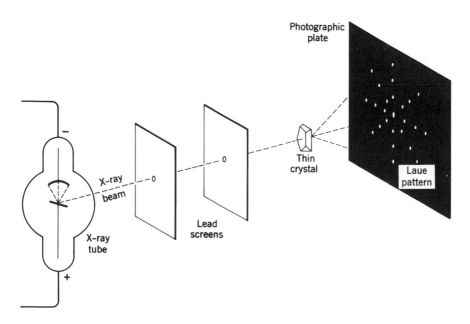

Fig. 27-2. Schematic arrangement for crystal diffraction by the Laue method.

183). Since Laue's time, several variations of the experimental technique have been developed. Laue used single crystals and polychromatic radiation, that is, X-ray beams of many wavelengths. Another method uses monochromatic X rays (only one wavelength) and powdered samples containing many tiny crystals randomly oriented. Still another method uses monochromatic X rays and single crystals which are rotated so that all sides are exposed consecutively to the beam. Choice of method is governed by information desired, the availability of large-enough single crystals, and the type of instrumentation available.

The diffraction patterns produced by these methods depend on the location and spacing of the atoms in the crystal. Accordingly, careful analysis of a diffraction pattern yields information about atom arrangements and distances in crystals.

This analysis is a difficult task. We usually try to construct a reasonable model of the crystal, based on available chemical and structural knowledge. We call it an "educated guess" based on "chemical intuition." Then we calculate the diffraction pattern produced by such a model, and compare that pattern with the observed pattern. If the comparison indicates that we have guessed the proper geometric arrangement, we adjust the interatomic distances in our model and recalculate the pattern, repeating this process until we obtain agreement between the calculated and observed patterns. This tedious process is facilitated nowadays by computer techniques.

At its best, X-ray diffraction yields a detailed "map" of the atoms in a crystal. Very accurate interatomic distances have been obtained with this technique, with errors commonly as small as 0.01 A—in distances of about 2 A (*see* Problem 13-8).

X-ray diffraction has one major limitation—it is not suitable for locating hydrogen atoms with high precision, because they do not scatter X rays sufficiently. Fortunately another method, neutron diffraction, fills that gap.

Neutron and Electron Diffraction

Neutron and electron diffraction methods are based on the wave nature of matter. In 1923 Louis de Broglie (1892–   ) suggested that matter might behave like waves. In 1927 the wave nature of electrons was verified through the electron-diffraction experiments of C. J. Davisson (1881–1958) and L. H. Germer (1896–   ) in the United States and G. P. Thomson (1892–   ) in Scotland (*see* Chapter 20). Neutron diffraction was first demonstrated in 1936, four years after the neutron's discovery by the English physicist James Chadwick (1891–   ); but it became a practical tool only after World War II, when neutron beams could be obtained from atomic piles.

Neutron and electron diffraction work on the same principle as X-ray diffraction. Neutron diffraction is particularly suitable for locating hydrogen atoms in crystals after the heavier atoms have been pinpointed by X-ray diffraction. The main disadvantage of neutron diffraction is the requirement of a nuclear reactor as a neutron source.

Electron diffraction has no such drawback—there are plenty of electrons available. Electron beams offer several additional advantages over neutron beams as well as over X rays:

1. Higher beam intensities are available; therefore shorter exposure times produce usable patterns on film—often seconds as opposed to hours.

2. Electron-beam wavelength can easily be altered by adjusting the accelerating voltage. (Recall the equation in Example 20-1 relating the wavelength, λ, in angstroms to voltage, V: $\lambda = 12.3 \sqrt{V}$.)

3. Electron beams are sensitive to electric and magnetic fields. These fields deflect and can thereby focus electron beams; as lenses focus light. (No such focusing has yet been developed for X rays.)

Electrons are so strongly scattered by atomic nuclei that electron-diffraction methods can be applied even to gases. The method has proved particularly useful when applied to small molecules. Its accuracy is comparable to that of X-ray diffraction.

MASS SPECTROMETRY

The mass spectrometer is an instrument which measures masses of ions, or, more precisely, the mass-to-charge ratios of ions. The basic function of the instrument is to smash molecules by electron bombardment and then to sort and count the fragments with the help of electric and magnetic fields. We may compare the technique to shattering a piece of ceramic or glassware, say a coffeepot, then sorting the pieces according to size into neat piles and counting the number of pieces in each pile (Table 27-1).

This is a simple analogy, but the basic action of the mass spectrometer *is* that simple. The molecules of a given sample are ionized and fragmented through collision with a beam of electrons. The positive ions are then attracted by a negatively charged electric plate and their paths are bent by a magnetic field. The observed path curvature is a direct measure of an ion's mass-to-charge ratio (Fig. 27-3).

The modern mass spectrometer is based on the discovery of canal rays in 1886 by the German physicist Eugen Goldstein (*see* Chapter 8, p.

Table 27-1. The Mass Spectrum of Methane, CH_4

Mass Number (a.m.u.)	Relative Number of Fragments	Fragment
16	100	CH_4 ("parent")
15	85	CH_3
14	9	CH_2
13	4	CH
12	1	C
1	3	H

132). In modified cathode-ray tubes Goldstein observed positively charged rays traveling in a direction opposite to that of the cathode rays. Eleven years later the Englishman J. J. Thomson, using magnetic and electric deflection, measured the mass-to-charge ratios of the particles in these canal rays and identified them as positive ions. These ions were created by the collisions of cathode-ray electrons with the molecules of the residual gas in the canal-ray tube. In 1913 Thomson, using a modified canal-ray tube, split a beam of neon ions into two beams, thus revealing the existence of two isotopes, neon-20 and neon-22 (see Chapter 11, p. 153). By 1919 F. W. Aston (1877–1945) in England and A. J. Dempster (Canadian, 1886–1950) in the United States had developed the canal-ray tube into precision mass spectrometers.

We cannot *predict* exactly how a particular coffeepot or molecule will shatter. Nevertheless, the results can be reproducible. That is, if we hit many coffeepots or molecules the same way each time, they will break more or less the same way each time, and we can recognize certain patterns of breakage. On examining the fragments of many coffeepots of a particular brand, we may find mostly unbroken handles and spouts. Another manufacturer's coffeepots, however, when treated similarly may consistently have its handles broken into, say, three pieces. We can accordingly use this method to distinguish different brands, even though the original pots appeared to be identical. Likewise, by merely looking at the numbers and types of molecular fragments we may be able to distinguish different chemical substances, even though they may otherwise appear to be quite similar and perhaps have nearly identical chemical properties. The mass spectrometer thus is a useful tool for identifying previously known substances.[3]

---

[3] This method of identification is analogous to matching a suspect's fingerprints with those in a police file, and we actually call it the *fingerprint method* of identification (see Chapter 14, p. 205).

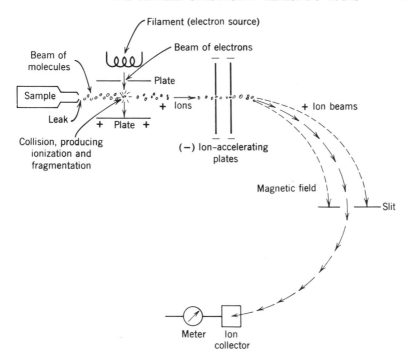

Fig. 27-3. Schematic diagram of a mass spectrometer. A small amount of sample vapor (upper left) is admitted through a pinhole into the evacuated interior of the instrument. An intersecting stream of electrons, emitted by a glowing filament and accelerated by application of a moderate voltage (perhaps 70 volts), collides with the molecules of the sample and ionizes and fragments them. Neutral and negatively charged fragments drift around aimlessly or are repelled by the negatively charged plates and are eventually pumped out of the system. Positively charged fragments are accelerated toward the negative plates and pass through a slit in these plates— just as Goldstein's canal rays passed through perforations in his cathodes.

The positive-ion beam then passes between the poles of a magnet and is bent by the magnetic field. The amount of bending depends on the ion-accelerating voltage (perhaps 2000 volts), the strength of the electromagnet, and the charge and mass of the ion. At a given voltage setting and magnetic field strength, there will be as many curved paths as there are fragments of different masses. The combined action of the electric and magnetic fields thus serves to sort the fragments according to their masses.

All that has to be done now is to count the number of ions following each path. Counting is usually done by neutralizing the positive fragments at a "collector" station. As each positive particle arrives it picks up an electron; the measured electron flow (current) indicates the arrival rate of ions at the collector. The measured electric current is directly proportional to the abundance of fragment-ions of a particular mass. In practice, there is usually only one ion-collector, monitoring only one path. By varying the applied accelerating voltage or the magnetic field, particles of different mass can be made to arrive at the collector. Thus, we end up knowing the mass numbers and abundances of the various fragments of a given sample.

Even if the substance has not previously been catalogued, mass spectrometry may yield quite a bit of information. By carefully examining the fragments, we can usually deduce much about the structure of the original molecules. Again, take coffeepot fragments. The presence of pieces of spouts and handles indicates that we are dealing with shattered coffeepots rather than cocktail glasses. With some luck we may be able to picture the original coffeepot, even when we are unable to identify all the pieces—very similar to an archeologist's picturing an extinct animal from a collection of its bones. Likewise, an experienced mass spectroscopist can often obtain a good idea of a molecule's structure from its fragments.

From the fragments we can also deduce where the weak points of a molecule lie—which bonds are most likely to break. More than that, we often can tell *how* weak the bonds are, by determining how much electron energy is required to break the bonds.

The potentialities of mass spectrometry were not fully exploited for quite some time. The first commercial instrument was built only in 1942, as a tool for quantitative analysis in the refinement of petroleum. The need for high-grade aviation fuels during World War II turned mass spectrometry into a practical analytical tool. It was not until the early 1950s however, that mass spectrometry was applied extensively to structural determinations.

The experimentally difficult part of mass spectrometry lies in obtaining reproducible patterns. We pointed out that fragmentation patterns may be reproducible, *provided* we hit the coffeepots or molecules the same way each time. This requirement is partly responsible for the complexity and expense of commercial instruments. Today's mass spectrometers vary in size from a few inches in length—useful for analyzing gases in interplanetary space—to a roomful of equipment. Prices of commercial instruments range from a few thousand dollars to more than a hundred-thousand dollars. Common instruments in chemical research laboratories cost about $50,000.

Problem 27-1. The observed mass numbers in the mass spectrum of ethane, C_2H_6, are 30, 29, 28, 27, 26, 25, 24, 15, 14, 13, 12, 1. Recall that the structure of ethane is

$$
\begin{array}{ccc}
 & H & H \\
 & | & | \\
H- & C- & C-H \\
 & | & | \\
 & H & H
\end{array}
$$

Indicate the fragments represented by each mass number (similar to Table 27-1).

Problem 27-2. The structure of ethylene, C_2H_4, is

$$\begin{matrix} H \\ \diagdown \\ & C = C \\ \diagup & & \diagdown \\ H & & H \end{matrix} \begin{matrix} H \\ \diagup \\ \\ \\ \end{matrix}$$

(a) List all fragments and their mass numbers (including the "parent") that can be produced by the rupture of one or more bonds.

(b) Which peaks may appear for ethane but not for ethylene?

(c) Which peaks may appear for ethylene but not for ethane, if any?

(d) A sample is known to be either ethane or ethylene. Merely on the basis of *presence* of peaks, could you identify ethane?

(e) Could you identify ethylene?

MOLECULAR SPECTROSCOPY

In Chapters 14 and 15 we introduced atomic spectroscopy and discussed its two principal areas of application. Chemists used it for chemical analysis and for the identification of new elements, such as the noble gases; physicists used it to gain insight into the structure of atoms.

We can also use spectroscopy to learn about molecules. Molecular spectroscopy is a particularly versatile tool. In chemical analysis, molecular spectra help us find out what substances are present in an unknown sample, how much of each, and also what substances are not present. From the spectrum of a new substance we can determine its functional groups and how they are arranged, the distances and angles between atoms, the strengths of bonds, how much energy it takes to rupture them, and the amount of heat we might obtain from burning the substance (for instance, as a rocket fuel).

Moreover, we are not limited to the examination of substances we can grasp with our hands. The universe itself is our laboratory. Astronomers, through spectral analysis, study compositions of heavenly bodies. Recall the discovery of helium in the sun in 1868, almost thirty years before it was found on earth. Spectroscopes carried aloft in balloons or rockets, or radioing back information from a journey millions of miles into space, have given us much information about our own atmosphere, as well as the atmospheres of our planetary neighbors and the possible existence of life thereon. Beyond the chemical aspects, astronomical spectroscopy yields information about the relative motions, distances, and ages of the stars, even about the size and age of the universe itself.

Components of a Spectroscope

The basic components of modern spectroscopes are the same as those used by Kirchhoff and Bunsen in the middle of the last century (Chap-

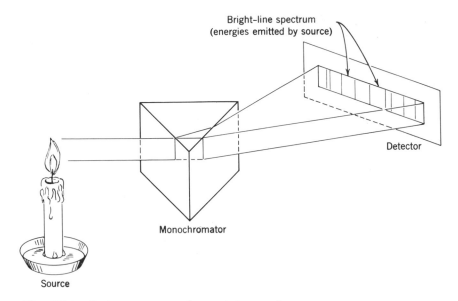

Fig. 27-4.   Basic components of an emission spectroscope.

ter 14)—namely, a radiation source, a monochromator, and a detector (Fig. 27-4).

In emission spectroscopy, the radiation source is the radiating sample. It may be a pinch of material heated by a Bunsen burner, or a fluorescent gas in a Crookes tube, or an incandescent astronomical body millions of light years away.

The monochromator disperses the polychromatic light into its constituent "colors," which may be visible colors or invisible ultraviolet, infrared, and other radiations. The dispersing element may be a simple prism, as in Newton's experiments, or various types of diffraction gratings (transmission or reflection), or combinations of several of these.

The detector may be very complex, such as the human eye, or as simple as a photographic plate, or perhaps some sort of photoelectronic detector that can amplify a signal. The recorded end product is usually either a series of lines on a photographic plate or a curve traced on a piece of chart paper.

In emission spectroscopy, the sample first is "excited" (raised to a higher energy level) by absorption of radiant, heat, or kinetic energy (for example through bombardment by electrons in a Crookes tube). When the atoms or molecules return to lower energy states they may radiate the excess energy. If the transitions are between discrete energy levels, such as for the hydrogen atom, the radiated energy will be packaged in discrete units—namely, photons associated with definite frequencies and wavelengths ($\nu = E/h$). These photons are admitted through a slit into the monochromator, where the photons of various frequencies are sorted out and sent on to the detector.

Absorption spectroscopy is quite similar to emission spectroscopy. However, instead of looking at the radiation *emitted* by a substance that has dropped from a higher energy level to a lower one, we determine what radiation has been *absorbed* by a substance in going from a lower energy level to a higher one. The processes of emission and absorption are really two sides of the same coin—transitions between energy levels. In absorption spectroscopy we irradiate the sample with a whole range of frequencies and see which are absorbed by the sample—in other words, which frequencies or energies correspond to an energy-level difference of that atom or molecule. The essential distinction between an emission spectroscope and an absorption spectroscope lies in the light source. In the former, the source of the radiation is the excited sample itself. In the latter the light source is distinct from the sample; it may be an incandescent bulb, sending out all frequencies of visible light, or perhaps a dull-red glowing piece of ceramic, yielding a range of infrared frequencies. This radiation is allowed to pass through a sample before entering the monochromator (Fig. 27-5).

The choice of obtaining an emission versus an absorption spectrum may be merely a matter of convenience or a question of available instrumentation, or it may be dictated by external circumstances. We can, for example, obtain both types of solar spectra. Usually we obtain an absorption (dark-line) spectrum. The surface of the sun is the source of con-

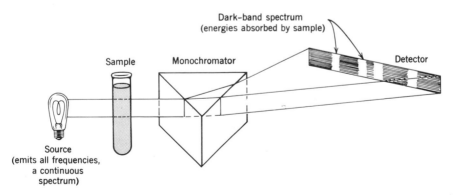

Fig. 27-5. Basic components of an absorption spectroscope.

tinuous (all frequencies) radiation, which then is selectively absorbed in passing through the gases of the sun's atmosphere. After a quick eight-minute trip of 93 million miles the radiation enters our earthbound monochromators. An emission (bright-line) spectrum of the sun's atmosphere can be obtained when the body of the sun is blacked out by the moon during a solar eclipse. It was from such bright-line spectra that Janssen and Lockyer discovered helium in 1868 (p. 207).

Spectral Regions and Energy-Transition Processes

From the energies of observed photons we try to make deductions about the energy levels in the atoms and molecules that emitted these photons. From these energy levels, in turn, we try to make further deductions about the structures of the atoms and molecules. Niels Bohr used this approach in devising his atom model from observed atomic spectra (Chapters 16 and 18).

Different processes in atoms and molecules result in the emission and absorption of various kinds of radiation.

In connection with Bohr's atom model, we discussed *electronic transitions*, especially in the hydrogen atom. In general, transitions involving loosely held outer electrons of atoms and molecules are associated with *visible* and *ultraviolet* radiation, while transition of tightly held inner electrons involve energies in the *far-ultraviolet* and *X-ray* regions of the spectrum. (Recall Moseley's X-ray spectrum, 1913, obtained by bombarding metals with high-speed electrons and resulting in the discovery of atomic numbers—Chapter 11). Absorption spectroscopy using ultraviolet and visible light was one of the earliest physical methods employed in the examination of molecular structure as well. It yields information about the electronic structures of molecules and is especially useful for the study of multiple bonds.

The atoms of all molecules, even at absolute zero temperature, are perpetually in a state of rapid oscillation. Like the particle in a box and in accordance with the Uncertainty Principle, atoms can never stand still (Fig. 27-6). Vibrational frequencies depend on bond strengths and the masses of the vibrating atoms—the stronger the bond and the lighter the

Fig. 27-6. A vibrating carbon monoxide molecule—6.40×10^{13} oscillations per second.

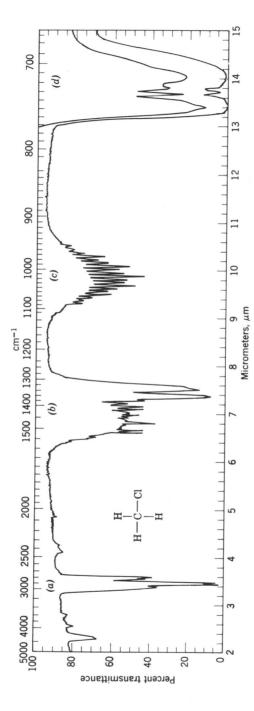

Fig. 27-7. Infrared spectrum of CH₃Cl. The upper abscissa scale reads wave numbers in cm⁻¹; the lower abscissa scale reads wave lengths in micrometers. The ordinate scale indicates percent transmittance of infrared radiation by the sample: 100% transmittance (no absorption) is at the top, 0% transmittance (complete absorption, therefore "dark bands") is at the bottom. Four absorption regions are clearly visible, caused by the following molecular vibrations: a–stretching of the C–H bonds; b–deformations of the CH₃ bonds; c–rocking of the CH₃ group as a whole; d–C–Cl stretching.

atoms, the higher the vibrational frequency. Typical vibrational frequencies are on the order of 10^{13} to 10^{14} oscillations per second. Consequently, transitions between *vibrational energy levels* result in spectral bands in the *middle-infrared* region (Fig. 27-7). From the frequencies of these bands we can deduce the vibrational frequencies within a molecule and can then calculate bond strengths and determine the structure of the molecule—e.g., is it linear like $BeCl_2$ or angular like H_2O; planar like BF_3 or pyramidal like NH_3; is it NNO or NON?

Example 27-1. We describe the elastic strength of a spring in terms of its force constant. The *force constant* of a spring is defined as the amount of force required to stretch or compress the spring a given distance. In the English system we often express force constants in pounds per inch. For molecules we usually use millidynes per angstrom. The force constant, k, of a diatomic molecule can be calculated from its vibrational frequency, ν, by the formula

$$k = 4\pi^2\mu\nu^2$$

where μ is called the "reduced mass" of the system and is calculated from the two atomic masses by the formula

$$\mu = \frac{m_1m_2}{m_1 + m_2}$$

Note that reduced mass has the dimension of mass. We previously mentioned reduced mass in Chapter 18, "Extension of Bohr's Theory to Other One-Electron Atoms."
(a) Calculate the reduced mass of the CO molecule in atomic mass units (amu).
(b) Express the reduced mass in grams (1 amu $= 1.67 \times 10^{-24}$ g).
(c) The vibrational frequency of the carbon monoxide molecule is 6.40×10^{13} cycles/sec. Calculate the force constant of the CO molecule.
(d) The force constant of the CO molecule is more than three times that of the HCl molecule (4.8 mdyn/A). Draw an electron-dot formula to indicate the bonding in the CO molecule, to explain why the CO bond is so much stronger than the HCl bond. In drawing the electron-dot formula, assume validity of the octet rule.

Solution.

(a) $\mu = \dfrac{m_C m_O}{m_C + m_O} = \dfrac{12 \times 16}{12 + 16} = \dfrac{192}{28} = 6.86$ amu

(b) 1.15×10^{-23} g.
(c) $k = 4\pi^2\mu\nu^2$
$= 4\pi^2(1.15 \times 10^{-23} \text{ g})(6.40 \times 10^{13} \text{ sec}^{-1})^2$
$= 17 \times 10^5 \text{ g-sec}^{-2} = 17 \times 10^5 \text{ dyne/cm}$
$= 17 \text{ mdyn/A}$
(Recall from $F = ma$ that a dyne is a g-cm-sec^{-2}).
(d) Carbon has four outer-shell electrons and oxygen has six, making a total of ten—$:C:::O:$. According to the octet rule, CO has a triple bond. This is confirmed by the force-constant calculation, based on the observed infrared vibrational frequency.

Molecules as a whole usually possess *rotational energy*. These rotational energies, too, are quantized; in other words, a particular molecule can have only certain fixed rotational frequencies (on the order of 10^{12} rotations per second). Transitions between rotational energy levels involve relatively small amounts of energy and are observed in the *far-infrared* and *microwave* regions of the spectrum. Rotational spectra yield information about molecular geometry and bond lengths, especially of small molecules.

Transitions involving the flipping-over in a magnetic field of *spinning nuclei* and *spinning electrons* result in the emission and absorption of low-energy radiation in the *radio* region of the electromagnetic spectrum (low frequencies, long wavelengths). These transitions are studied through the techniques of nuclear magnetic resonance spectroscopy (NMR) and electron spin resonance spectroscopy (ESR), also known as electron paramagnetic resonance spectroscopy (EPR). These types of molecular spectroscopy also yield information about molecular structure.

Fig. 27-8 correlates these various energy-transition mechanisms with an overall view of the electromagnetic spectrum.

The History of Infrared Spectroscopy

The chronology of infrared spectroscopy is outlined here because it illustrates again how developments in science are interrelated and how much these developments depend on other factors, such as political happenings.

1. Newton in 1666 uses prisms and a round hole to disperse sunlight into a spectrum (Chapter 12).
2. William Herschel in 1800, discovers infrared radiation while attempting to measure the heating power of the various colors of sunlight (Chapter 13).
3. Wollaston in 1802 and Fraunhofer in 1814 use slits instead of round

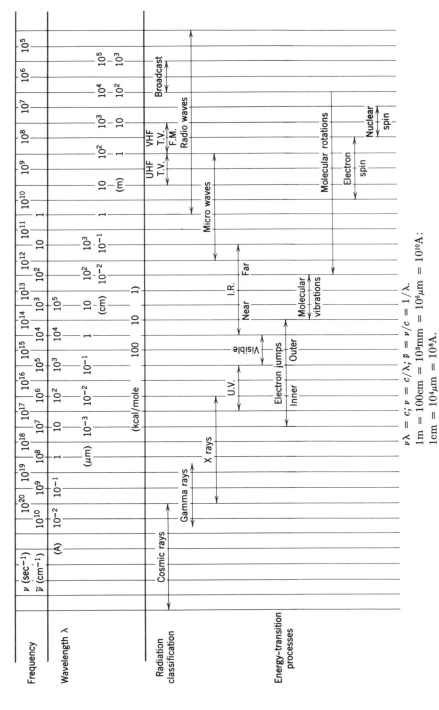

Fig. 27-8. Energy-transition processes in relation to the regions of the electromagnetic spectrum.

$\nu\lambda = c;\ \nu = c/\lambda;\ \bar{\nu} = \nu/c = 1/\lambda.$
$1\,m = 100\,cm = 10^3\,mm = 10^6\,\mu m = 10^{10}\,\text{Å}:$
$1\,cm = 10^4\,\mu m = 10^8\,\text{Å}.$

holes and discover dark lines ("Fraunhofer lines") in the visible solar spectrum (Chapter 14).

4. Fraunhofer in 1820 invents the transmission grating for diffraction of light (Chapter 12).

5. John F. W. Herschel, son of William Herschel, in 1840 records the first infrared spectrum.

6. Kirchhoff and Bunsen in 1859 build the first practical spectroscope and use it for chemical analysis. They discover several new elements and, via the Fraunhofer lines, demonstrate the presence of many known elements in the sun (Chapter 14). Their work is based on the discovery that the spectrum of each element is characterized by a definite set of lines. Identification of elements thus becomes a task of matching observed spectral lines with previously catalogued lines. This method is known as the *fingerprint method* of spectroscopic identification.

7. Julius in 1892 relates the presence of *infrared absorption bands* to the presence of certain functional groups in molecules. (These bands in infrared spectra are analogous to Fraunhofer lines in the visible part of the spectrum.) Specifically, Julius observed that a band at 2900 cm^{-1} in the spectrum of a compound usually indicates the presence of a methyl group ($-CH_3$) in that compound. In 1895 Aschkinass and Ransohoff assign a similar band at 3300 cm^{-1} to the hydroxyl group ($-OH$). Spectral bands that are associated with particular functional groups are known as *characteristic bands,* or *characteristic frequencies* (*see* Fig. 27-7). Matching observed bands in the infrared spectrum of an unknown substance with previously catalogued frequencies to determine the presence of functional groups is an extension of the fingerprint method. To be widely useful, this method requires the prior cataloguing of a large number of spectra of known compounds. In the visible region, this was done notably by Ångström.

8. Between 1905 and 1908 Coblentz examines the infrared spectra of more than one hundred compounds and issues the first major catalogue of infrared spectra, also listing several new characteristic frequencies.

9. Advances in instrumentation: (a) development of heat detectors for infrared radiation in the latter half of the nineteenth century; (b) achievement of high resolution—that is, the ability to separate two closely spaced spectral lines and thus to detect fine structure in the spectrum. (Recall that high resolution enabled Paschen to observe fine structure in the spectrum of atomic hydrogen; as a result, Sommerfeld in 1915 introduced elliptical orbits to explain this fine structure—*see* p. 251).

10. The discovery in 1913 of fine structure in the infrared spectrum of water vapor, subsequently ascribed to rotation of the molecule.

11. Bjerrum in 1914 introduces the concept of the vibrating molecule—atoms held together by elastic springs—to account for observed infrared spectra and measured heat capacities.

12. Construction of a few infrared spectrometers for industrial chemical applications, in Germany in the 1920s and in the United States in the 1930s. Automated recording of spectra (versus laborious and time-consuming point-by-point recording) is introduced to infrared spectroscopy in 1939. (It had already been used for visible spectra by Ångström in 1895.) The first commercially sold instrument is designed about 1940. Further construction of industrial instruments is delayed by World War II.

13. Demonstration of the great utility of infrared spectroscopy as an analytical tool during World War II (a) in the preparation of high-test aviation gasoline; (b) in the synthesis of penicillin to treat battlefield casualties; (c) in the manufacture of synthetic rubber after the Allies were cut off from their supply of natural rubber through the Japanese occupation of Malaya.

14. Postwar development of reasonably priced, practical commercial instruments, using the double-beam principle and electronic instead of photographic techniques—made possible by World War II research in electronics.

15. The availability since 1956 of low-cost ($5000) spectrometers, making the infrared method available to laboratories that could not previously afford the higher-priced ($11,000-20,000) precision instruments, and also allowing the chemist to have such an instrument at his elbow for personal use.

16. Extension in the 1960s of the usable spectral range into the far-infrared region, making it possible to observe low-frequency vibrations and rotations. In early infrared spectrometers rock salt was used as material for windows and prisms because glass is not transparent to infrared radiation. At wavelengths beyond 16 micrometers (below 650 cm^{-1}—see Fig. 27-8) NaCl also becomes opaque and other materials must be used, such as KBr and CsI.

ADDITIONAL READING FOR PART FIVE

Amoore, J. E., J. W. Johnston, Jr., and M. Rubin. "The Stereochemical Theory of Odor." *Scientific American,* February 1964; Freeman Reprint No. 297.

Barrow, Gordon M. *The Structure of Molecules, An Introduction to Molecular Spectroscopy.* New York: W. A. Benjamin, 1964 (paperback). Chapters 1 and 2.

Benfey, Otto Theodor. *From Vital Force to Structural Formulas.* Boston: Houghton Mifflin, 1964 (paperback). The development of organic structural theory.

———. *The Names and Structures of Organic Compounds.* New York, John Wiley, 1966 (paperback). Programmed learning.

Bragg, Lawrence. "X-Ray Crystallography." *Scientific American,* July 1968.

Brey, Wallace S. *Physical Methods for Determining Molecular Geometry.* New York: Reinhold, 1965 (paperback).

Buswell, A. M., and W. H. Rodebush. "Water." *Scientific American,* April 1956; Freeman Reprint No. 262.

Grundy, J. *Stereochemistry,* London: Butterworth, 1964.

Mahan, Bruce. *University Chemistry.* 2nd ed. Reading, Mass.: Addison-Wesley, 1969.

Pake, George E. "Magnetic Resonance." *Scientific American,* August 1958; Freeman Reprint No. 233.

Pauling, Linus. *The Nature of the Chemical Bond.* Ithaca, N. Y.: Cornell University Press, 1960.

Ryschkewitsch, G. E. *Chemical Bonding and the Geometry of Molecules,* New York: Reinhold, 1963 (paperback).

Schwarz, J. C. P. *Physical Methods in Organic Chemistry.* San Francisco: Holden-Day, 1964.

Sebera, Donald K. *Electronic Structure and Chemical Bonding.* New York: Blaisdall Publishing Company, 1964 (paperback).

Simon, Ivan. *Infrared Radiation.* Princeton, N. J.: Van Nostrand, 1966 (paperback).

Sonnessa, Anthony J. *Introduction to Molecular Spectroscopy.* New York: Reinhold, 1966 (paperback).

Silverstein, R. M., and G. C. Bassler. *Spectrometric Identification of Organic Compounds.* 2nd ed. New York: John Wiley, 1969.

Yates, Peter. *Structure Determination.* New York, W. A. Benjamin, 1967 (paperback). Mostly on "wet chemistry."

Appendix 1.

A QUICK REFRESHER ON ARITHMETIC PROCEDURES

> *"True philosophy expounds nature to us; but she can be understood only by him who has learned the speech and symbols in which she speaks to us. This speech is mathematics, and its symbols are mathematical figures."*
>
> Galileo Galilei (1564–1642)

A. EXPONENTIAL NOTATION

Scientists deal with very large and very small numbers—for example, the numbers of atoms in a gram of hydrogen is 602,000,000,000,000,000,-000,000 and the weight of a single hydrogen atom is 0.000000000000000-0000000017 gram. These numbers can be expressed more conveniently, and as a result are better visualized, through exponential notation—6.02×10^{23} and 1.7×10^{-24} grams, respectively. The exponent in each case tells us how far to the right (positive index) or to the left (negative index) the decimal point should be moved.

Moving the decimal place to the left corresponds to division by powers powers of 10. To multiply by 10 (10^1) move the decimal one place to the right; to multiply by 100 (10^2), move it two places to the right, etc.

Example: $2.54 \times 1,000 = 2.54 \times 10^3 = 2,540$

Moving the decimal place to the left corresponds to division by powers of 10.

403

Example: $4.54 \div 100 = 4.54 \div 10^2 = 4.54 \times 10^{-2} = 0.0454$

Powers of ten are expressed as follows:

$$1 \quad = 10^0 \text{ (by definition)}$$
$$10 \quad = 10^1$$
$$100 \quad = 10^2 \ (10 \times 10)$$
$$1000 \quad = 10^3 \ (10 \times 10 \times 10)$$

etc.

$$1 \quad = 10^0$$
$$0.1 \quad = 10^{-1} \ (1 \div 10)$$
$$0.01 \quad = 10^{-2} \ (1 \div 100)$$
$$0.001 = 10^{-3} \ (1 \div 1000)$$

etc.

With exponential notation,

$$2{,}540 \text{ becomes } 2.54 \times 10^3$$
$$0.0454 \text{ becomes } 4.54 \times 10^{-2}$$

Similarly,

$$735 \text{ becomes } 7.35 \times 10^2$$
$$1{,}071 \text{ becomes } 1.071 \times 10^3$$
$$0.0036 \text{ becomes } 3.6 \times 10^{-3}$$

(Count the number of places the decimal point is moved in each example.)

The number 2,540 can be expressed equally well as 254×10 or 25.40×10^2. Conventionally, however, the form with one digit to the left of the decimal point is preferred: 2.54×10^3. The decimal point may be shifted if a compensating change is made in the exponent. For example:

$$2.54 \times 10^3 \ = 25.4 \times 10^2 \ = 254 \times 10^1, \text{ etc.}$$
$$4.54 \times 10^{-4} = 45.4 \times 10^{-5} = 454 \times 10^{-6}, \text{ etc.}$$

This procedure sometimes is useful in performing arithmetic operations, since exponential numbers cannot be added or subtracted directly unless the exponents are the same:

$$(1.2 \times 10^4) + (3.45 \times 10^5) = (1.2 \times 10^4) + (34.5 \times 10^4)$$
$$= 35.7 \times 10^4 = 3.57 \times 10^5$$

B. MULTIPLICATION AND DIVISION OF EXPONENTIAL NUMBERS

To *multiply* exponential numbers, add the exponents:

(a) $10^2 \times 10^3 = 10^{2+3} = 10^5$

$100 \times 1000 = 100,000$

(b) $10^5 \times 10^{-2} = 10^{5-2} = 10^3$

$100,000 \times 0.01 = 1,000$

To *divide*, subtract the exponents:

(a) $10^5 \div 10^3 = 10^{5-3} = 10^2$

$100,000 \div 1,000 = 100$

(b) $10^3 \div 10^{-2} = 10^{3-(-2)} = 10^{3+2} = 10^5$

$1,000 \div .01 = 1,000 \div 1/100 = 1,000 \times 100 = 100,000$

Note that *squaring* (or *cubing*) is equivalent to multiplying the number by itself a total of two (or three) times. For example:

$$(10^6)^2 = 10^6 \times 10^6 = 10^{6+6} = 10^{6 \times 2} = 10^{12} \text{ (not } 10^8)$$

$$(10^5)^3 = 10^{5 \times 3} = 10^{15}$$

If raising exponential numbers to a certain power is accomplished by multiplying the indices, then obviously taking *roots* is accomplished by dividing the indices. For example:

(a) $\sqrt{10^{12}} = 10^{12 \div 2} = 10^6$

Since $10^{12 \div 2} = 10^{12 \times 1/2}$, the root is often indicated by a fractional exponent:

$\sqrt{10^{12}} = (10^{12})^{1/2}$; or, in general,

$\sqrt{N} = N^{1/2}$

(b) $\sqrt[3]{10^{15}} = (10^{15})^{1/3} = 10^5$

In *multiplying and dividing mixed numbers,* the numbers and the exponents are multiplied separately. For example:

(a) $(2 \times 10^3) \times (4.6 \times 10^{-1}) = (2 \times 4.6) \times (10^3 \times 10^{-1})$

$$= 9.2 \times 10^2$$

(b) $(1.6 \times 10^{-24}) \times (3 \times 10^{10})^2 = (1.6 \times 3^2) \times (10^{-24} \times 10^{20})$

$$= 14.4 \times 10^{-4} = 1.44 \times 10^{-3}$$

(Imagine doing this calculation in the ordinary decimal notation!)

(c) $\sqrt{2.5 \times 10^{11}} = (2.5 \times 10^{11})^{1/2} = (25 \times 10^{10})^{1/2} =$

$$= \sqrt{25} \times (10^{10})^{1/2} = 5 \times 10^5$$

(Note the trick of expressing 2.5×10^{11} as 25×10^{10}, so that the exponent becomes divisible by 2.)

(d) $\sqrt[3]{1.25 \times 10^{14}} = (125 \times 10^{12})^{1/3} = 5 \times 10^4$

C. SIGNIFICANT FIGURES

Significant figures are those digits of a number that are judged to be reasonably trustworthy. When expressing a physical quantity, we generally retain only one doubtful digit. For example, if a chemical balance is capable of weighing to within two milligrams (± 0.002 g), we would record a typical measured weight as 10.453 g (not 10.4530). That is, the result has five significant figures, with the last digit, 3, being slightly uncertain.

Zeros used merely to indicate the position of the decimal point are not significant figures. Thus 0.00345 has only three significant figures. On the other hand, when we write 2.100, we wish to indicate that only the third decimal place is slightly uncertain; thus we have four significant figures.

When we write 4000, however, it is not apparent where the uncertainty lies. Do we have four, three, two, or only one significant figure? Hence, it is better to use exponential notation. 4.0×10^3 clearly indicates two significant figures.

In a computed result the number of significant figures depends on the propagation of the individual uncertainties. Unnecessary labor usually can be saved by properly rounding off the individual figures used in the computation. It is a general rule, however, that for values used for further computation, one extra digit should be retained.

In rounding off numbers to retain the proper number of significant figures, round off to the nearest whole number when the digit to be dropped is not 5:

$$32.64 \rightarrow 32.6$$

$$32.66 \rightarrow 32.7$$

When the last digit is 5, round off to the nearest even digit:

$$32.65 \rightarrow 32.6$$

$$32.75 \rightarrow 32.8$$

In addition and subtraction, do not carry the *final* result beyond the first column that contains a doubtful figure (be sure to line up the decimal points):

Given	Correct Rounding Off	Incorrect (rounded off too soon)
72.95	72.95	72.95
1.054	1.054	1.05
0.43421	0.434	0.43
+0.21011	+0.210	+0.21
	74.648 $\rightarrow$ 74.65	74.64

Suppose, for example, that your weight is approximately 140 pounds. If you put on a 2-ounce (1/8, or 0.125-pound) wristwatch, you would certainly not say that your weight is now 140.125 pounds.

$$140 + 0.125 \rightarrow 140 + 0.1 = 140.1 \rightarrow 140$$

In multiplication and division the *relative* uncertainty is of importance. The uncertainty in the product is about the same as the uncertainty in the least precise figure entering the calculation:

$$142.7 \times 0.081 = 11.5587 \rightarrow 11.6$$

approximate relative uncertainty $\left\{ \begin{array}{} \uparrow \quad \uparrow \\ 1/1400 \quad 1/80 \end{array} \right.$ $\begin{array}{} \uparrow \\ 1/120 \end{array}$

(An answer of 11.56 would imply an uncertainty of only one part in 1200, which would be incorrect since one of the figures entering the calculation has a much greater uncertainty, namely one part in 80.)

Averages may carry one more significant figure than the least precise contributing number:

$$\frac{13.2 + 13.3}{2} = 13.25$$

D. APPROXIMATE CALCULATIONS

Approximate calculations are a necessary adjunct to the use of a slide rule, because the slide rule does not tell us where to place the decimal point. Beyond that, approximate calculations should almost always be made as a matter of routine to catch gross errors in the more precise calculation.

The first step in solving any problem should be a logical analysis of the question asked and the data available, to determine whether the data are sufficient to answer the question. (A negative answer to this question often helps to point the scientist toward additional observations or experimentation.)

After analyzing the problem, we may decide to set up the problem in the form of one or more mathematical equations. (Some problems can be solved directly, without the use of equations.)

The next step as a rule should be an approximate calculation. Is the approximate answer a reasonable one? If not, is the error in the arithmetic or in our setting up of the problem? If the latter is the case, there certainly is no point in doing a laborious precise calculation. Various tricks may be employed for approximate calculations. For example, in a series of multiplications and divisions, it is generally useful to express the numbers in conventional exponential notation and then to group the numbers (perhaps only mentally) to make simple relationships apparent. The following problem might be done in these steps:

$$\frac{0.00127 \times 0.082 \times 760 \times 511}{737 \times 0.0532}$$

$$= \frac{(1.27 \times 10^{-3}) \times (8.2 \times 10^{-2}) \times (7.60 \times 10^2) \times (5.11 \times 10^2)}{(7.37 \times 10^2) \times (5.32 \times 10^{-2})}$$

$$= \frac{1.27 \times 8.2 \times 7.60 \times 5.11}{7.37 \times 5.32} \times \frac{10^{-3} \times 10^{-2} \times 10^2 \times 10^2}{10^2 \times 10^{-2}}$$

Now we recognize at a glance that $7.60/7.37$ is a little greater than one, while $5.11/5.32$ is a little less than one; thus these four numbers yield about one. 1.27×8.2 is approximately equal to 8.2 multiplied by one and a quarter; $8.2 + (\frac{1}{4} \times 8.2) \approx 8.2 + 2$, or about 10.2. The exponents add up to 10^{-1}. Our approximate answer therefore is 10.2×10^{-1}, or 1.02. A slide-rule calculation yields 1.032, or 1.03.[1]

Other useful tricks in the manipulation of numbers:

(a) $N \times 5 = N \times 10 \div 2 = \frac{1}{2}(10N)$ (that is, move the decimal point to right and take half).

---

[1] For most purposes in this text, approximate answers with two, perhaps three, significant figures are adequate.

(b) $N \times 25 = \frac{1}{4}(N \times 100)$.

(c) $N \times 9 = 10N - N$.

(d) $N \times 49 = N(50 - 1) = \frac{1}{2}100N - N$.

(e) $N \times 18 = 20N - 0.1$ of $20\ N$.

(f) $N \times 450 = 500N - 0.1$ of $500N$.

(g) $0.12 \sim 1/8$.

(h) $0.14 \sim 1/7$.

E. THE USE OF UNITS IN CALCULATIONS AND THE CONVERSION OF UNITS

Most quantities in scientific calculations are the result of measurements and are expressed in some sort of unit—inches, feet, centimeters, seconds, kilograms, calories, miles per hour, foot-pounds, etc. It is generally advisable to carry the units through the entire calculational process.

If our answer has inappropriate units, this is a good indication that we have set up the problem improperly. *If*, for example, the question calls for volume and our answer is in inches, we have good reason to be suspicious. Perhaps we divided two quantities instead of multiplying them.

Often the units used give a hint of the formula needed in a calculation. To illustrate with a trivial example: Given a triangle whose area is 150 cm² and whose base is 30 cm, calculate its height. Everyone of course knows the formula for the area of a triangle, but let us assume, for purpose of illustration, that we do not. Let's try adding the two numbers: But we can't add square centimeters to centimeters. Too bad! This also kills the idea of subtracting. What about multiplication? But cm² × cm yield cm³, which represents volume, not length. What is left? Division! cm²/cm is cm, which is what we need. The units of the problem have correctly indicated that the problem calls for division. Caution: This type of "dimensional analysis" does not tell us that the factor ½ should be used—area of triangle = ½ base × height.[2]

Dimensional analysis is particularly useful in the derivation of conversion formulas. For example, how do we convert miles per hour to feet

---

[2] Dimensional analysis is not recommended as a substitute for studying, but if applied properly it can be useful. Dimensional considerations helped Bohr to develop his quantum condition (1913): The angular momentum of circling electrons is quantized in units of h/2π, or $mvr = nh/2\pi$. His reasoning was based in part on his recognition that Planck's constant h has the dimensions of angular momentum. On the basis of this formula, Bohr then devised his model of the hydrogen atom with its discrete orbits. Dimensional analysis, as in our triangle problem, did not tell Bohr about the constant 2π; that had to be obtained from the physics of the problem.

per second? It is quite simple, if we proceed systematically. Multiplying both the numerator and the denominator of a fraction by the same quantity leaves its value unchanged. One mile equals 5280 feet. Hence, to convert 1 mile/hour to n feet/second, we multiply the denominator by 1 mile (this will cancel the miles) and the numerator by 5280 feet. (Multiplying by 5280 feet/1 mile leaves the value of the original fraction unchanged):

$$1 \, \frac{\text{mile}}{\text{hour}} = \left(\frac{1 \, \cancel{\text{mile}}}{\text{hour}} \right)\left(\frac{5280 \text{ feet}}{\cancel{\text{mile}}} \right) = 5280 \, \frac{\text{feet}}{\text{hour}}$$

Then, since 1 hour equals 3600 seconds, we multiply the numerator by 1 hour (to cancel hours) and the denominator by 3600 seconds:

$$\frac{5280 \text{ feet}}{\text{hour}} = \left(\frac{5280 \text{ feet}}{\cancel{\text{hour}}} \right)\left(\frac{\cancel{\text{hour}}}{3600 \text{ seconds}} \right)$$

$$= \left(\frac{5280}{3600} \right)\left(\frac{\text{feet}}{\text{second}} \right) = 1.47 \, \frac{\text{feet}}{\text{second}}$$

(We could have done all this in one sequence, of course.)

One mph, then, is slightly less than 1.5 feet per second. On this basis, 60 mph is a little less than 90 feet per second. An exact calculation yields:

$$60 \text{ mph} = 60 \, \frac{\cancel{\text{miles}}}{\cancel{\text{hour}}} \, \frac{5280 \text{ feet}}{\cancel{\text{mile}}} \, \frac{\cancel{\text{hour}}}{60 \, \cancel{\text{minutes}}} \, \frac{\cancel{\text{minutes}}}{60 \text{ seconds}}$$

We can cancel the 60s before we multiply. (This is the advantage of setting up all members of an expression in one line *before* doing any actual calculating.)

$$60 \text{ mph} = \frac{5280}{60} \text{ feet/second} = \frac{528}{6} \text{ feet/second}$$

$$= 88 \text{ feet/second (exactly)}$$

Problem. Using the foregoing approach, convert 1 mph to meters per minute. (Is your answer reasonable? To check yourself, mentally convert 1.5 ft/sec to meters/minutes, by approximating 3 feet/meter.)

Appendix 2

THE METRIC SYSTEM

LENGTH

The meter was defined in France during the French Revolution (ca. 1790) as one ten-millionth of the distance from the North Pole to the Equator. (Lavoisier was one of the members of the commission on metric units.) On this basis, a standard meter was constructed—the distance between two scratches on a platinum bar. The original bar is kept in a vault in Sèvres, near Paris; a copy is at the National Bureau of Standards in Washington, D. C. Almost all countries of the world now use the metric system, with the notable exception of Great Britain and the United States, although Britain is in the process of adopting it.

In the United States there is no standard yard; instead, the yard is defined as 3600/3937 of the standard meter. That is, 1 m = 39.37 inches = 1.1 yard; 1 yard = 0.9144 m (exactly); 1 inch = 2.54 cm (exactly).

After the adoption of the standard meter it was found that the earth's quadrant is approximately 0.2% longer than the original 10 million meters assigned to it; but this is of no consequence in view of the existence of the standard-meter bar. Recent developments in the techniques of measurement, however, have necessitated a more precise definition of the meter. It has now been redefined by international agreement (1960) as a multiple of the wavelength of the orange-red spectral line at 6056 A emitted under specified conditions by the gas krypton-86.

To convert units in the English system is a complex task. One mile equals, for example, 5280 feet or 1728 yards; 1 yard equals 3 feet; 1 foot equals 12 inches. (How many inches to a mile?) In the metric system, on the other hand, conversion factors are simply powers of ten: One

411

meter (m) equals 10 decimeters (dm) or 100 centimeters (cm) or 1000 millimeters (mm); 1000 meters equal 1 kilometer (km), etc.

Table A2-1 summarizes the prefixes of the metric system. Table A2-2 gives some conversion factors between metric and English units of measurement.

VOLUME

In the metric system most other units (except those of time and temperature) are derived from the meter. The units of volume are the cubic centimeter (cc) and the liter (l). The liter is equal to 1000 cc or 1000 milliliters (ml); it is just a little larger than the quart (*see* Table A2-2).[1]

MASS

Mass units are defined with the help of water. A gram is the mass of one cubic centimeter of water (at a temperature of 4°C). Thus, a liter of water weighs exactly 1 kilogram. A U.S. one-cent coin weighs approximately 3 g; a five-cent piece weighs 5 g.

Again, there is no standard pound. The pound is simply defined in terms of the standard kilogram kept at the Bureau of Standards in Washington: 1 pound = 0.45359237 kg (exactly). For other conversions *see* Table A2-2. A useful fact to remember is that a pint of water weighs nearly a pound.

TIME

The sources of our units of time can be traced back to prehistoric times. Until recently, the second was defined as 1/86,400 of a day, the day being the time required for the earth to complete one rotation on its own axis. For the sake of greater precision, the second was redefined in 1956 as a fractional part of a year, the year being the time required for the earth to complete a revolution around the sun.

In 1967 the second was defined by international agreement in terms of an "atomic clock." It is 9,192,631,770 times the duration of the radiation pulse emitted when an atom of cesium-133 makes a transition between two specified energy levels. The new second was chosen to be identical

---

[1] Originally there was a slight discrepancy between the milliliter and the cubic centimeter. Redefinition of the liter in 1964 as exactly one cubic decimeter has eliminated that discrepancy.

with the second based on the earth's rotation. The major advantage of the new second is that it can be obtained accurately from commercially available cesium "atomic clocks."[2]

TEMPERATURE

Our present temperature scales also predate the metric system. Galileo invented an air thermometer in 1597. The first mercury thermometer was used in 1643. About 1714 Fahrenheit proposed the scale now named after him; this was followed in 1730 by the Réaumur scale and in 1742 by the Celsius scale (known also as the Centrigrade scale). The latter is the scale most commonly used today, especially in scientific work.

To establish a temperature scale we must either define the scale interval (i.e., the size of the degree) and also a fixed point from which to begin measuring; or else we must define two fixed points and divide the interval between them into an arbitrary number of degrees. Both Fahrenheit and Celsius chose the latter approach, and both used the freezing and boiling points of water as their fixed points. Fahrenheit arbitrarily chose 180 scale divisions in the interval 32–212 degrees.[3] Celsius chose larger divisions, using only 100 divisions in the interval 0–100 degrees. In other words, the freezing point of pure water is *by definition* 0°C or 32°F; the boiling point is *by definition* 100°C or 212°F. You now have enough information to work out the conversion formulas between the two scales, shown in Table II-2.

For certain scientific work an "absolute" scale is needed. Zero degrees on all absolute scales is the lowest temperature that is theoretically attainable; at this temperature gases would shrink to nothing (if they did not liquify or freeze, and if we could actually reach the temperature of absolute zero). The absolute scale most commonly used is the Kelvin scale, named in honor of Lord Kelvin, who first proposed an absolute scale in 1848. The scale intervals on the Kelvin scale—simply called kelvin, K—are the same as on the Celsius scale. The freezing point of water is at 273.15 K, the boiling point at 373.15 K. The accompanying figure compares the Fahrenheit, Celsius, and Kelvin scales.

[2] For additional reading, *see* H. Lyon, "Atomic Clocks," *Scientific American*, February 1957; available as offprint No. 225 from W. H. Freeman and Company, San Francisco, California.

[3] At first, Fahrenheit had used the lowest temperature obtainable by an ice-ammonia mixture as his zero, and the human body temperature as 96°F (a convenient 8×12). On the modern Fahrenheit scale, our body temperature is 98.6°F. The discrepancy is perhaps due to nonlinearity of Fahrenheit's original scale.

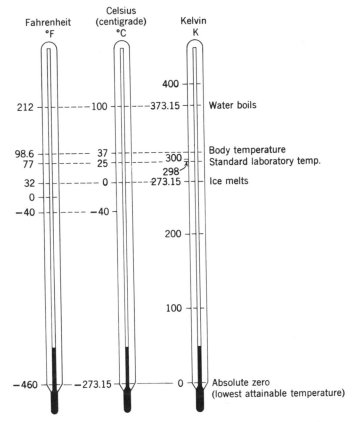

Comparison of the Fahrenheit, Celsius, and Kelvin scales of temperature.

ENERGY

We can now define energy units. In one scheme heat-energy units are defined as the energy needed to warm a certain amount of water through an agreed-upon temperature interval: The calorie is the heat energy required to warm up 1 gram of water 1 degree, from 14.5°C to 15.5°C. (The calorie used in dieting and food evaluation is actually equal to 1000 calories, that is, a kilocalorie, also known as a "large calorie.") The British thermal unit (Btu) is the heat energy required to raise the temperature of 1 pound of water 1 degree Fahrenheit.

Mechanical energy, or work, is defined in terms of forces and distances, yielding such units as the foot-pound, the erg, and the joule (*see* Chapter 6). In 1845 James Joule demonstrated that mechanical energy can be converted to heat energy (and vice versa) and measured the amount of heat energy produced by a given amount of mechanical energy (p. 101).

Still other energy units, such as the electron volt (ev), are based on electrical measurements. Conversion factors for various energy units are included in Table A2-2.

For convenience, Table A2-3 lists some miscellaneous constants discussed in various chapters of the text.

Table A2-1. The Metric-System Prefixes[a]

Factor	Prefix	Symbol	Examples
10^{12}	tera-	T	
10^{9}	giga-	G	
→10^{6}	mega-	M	1 megaton = 10^6 tons
→10^{3}	kilo-	k	1 km = 10^3 m
10^{2}	hecto-	h	
10^{1}	deka-	da	
10^{-1}	deci-	d	
→10^{-2}	centi-	c	1 cm = 10^{-2} m (1 m = 100 cm)
→10^{-3}	milli-	m	1 mm = 10^{-3} m (1 m = 1000 mm)
→10^{-6}	micro-	μ	1 μm = 10^{-6} m (1 m = 10^6 μm)
10^{-9}	nano-	n	
10^{-12}	pico-	p	

[a] The most commonly used prefixes are indicated by arrows.

Table A2-2. Conversion Factors

Length

1 kilometer (km) = 1,000 meters (m) = 0.621 mile $\approx$ 3/5 mile
1 meter (m) = 100 centimeters (cm) = 1,000 millimeters (mm) = 1.1 yards
    = 3.3 feet = 39.37 inches
1 centimeter (cm) = 10 millimeters (mm) = 10^8 angstroms (A) = 0.394 inch
1 angstrom (A) = 10^{-8} centimeter (cm) = 3.94×10^{-9} inch
1 mile = 5280 feet = 1728 yards = 1.609 km
1 yard = 0.9144 m (exactly)
1 foot = 12 inches = 30.48 cm (exactly)
1 inch = 2.54 cm (exactly)

Volume

1 liter (l) = 1,000 milliliters (ml) = 1,000 cubic centimeters (cc)
    = 1.06 quarts (U.S.)
1 milliliter (ml) = 1 cubic centimeter (cc) (exactly)
1 gallon (U.S.) = 3.8 liters
1 quart (U.S.) = 0.95 liter

Table A2-2. (continued)

1 fluid ounce (U.S.) = 29.57 cc
1 cubic foot = 28.32 liters

Mass (Weight)[a]

1 kilogram (kg) = 1,000 grams (g) = 2.20 pounds
1 gram (g) = 0.035 ounce, avoirdupois
1 pound = 454 grams
1 ounce, avoirdupois = 28.35 grams

Force[b]

1 newton = 1 × 10⁵ dynes = 0.225 pound
1 pound = 4.45 newtons = 4.45 × 10⁵ dynes

Temperature

°C = 5/9(°F − 32) = K − 273.15
°F = 9/5 °C + 32
K = °C + 273.15

Energy

1 calorie = 3.97×10^{-3} Btu = 4.18 joules = 3.09 foot-pounds = 2.61×10^{19} *ev*
1 British thermal unit (Btu) = 252 calories = 778 foot-pounds = 1,055 joules
1 joule = 10^7 ergs = 0.74 foot-pound = 0.239 calorie
1 erg = 10^{-7} joule = 6.24×10^{11} *ev*
1 electron volt (*ev*) = 1.60×10^{-12} erg = 3.83×10^{-20} cal
1 *ev* per molecule = 23 kcal per mole

Electric Charge[c]

1 coulomb = 2.998×10^9 statcoulombs (the cgs unit of charge, also called esu)
          = 6.28×10^{18} electron charges
1 faraday = 96,500 coulombs = 6×10^{23} electron charges

[a] The distinction between mass and weight is discussed in Chapter 6.
[b] Defined in Chapter 6.
[c] Defined in Chapter 7.

Table A2-3. Miscellaneous Constants

speed of light, c	3.00×10^{-10} cm/sec
charge of the electron, e	1.60×10^{-19} coulomb
	4.80×10^{-10} statcoulomb
mass of the electron, m	9.11×10^{-28} gram
	0.00055 atomic mass units (amu)
atomic mass unit, amu	1.67×10^{-24} gram
Avogadro's number	6.02×10^{23}
Planck's constant, h	6.63×10^{-27} erg-sec

Appendix 3

NOBEL PRIZE WINNERS IN CHEMISTRY

1901	Jacobus Van't Hoff	Neth.	Laws of chemical dynamics and osmotic pressure.
1902	Emil Fischer	German	Work on sugar and purine syntheses.
1903	Svante Arrhenius	Swedish	Theory of electrolytic dissociation.
1904	Sir William Ramsay	English	Discovery of inert gaseous elements and determination of their places in the periodic system.
1905	Adolf von Baeyer	German	Work on organic dyes, hydroaromatic compounds.
1906	Henri Moissan	French	Isolation of fluorine; introduction of Moissan furnace.
1907	Eduard Buchner	German	Discovery of noncellular fermentation.
1908	Lord Rutherford	English	Investigations into the disintegration of elements and the chemistry of radioactive substances.
1909	Wilhelm Ostwald	German	Pioneer work on catalysis, chemical equilibrium and reaction velocities.
1910	Otto Wallach	German	Pioneer work in alicyclic combinations.
1911	Marie Curie	French	Discovery of radium and polonium; isolation of radium.
1912	Victor Grignard	French	Discovery of the so-called Grignard reagents.

417

	Paul Sabatier	French	Method of hydrogenating organic compounds in the presence of finely powdered metals.
1913	Alfred Werner	Swiss	Work on the linkage of atoms in molecules.
1914	Theodore Richards	U. S.	Accurate determination of the atomic weights of numerous elements.
1915	Richard Willstätter	German	Pioneer researches on plant pigments, especially chlorophyll.
1916	(No award)		
1917	(No award)		
1918	Fritz Haber	German	Synthesis of ammonia from its elements.
1919	(No award)		
1920	Walther Nernst	German	Work in thermochemistry.
1921	Frederick Soddy	English	Chemistry of radioactive substances; occurrence and nature of isotopes.
1922	Francis Aston	English	Work with mass spectrograph; whole-number rule.
1923	Fritz Pregl	Austrian	Method of microanalysis of organic substances.
1924	(No award)		
1925	Richard Zsigmondy	Austrian	Elucidation of the heterogeneous nature of colloidal solutions.
1926	Theodor Svedberg	Swedish	Work on disperse systems.
1927	Heinrich Wieland	German	Researches into the constitution of bile acids.
1928	Adolf Windaus	German	Constitution of sterols and their connection with vitamins.
1929	Sir Arthur Harden	English	Investigations on the fermentation of sugars and the enzymes acting in this connection.
	H. von Euler-Chelpin	Swedish	
1930	Hans Fischer	German	Hemin, chlorophyll research; synthesis of hemin.
1931	Karl Bosch	German	Invention and development of chemical high-pressure methods.
	Friedrich Bergius	German	
1932	Irving Langmuir	U. S.	Discoveries and investigations in surface chemistry.
1933	(No award)		
1934	Harold Urey	U. S.	Discovery of heavy hydrogen.
1935	Frédéric Joliot	French	Synthesis of new radioactive elements.
	Irène Joliot-Curie	French	

1936	Peter Debye	Neth.	Studies of dipole moments and the diffraction of X rays and electrons in gases.
1937	Sir Walter Haworth	English	Research on carbohydrates and vitamin C.
	Paul Karrer	Swiss	Research on carotenoids, flavins and vitamins.
1938	Richard Kuhn	German	Carotenoid and vitamin research (declined).
1939	Adolf Butenandt	German	Work on sexual hormones (declined).
	Leopold Ruzicka	Swiss	Work on polymethylenes and higher terpens.
1940– 1942	(No award)		
1943	George de Hevesy	Hung.	Use of isotopes as tracers in chemical research.
1944	Otto Hahn	German	Discovery of the fission of heavy nuclei.
1945	Artturi Virtanen	Fin.	Invention of fodder preservation method.
1946	James Sumner	U. S.	Discovery of enzyme crystallization.
	John Northrop	U. S.	Preparation of enzymes and virus
	Wendell Stanley	U. S.	proteins in pure form.
1947	Sir Robert Robinson	English	Investigations on alkaloids and other plant products.
1948	Arne Tiselius	Swedish	Researches on electrophoresis and adsorption analysis; researches on the serum proteins.
1949	William Giauque	U. S.	Behaviour of substances at extremely low temperatures.
1950	Otto Diels	German	Discovery and development of
	Kurt Alder	German	diene synthesis.
1951	Edwin McMillan	U. S.	Discovery of and research on trans-
	Glenn Seaborg	U. S.	uranium elements.
1952	Archer Martin	English	Method of identifying and separat-
	Richard Synge	English	ing chemical elements by chromatography.
1953	Hermann Staudinger	German	Work on macromolecules.
1954	Linus Carl Pauling	U. S.	Study of the nature of the chemical bond.
1955	Vincent Du Vigneaud	U. S.	First synthesis of a polypeptide hormone.
1956	Nikolai Semenov	Russian	Work on the kinetics of chemical
	Sir Cyril Hinshelwood	English	reactions.

1957	Sir Alexander Todd	English	Work on nucleotides and nucleotide coenzymes.
1958	Frederick Sanger	English	Determination of the structure of the insulin molecule.
1959	Jaroslav Heyrovsky	Czech.	Discovery and development of polarography.
1960	Willard Libby	U. S.	Development of radiocarbon dating.
1961	Melvin Calvin	U. S.	Study of chemical steps that take place during photosynthesis.
1962	John C. Kendrew Max F. Perutz	English English	Determination of the structure of hemoproteins.
1963	Giulio Natta Karl Ziegler	Italian German	Structure and synthesis of polymers in the field of plastics.
1964	Dorothy M. C. Hodgkin	English	Determining the structure of biochemical compounds essential in combating pernicious anemia.
1965	Robert B. Woodward	U. S.	Synthesis of sterols, chlorophyll and other substances once thought to be produced only by living things.
1966	Robert S. Mulliken	U. S.	Work concerning chemical bonds and the electronic structure of molecules by the molecular orbital method.
1967	Manfred Eigen Ronald G. W. Norrish George Porter	German English English	Studies of extremely fast chemical reactions.
1968	Lars Onsager	U. S.	Contributions to the theory of the thermodynamics of irreversible processes.
1969	Derek H. R. Barton Odd Hassel	English Norway	Work in determining actual three-dimensional shape of certain organic compounds.
1970	Luis F. Leloir	Arg.	Discovery of sugar nucleotides and their role in the biosynthesis of carbohydrates.
1971	Gerhard Herzberg	Canadian	Contribution to knowledge of electronic structure and geometry of molecules, particularly free radicals.

Appendix 4

NOBLE PRIZE WINNERS IN PHYSICS[a]

1901	Wilhelm Konrad Röntgen	German	Discovery of X-rays.
1902	Hendrik Antoon Lorentz	Dutch	Influence of magnetism on
	Pieter Zeeman	Dutch	the phenomena of radiation.
1903	Henri Becquerel	French	Discovery of the radioac-
	Pierre Curie	French	tive elements of radium
	Marie Curie	French	and polonium.
1904	Baron Rayleigh	English	Discovery of argon.
1905	Philipp Lenard	German	Research in cathode rays.
1906	Sir Joseph John Thomson	English	Conduction of electricity through gases.
1907	Albert A. Michelson	U. S.	Spectroscopic and metrological investigations.
1908	Gabriel Lippmann	French	Photographic reproduction of colors.
1909	Guglielmo Marconi	Italian	Development of wireless
	Karl Ferdinand Braun	German	telegraphy.
1910	Johannes Diderik van der Walls	Dutch	Equations of state of gases and fluids.
1911	Wilhelm Wien	German	Laws of heat radiation.
1912	Nils Gustaf Dalen	Swedish	Coast lighting.
1913	Heike Kamerlingh-Onnes	Dutch	Properties of matter at low temperatures; production of liquid helium.
1914	Max von Laue	German	Diffraction of X-rays in crystals.
1915	Sir William Henry Bragg	English	Study of crystal structure

[a] From the *Encyclopaedia Britannica*.

	William Lawrence Bragg	English— his son	by means of X-rays.
1916	(No award)		
1917	Charles Glover Barkla	English	Discovery of the characteristic Röntgen radiation of elements.
1918	Max Planck	German	Discovery of the elemental quantum.
1919	Johannes Stark	German	Discovery of the Doppler effect in canal rays and the division of spectral lines in the electric field.
1920	Charles Edouard Guillaume	Swiss	Discovery of the anomalies of nickel-steel alloys.
1921	Albert Einstein	German	Founder of theory of relativity and discoverer of law of photoelectric effect.
1922	Niels Bohr	Danish	Study of structure and radiations of atoms.
1923	Robert Andrews Millikan	U. S.	Work on elementary electric charge and the photoelectric effect.
1924	Karl Manne Siegbahn	Swedish	Discoveries in the area of X-ray spectra.
1925	James Franck Gustav Hertz	German German	Laws governing collision between electron and atom.
1926	Jean Perrin	French	Discovery of the equilibrium of sedimentation.
1927	Arthur H. Compton	U. S.	Discovery of the dispersion of X-rays.
	Charles T. R. Wilson	English	Method of rendering discernible the courses of electrically charged particles by water condensation.
1928	Sir Owen Williams Richardson	English	Discovery of the law known by his name (the dependency of the emission of electrons on temperature).
1929	Louis-Victor de Broglie	French	Wave nature of electrons.

1930	Sir Chandrasekhara Raman	Indian	Works on the diffusion of light and discovery of the effect known by his name.
1931	(No award)		
1932	Werner Heisenberg	German	Creation of quantum mechanics.
1933	Paul Adrien Maurice Dirac	English	Discovery of new fertile forms of the atomic theory.
	Erwin Schrödinger	Austrian	
1934	(No award)		
1935	James Chadwick	English	Discovery of the neutron.
1936	Victor Hess	Austrian	Discovery of cosmic radiation.
	Carl David Anderson	U. S.	Discovery of the positron.
1937	Clinton Joseph Davisson	U. S.	Discovery of diffraction of electrons by crystals.
	George P. Thomson	English	
1938	Enrico Fermi	Italian	Artificial radioactive substances.
1939	E. O. Lawrence	U. S.	Invention of the cyclotron.
1940–1942	(No award)		
1943	Otto Stern	U. S.	Detection of magnetic movements of protons.
1944	Isidor Isaac Rabi	U. S.	Studies of atom's nucleus.
1945	Wolfgang Pauli	Austrian	Exclusion principle.
1946	Percy Williams Bridgman	U. S.	High-pressure physics.
1947	Sir Edward Appleton	English	Discovery of Appleton layer.
1948	Patrick Maynard Stuart Blackett	English	Discoveries in cosmic radiation.
1949	Hideki Yukawa	Japanese	Theoretical work on meson.
1950	Cecil Frank Powell	English	Photographic method of studying atomic nuclei; discoveries about mesons.
1951	Sir John Douglas Cockcroft	English	Transmutation of atomic nuclei. by artificially accelerated atomic particles.
	Ernest Thomas Sinton Walton	Irish	
1952	Felix Bloch	U. S.	Measure of magnetic fields in atomic nuclei.
	Edward Mills Purcell	U. S.	
1953	Frits Zernike	Dutch	Introduction of phase contrast microscopy.

1954	Max Born	English[b]	Work in mathematics which enabled physicists to understand atom behavior.
	Walther Bothe	German	Analysis of cosmic radiation; "the coincidence method."
1955	Willis E. Lamb, Jr.	U. S.	Work with atomic measurements.
	Polykarp Kusch	U. S.	
1956	John Bardeen	U. S.	Invention and development of transistor.
	Walter H. Brattain	U. S.	
	William B. Shockley	U. S.[c]	
1957	Chen Ning Yang	China[d]	Overthrow of principle of conservation of parity.
	Tsung Dao Lee	China[d]	
1958	Pavel A. Cerenkov	Russian	Discovery and interpretation of Cerenkov effect.
	Ilya M. Frank	Russian	
	Igor Y. Tamm	Russian	
1959	Owen Chamberlain	U. S.	Discovery of the antiproton.
	Emilio Gino Segré	U. S.[e]	
1960	Donald A. Glaser	U. S.	Invention of bubble chamber.
1961	Robert L. Hofstadter	U. S.	Electromagnetic structure of nucleons from high-energy electron scattering.
	Rudolf L. Mössbauer	German	Discovery of recoilless resonance absorption of gamma rays in nuclei.
1962	Lev D. Landau	Russian	Contributions to the understanding of condensed states of matter (superfluidity in liquid helium).
1963	J. Hans D. Jensen	German	Development of shall model theory of the structure of atomic nuclei.
	Maria Goeppert Mayer	U. S.	
	Eugene Paul Wigner	U. S.	Principles governing mechanics and interaction of protons and neutrons in the atomic nucleus.

[b] Born in Germany; naturalized British citizen.
[c] Born in England; naturalized U. S. citizen.
[d] Both have permanent U. S. resident status.
[e] Born in Italy; naturalized U. S. citizen.

1964	Charles H. Townes	U. S.	Work in quanum electronics leading to construction of instruments based on maser-laser principles.
	Nikolai G. Basov	Russian	
	Aleksandr M. Prokhorov	Russian	
1965	Julian S. Schwinger	U. S.	Basic principles of quantum electrodynamics.
	Richard P. Feynman	U. S.	
	Shin'ichirō Tomonaga	Japanese	
1966	Alfred Kastler	French	Discovery and development of optical methods for studying Hertzian resonances in atoms.
1967	Hans A. Bethe	U. S.	Discoveries concerning the energy production of stars.
1968	Luis W. Alvarez	U. S.	Work with elementary particles, including the discovery of resonance states.
1969	Murray Gell-Mann	U. S.	Discoveries concerning classification of elementary particles and their interactions.
1970	Hannes Alfvén	Swedish	Work in magnetohydronamics leading to applications in plasma physics.
	Louis Éugene Félix Néel	French	Work in antiferromagnetism and ferrimagnetism.
1971	Dennis Gabor	English[f]	Work in holography.

[f] Born in Hungary.

GLOSSARY

absorption spectrum—a spectrum containing dark lines or bands on a light background, produced when white light has been passed through a substance. The dark lines or bands represent wavelengths, or energies, absorbed by the substance.

acceleration—change of speed and/or direction (i.e., change of velocity) per unit of time. In scientific usage acceleration is a vector quantity, embodying both magnitude and direction. Thus an orbiting body traveling at constant speed is considered to be accelerating because it is constantly changing its direction.

acid—a substance that produces hydrogen ions in solution; usually characterized by sour taste, the ability to turn litmus red, etc.

actinide—one of 14 (or 15 if Ac is included) elements between actinium (no. 89) and element 104.

alkali metal—a member of the lithium family: Li, Na, K, Rb, Cs, Fr.

alkaline earth metal—a member of the beryllium family: Be, Mg, Ca, Sr, Ba, Ra.

alkane, alkene, alkyne—see *hydrocarbon*.

alpha rays (or particles)—helium nuclei emitted by atomic nuclei during radioactivity.

amalgam—an alloy of mercury with other metals.

ampere—a unit of electric current. An ampere is the unvarying current that, if present in each of two parallel conductors of infinite length and 1 meter apart in empty space, causes each conductor to experience a force of exactly 0.02 dyne per meter of length. (An ampere amounts to a current of one coulomb per second.)

amplitude—the height of a wave.

amu—abbreviation for atomic mass unit (a dimensionless quantity). One amu is equivalent to 1.66×10^{-24} g.

angstrom—a unit of length, equal to 10^{-8} cm. Abbreviation, A.

426

angular momentum (symbol *L*)—the product of a body's moment of inertia (*I*) and its angular velocity (*ω*): $L = I\omega$. Angular momentum
tends to maintain a body's rate and direction of rotation.

angular velocity—angular distance divided by time. The earth's angular
velocity, for example, is 360 degrees per day, or 15 degrees per hour.

anode—the positive electrode (in electrolysis or cathode-ray tubes).

anion—a negatively charged ion—e.g., Cl^-, $SO_4^=$.

atom—the smallest particle of an element that enters into chemical combination. At present we recognize atoms of 104 different elements.
Originally atoms were thought to be indivisible, but today we realize
that atoms contain many subatomic particles such as electrons, protons, neutrons, positrons, mesons.

atomic mass—synonymous with *atomic weight*.

atomic number—the number of protons in the nucleus of an element. Or,
the number of electrons in a neutral atom. Each element is characterized by a definite atomic number; these range from 1 (for hydrogen) to 104.

atomic weight—the relative weight of the atom of an element on a scale
on which the hydrogen atom equals 1 (more precisely, a scale on
which carbon-12 equals 12). Atomic weight is a dimensionless quantity (since it represents a ratio of two weights); we sometimes speak
of *atomic mass units* (amu). For example the atomic weight of
oxygen is 16, or 16 amu.

Avogadro's number—6.02×10^{23}. It is the number of atoms or molecules
in a mole. Precisely, it is the number of carbon atoms in 12 grams of
carbon-12.

base—a hydrogen-ion acceptor, such as a metal oxide. Bases react with
acids to form water.

beta rays or particles—a stream of high-speed electrons ejected by atomic
nuclei during radioactivity.

black body—a body that absorbs all radiation incident upon it (and re
flects none).

blackbody radiation—the type of radiation emitted by a black body or
perfectly black surface. The equivalent effect can be obtained from
a specially designed cavity (*see* Fig. 17-1).

Bohr radius—the radius of the smallest electron orbit in Bohr's model of
the hydrogen atom: 0.53×10^{-8} cm, or 0.53 A.

bond (chemical)—the attraction between two or more atoms within a
molecule.

bond energy—the energy required to rupture a bond.

Bremsstrahlung—X rays produced by the deceleration of electrons.

bright-line spectrum—an emission spectrum consisting of discrete lines.

The spectral lines represent energies emitted by atoms or molecules making transitions from higher to lower energy levels.

British thermal unit (Btu)—the amount of heat energy required to raise the temperature of 1 pound of water 1 degree Fahrenheit. 1 Btu = 252 calories

calorie—an energy unit; the heat energy required to raise the temperature of 1 gram of water 1 degree Celsius; 1 calorie = 4.18 joules.

canal ray—a stream of positive ions in cathode-ray tubes.

cathode—the negative electrode (in electrolysis or in cathode-ray tubes).

cathode ray—a stream of electrons (in a cathode-ray tube).

cation—a positively charged ion, e.g., H^+, Mg^{++}, NH_4^+.

cavity radiation—see *blackbody radiation.*

centrifugal force—the apparent outward force acting on an orbiting body.

centripetal force—the inward-directed force that must be exerted to keep an orbiting body from flying off at a tangent.

cgs—abbreviation for *centimeter-gram-second.*

combining power—the number of bonds an atom can form. Some authors use the word *valence* synonymously with combining power.

compound—a substance of characteristically constant composition that can be broken down into two or more simpler substances. Or, a combination of two or more elements in definite proportions. (But also see *nonstoichiometric compounds.*)

conductor—a substance through which electric current can flow readily.

constructive interference—see *interference.*

continuous spectrum (or radiation)—a spectrum (or radiation) containing all frequencies or wavelengths of radiation.

coulomb—a unit of charge. It is the amount of charge that flows in 1 second through a wire carrying a current of 1 ampere; 1 coulomb = 3×10^9 statcoulombs.

covalent bond—a chemical bond resulting from the sharing of electrons, usually in pairs.

current (electric)—electric charges in motion.

dark-line spectrum—an absorption spectrum consisting of dark lines superimposed on a continuous spectrum. The dark spectral lines represent energies absorbed by atoms or molecules making transitions from lower to higher energy levels.

deceleration—a slowing-down (see *acceleration*).

density—weight or mass per unit volume.

destructive interference—see *interference.*

deuterium—an isotope of hydrogen of atomic weight 2.

dew point—the temperature below which vapors tend to condense into liquids.

diffraction—the bending of light by the edges of obstacles, such as a knife edge or the edges of a hole or slit.

diffraction grating—a device for dispersing radiation into a spectrum by diffraction and interference. See *reflection grating* and *transmission grating*.

dipole moment—an effect due to separation of electric charges. Dipole-moment measurements provide valuable information about molecular structure.

dispersion—the separation of polychromatic light into its constituent colors, through refraction or diffraction and interference. (Monochromatic light cannot be dispersed.)

double bond—a bond between two atoms in which two electron pairs are shared.

doublet—two closely spaced spectral lines.

dyne—the cgs unit of force; a g-cm/sec². It is the force required to accelerate a mass of 1 gram at the rate of 1 cm/sec².

elastic collision—a collision in which the total kinetic energy is conserved (in contrast to an inelastic collision). Collisions between billiard balls, for example, are elastic collisions.

electrolysis—the decomposition of solutions and molten salts by passage of electric current.

electromagnetic radiation (or wave)—a varying electric field moving through space with an associated varying magnetic field. Examples: light, infrared and ultraviolet radiation, radio waves, X rays.

electromagnetic spectrum—the range of electromagnetic radiations. (*See* Fig. 13-4.)

electron—the smallest unit of negative electric charge. Also, the particle possessing that charge.

$$\text{mass} = 9.1 \times 10^{-28}\, \text{g} = 1/1840\, \text{amu}$$

$$\text{charge} = 1.6 \times 10^{-19}\, \text{coulomb}$$

$$= 4.8 \times 10^{-10}\, \text{statcoulomb, or esu}$$

electron-dot formula—*see* pp. 261 and 341.

electron shell—in atoms or molecules, a grouping of electrons having similar energies. The innermost shell may be occupied by two electrons; the others, respectively, by 8, 18, 32, or 54 electrons. In the quantum-mechanical picture, orbitals having the same principal quantum number (n) are said to belong to the same shell, also called *principal shell*. Within a shell are *subshells*, distinguished by different values of the angular momentum quantum numbers (l).

electron volt—an energy unit equal to 1.60×10^{-12} erg.

electrostatic—pertaining to electricity not in motion.

electrostatic unit of charge (esu)—see *statcoulomb*.

element—one of a hundred-odd substances that cannot be broken down by chemical means into two or more substances. Each element is characterized by an individual atomic number. (*See* endpaper for a listing of the known elements.)

emission spectrum—a spectrum resulting from the emission of radiation by substances. A bright-line spectrum is one type of emission spectrum. (*See* Fig. 27-4.)

empirical formula—see *formula*.

energy—the ability to do work. We distinguish various kinds of energy such as kinetic, potential, and heat energy. The cgs unit of energy is the erg.

energy level—a discrete energy state.

equation (chemical)—a combination of formulas that indicates the molecules that participate in a given chemical reaction, e.g., $2H_2 + O_2 \rightarrow 2H_2O$.

erg—the cgs unit of work, or energy; a dyne-cm. It is the work done by a force of one dyne moving through a distance of one cm.

esu—abbreviation for electrostatic unit (of charge). See *statcoulomb*.

excitation energy—the energy required to promote an atom, molecule, or other substance from its ground state to the next energy level.

exponent—the superscript n in the expression y^n, which indicates the y is to be multiplied by itself n times; e.g., $y^3 = y \times y \times y$.

family (chemical)—the group of chemically similar elements in a vertical column of the periodic table—e.g., He, Ne, Ar, etc.

faraday—a quantity of charge equal to 96,500 coulombs. A mole of electrons equals exactly one faraday of negative charge.

fermentation—the process whereby certain minute organisms convert sugar into alcohol and carbon dioxide.

fine structure—Spectral "lines" or bands that are found upon higher resolution to consist of several closely spaced lines are said to have "fine structure."

first ionization energy—see *ionization energy*.

fluorescence—the emission of light by a substance under illumination. Also see *phosphorescence*.

force—an influence that changes, or tends to change, the state of rest or motion of a body. The cgs unit of force is the dyne.

force constant—the force needed to compress or stretch a spring or bond a unit distance (expressed in units like pounds per inch, dynes per cm, and millidynes per angstrom).

formula (chemical)—the combination of element symbols that describes the composition of a compound or molecule. *Empirical formulas* indicate the relative number of atoms of each element; *molecular formulas* express the actual number of atoms in a molecule; *structural formulas* show the spatial arrangement of the atoms within a molecule. For example, the molecular formula of ethylene is C_2H_4; its empirical formula is CH_2 (see p. 362).

Fraunhofer lines—dark lines in the solar spectrum due to absorption of radiation in the sun's atmosphere.

frequency (of a wave)—the number of waves arriving at or passing a given point in a given interval of time. Symbol: ν Common unit: cycles per second (abbreviated *cps*, sometimes written as sec^{-1}; also called hertz; abbreviated Hz).

functional group—a recognizable group of atoms that occurs in many molecules and that usually confers upon them characteristic properties. Examples: $-OH$, $-NO_2$, $C=C$.

fundamental frequency—the lowest frequency of a standing wave.

gamma rays—high-energy electromagnetic radiation emitted by atomic nuclei during radioactivity. (Gamma rays have shorter wavelengths and higher energies than ordinary X rays.)

gram—a metric unit of mass, equal to the mass of 1 cubic centimeter of water at 4° C; 1 gram equals 0.035 ounce.

gram-atomic weight—the atomic weight of an element expressed in grams. The GAW of oxygen, for example, is 16 g.

gram-molecular weight—the molecular weight expressed in grams. The GMW of oxygen, for example, is 32 g.

grating—see *reflection grating* and *transmission grating*.

ground level—the lowest energy state of an atom, molecule, or other substance.

halogen—a member of the fluorine family: F, Cl, Br, I, At.

harmonic frequencies (1st, 2nd, 3rd, etc. harmonics)—the possible frequencies of a standing wave, all integral multiples of the fundamental frequency, or first harmonic.

heat—a form of energy that depends on a body's temperature, its mass, and the kind of material of which it is composed. A common unit of heat is the calorie.

heat capacity (of an element)—the product of specific heat and atomic weight. In effect, it is the amount of heat, in calories, required to raise the temperature of one gram-atomic weight of an element by one degree Celsius (*see* p. 43).

hertz (Hz)—a frequency unit, equal to one (cycle) per second.

hydrocarbon—a compound containing only carbon and hydrogen. *Satu-*

rated hydrocarbons contain only single bonds and are known as *alkanes.* *Unsaturated hydrocarbons* contain double bonds (*alkenes*) or triple bonds (*alkynes*).

impulse—the product of force and the time through which it acts, producing a change of momentum. (Impulse = $F\Delta t = \Delta p$).

Induction (electric)—the production of charge separation by a nearby electric charge.

inelastic collision—a collision in which kinetic energy is converted to other forms of energy, or vice versa. Automobile collisions usually are inelastic collisions—the automobiles come to a standstill, and their kinetic energy is converted mostly into heat.

inert gas—a member of the relatively unreactive helium family; preferred term, *noble gas.*

inertia—the innate tendency of a body to persist in its state of rest or uniform motion. (Inertia depends only on the mass of a body.)

interference—the mutual reinforcing (*constructive interference*) or weakening (*destructive interference*) of two or more beams of electromagnetic radiation. Constructive interference is due to addition of simultaneously arriving crests (or troughs); destructive interference is due to partial or complete cancellation of a crest by a simultaneously arriving trough.

ion—an atom or group of atoms carrying one or more electric charges because of loss or acquisition of electrons. Examples: H^+, Cl^-, $S^=$, Al^{+++}, NO_3^-, NH_4^+, H_2^+.

ionic bond—a chemical bond due to electrostatic attraction between ions.

ionization—the production of ions by loss or gain of electrons.

ionization energy—the minimum energy required to remove an electron from an atom or molecule in its ground state. That, actually, is the *first ionization energy.* The second ionization energy is the energy required to subsequently remove the second electron, etc.

isomers—two or more compounds having identical chemical compositions and molecular weights but different properties. Or, two or more molecules having the same number and kinds of atoms but different arrangements. (For classification of isomers (*see* p. 371).

isotopes—two or more atoms of an element that have the same atomic number (same number of protons) but different atomic weights (different numbers of neutrons). The isotopes of an element possess nearly identical chemical properties.

joule—the mks unit of work, or energy; a newton-meter. (1 joule = 10^7 ergs.)

kilogram—a metric unit of mass, equal to the mass of a cubic decimeter of water. One kilogram equals 1000 grams, or 2.20 pounds.

kinetic energy—energy of motion. It is defined as one half of the mass times the square of the velocity: ½ mv^2.

lanthanide—one of 14 (or 15 if La is included) elements between lanthanum (no. 57) and hafnium (no. 72). Also called *rare earth*.

law of nature—a descriptive generalization based on experiments or observations that summarizes some phenomenon of nature. Laws are expressions of our factual knowledge and must not be confused with theories.

light-year—the distance light travels in one year: 9.5 $\times$ 10^{12} km, or 6 $\times$ 10^{12} miles.

linear momentum—see *momentum*.

liter—a metric unit of volume, equal to a cubic decimeter (1000 cc); 1 liter equals 1.06 quarts.

mass—the quantitative measure of a body's inertia. In layman's—but not Lehmann's—language mass is the amount of matter in a body. Sometimes, but erroneously, "mass" is used as a synonym for "weight.")

melting point—the temperature at which a solid liquefies.

metal—one of about 76 elements distinguished by metallic luster, malleability, electrical conductivity, etc. Metals are found on the left side of the periodic table.

meter—the basic unit of length in the metric system. One meter equals approximately 1.1 yards (*see* Appendix 2).

mixture—a combination of two or more substances in variable proportion; each component usually retains its own properties. Examples: rocks, dirt, seawater, air, sand-and-salt.

mks—abbreviation for *meter-kilogram-second*.

molar volume—the volume occupied by a mole of gas, approximately 22.4 liters.

mole—the number of atoms (or molecules) in gram-atomic (or gram-molecular) weight of a substance; Avogadro's number of particles.

molecular formula—see *formula*.

molecular weight—the weight of a molecule relative to that of the hydrogen atom. The sum of the atomic weights of all the atoms in a molecule. The MW of H_2O, for example, is 18.

molecule—a group of atoms held together strongly enough to be considered a unit. Examples: O_2, H_2O, C_2H_5OH.

moment of inertia (symbol, I)—the product of a rotating body's mass and the square of its distance from the axis of rotation: $I = mr^2$.

momentum (symbol, p)—the product of mass and velocity: $p = mv$. Momentum tends to keep a body moving with unchanged velocity; like velocity, it is a vector quantity.

monochromatic light—light having a single wavelength or frequency; that is, one color of the spectrum.

multiple bond—a double or triple bond.

potential energy—energy due to position.

neutrino—a subatomic particle without charge or mass.

neutron—a heavy neutral particle in atomic nuclei. Its mass is approximately equal to that of the proton (1 amu), or 1840 times that of the electron.

newton—the mks unit of force; a kg-m/sec^2. It is the force required to accelerate a mass of one kilogram at the rate of 1 m/sec^2. (1 newton = 10^5 dynes. 4.45 newtons = 1 pound.)

noble gas—a member of the relatively unreactive helium family: He, Ne, Ar, Kr, Xe, Rn (formerly called *inert gases*).

noble-gas configuration—an electron configuration resembling that of the noble gases, with eight electrons in the outer shell (only two for helium).

node—the position at which a standing wave has zero amplitude (*see* Fig. 22-3).

nonstoichiometric compound—a compound that may have slightly variable composition. Nonstoichiometric compounds represent an exception to the Law of Constant Composition. (*See* p. 18.)

nucleus (atomic)—the small heavy core of an atom, containing all of the atom's positive charge (protons) and most of its mass (protons and neutrons.) Nuclear diameters are of the order of 10^{-12} cm, or 10^{-4} A. (Atomic diameters range from 1 to 5 A.)

nuclide—an atom characterized by a specific atomic number and mass. For example, boron (element 5) is composed of nuclides of masses 10 and 11 amu. *Nuclides* is often used synonymously with *isotopes*.

operational definition—a definition that prescribes, directly or indirectly, an operation, experimental procedure, or measurement.

orbital—the spatial distribution of electron density in an atom or molecule. According to the Exclusion Principle not more than two electrons can occupy an orbital. (Quantum-mechanical orbitals have replaced Bohr's concept of electron orbits.)

organic compounds—compounds containing carbon (with a few exceptions such as CO_2, $NaCO_3$, etc.).

overtones—the 1st, 2nd, 3rd, etc., overtones are synonymous with the 2nd, 3rd, 4th, etc., harmonics (see *harmonic frequencies*).

period—a horizontal row of the periodic table.

phosphorescence—the glow produced by some substances after previous exposure to light. Phosphorescence, in contrast to fluorescence, continues for a time after removal of the exciting light or other radiation.

photon—a pulse or "packet" of electromagnetic radiation. Its energy is related to the wave frequency by the formula $E = h\nu$.

Planck's constant—a proportionality constant appearing in several fundamental formulas, equal to 6.626×10^{-27} erg-sec.

positron—a subatomic particle having the identical mass but a charge opposite that of the electron.

potential energy—energy due to position.

prism—a transparent body bounded in part by two plane surfaces that are not parallel, usually with a triangular base; used for dispersing light into a spectrum by refraction.

proton—the hydrogen nucleus; also, the heavy positive particles in other atomic nuclei. An element's atomic number equals the number of protons in its atoms. Protons are approximately 1840 times as heavy as electrons and have a charge equal but opposite in sign to that of the electron.

polychromatic light—light containing many wavelengths or frequencies; that is, many colors of the spectrum. White light is polychromatic light.

principal shell—see *electron shell*.

quantum—a pulse or "packet" of energy. A *photon* is a quantum of electromagnetic radiation energy.

radioactivity—the spontaneous disintegration of unstable atomic nuclei, accompanied by the emission of radiation (mainly alpha, beta, or gamma rays).

rare earth—one of 14 (or 15 if La is included) elements between lanthanum (no. 57) and hafnium (no. 72). Also called *lanthanide*.

reduced mass (symbol μ)—a mathematical quantity found convenient in some types of calculation:

$$\mu = \frac{m_1 m_2}{m_1 + m_2}$$

where m_1 and m_2 are the masses of two bodies.

reflection grating—a reflecting surface, such as polished metal or a mirrored piece of glass, on which many thousands of fine lines have been ruled. Radiation reflected from the mirrored surfaces between the scratches yields a spectrum due to interference effects.

refraction—the bending of light when it passes obliquely from one transparent medium to another. (Prisms refract light.)

Rydberg constant—the proportionality constant in the Rydberg-Ritz formula for atomic spectra. For hydrogen, $R = 109,677.58$ cm^{-1} or $3.288 \ldots 10^{15}$ sec^{-1}.

saturated compound—a compound containing only single bonds between atoms.

second—a unit of time, defined in terms of an "atomic clock" (*see* Appendix 2).

second ionization energy—see *ionization energy*.

shell—see *electron shell*.

significant figures—those digits of a number which are judged to be reasonably trustworthy (*see* Appendix 1).

solute—the dissolved substance in a solution—e.g., salt in a saline solution.

solution—a homogeneous mixture (of variable composition) of two or more substances. Or, a random distribution of the molecules, atoms, or ions of two or more substances. Solutions may be liquid (salt or alcohol in water), solid (amalgams), or gaseous (any two or more gases).

solvent—the dissolving medium in a solution—e.g., water in a saline solution.

specific heat—the amount of heat, in calories, required to raise the temperature of one gram of a substance by one degree Celsius.

spectroscopy—the study of the constituents of light, or radiation. The study of the interaction of radiation and matter.

spectrum—the band of colors produced by the dispersion of polychromatic light. The colors of the visible spectrum are violet (shortest wavelength—4000 A), blue, green, yellow, orange, and red (longest wavelength—8000 A). The term *spectrum* has been extended to include invisible radiation such as gamma rays, X rays, ultraviolet radiation, infrared radiation, and radio waves (*see* Fig. 13-4).

speed—distance traveled per unit time.

standing waves—waves prevented from moving along, for example, sound waves in an organ pipe and vibrations of a violin string.

statcoulomb—the cgs electrostatic unit of charge, also known as *esu*. It is the amount of charge which, when placed at a distance of 1 cm from a charge of the same sign and magnitude, will repel the latter charge with a force of 1 dyne.

structural formula—a formula showing how atoms are arranged in a molecule.

subshell—see *electron shell*.

symbol (chemical)—the one- or two-letter abbreviation for each element—e.g., H for hydrogen, He for helium, Na for sodium.

temperature scales—see Appendix 2.

term (spectroscopic)—the expression R/n^2, representing frequency or wave number in the Rydberg-Ritz formula for atomic spectra.

theory—a proposed explanation. A group of general propositions attempting to explain observed phenomena; e.g., Newton's corpuscular theory of light attempts to explain such phenomena as reflection and refraction.

transition element (or metal)—one of the elements in the middle of the periodic table between the beryllium and boron families.

transmission grating—a transparent material, often a flat piece of glass, on which many thousands of fine lines have been ruled. Radiation passing through the clear glass between the scratches is dispersed into a spectrum (due to diffraction and interference effects.)

traveling waves—propagating waves such as ocean waves and light waves.

triple bond—a bond between two atoms in which three electron pairs are shared.

tritium—an isotope of hydrogen of atomic weight 3.

valence—see *combining power.*

vector—an arrow used to indicate the direction and magnitude (by its length) of a quantity such as force or velocity.

vector quantity—a quantity that has both magnitude and direction, such as velocity, force, momentum.

velocity—distance traveled per unit time in a given direction. (Velocity is a vector quantity.)

volt—a unit of electric potential.

wave front—the tangent drawn to the foremost points of propagating waves (see Figs. 12-22 and 12-23).

wavelength—the distance between adjacent crests (or adjacent troughs) of a wave.

wave number—the number of waves in a given distance interval (e.g., so many waves per centimeter, written as cm^{-1}, called reciprocal centimeters).

weight—the gravitational force acting on a body. (Sometimes, but erroneously, used as a synonym for mass.)

wet chemistry—the kind of chemistry wherein most reactions are carried out in solution (a rather loose term).

white light—a mixture of all the colors of the visible spectrum.

work—the product of a force and the distance through which it acts. Common force units are the foot-pound, the erg, and the joule.

X *ray*—high-frequency electromagnetic radiation, of wavelength from approximately 0.01 to 100 A.

Name Index

Anderson, Carl David (1905-   ), 161
Angström, Anders Jonas (1814-1874), 144, 399
Aristotle (384-322 B.C.), 3, 206
Arrhenius, Svante (1859-1927), 118
Aschkinass, 399
Aston, Francis W. (1877-1945), 49, 153, 388
Avogadro, Amedeo (1776-1856), 30, 265

Balmer, Johann (1825-1898), 210, 223
Bartlett, Neil (1932-   ), 209
Becquerel, Henri Antoine (1852-1908), 136
Bernoulli, Daniel (1700-1782), 239
Berthollet, C. L. (1748-1822), 16
Berzelius, Jöns (1779-1848), 32, 36n, 39, 118, 335, 378
Bjerrum, N. J., (1879-1958), 400
Bohr, Niels (1885-1962), 157n, 218, 223-224, 233, 235
Born, Max (1882-1970), 294-295
Boyle, Robert (1627-1691), 4-5
Brackett, 214
Bradley, James (1692-1762), 270
Bragg, William H. (1862-1942), 286n
Brahe, Tycho (1546-1601), 104
Broglie, de, Louis (1892-   ), 275, 386
Bunsen, Robert Wilhelm (1811-1899), 205, 399
Butlerov, A. M., (1828-1886), 378

Cannizzaro, Stanislao (1826-1910), 39-40
Carlisle, Anthony (1768-1840), 117
Cassini, G. D. (1625-1712), 173n
Cavendish, Henry (1731-1810), 48, 114, 207
Celsius, Anders (1701-1744), 413
Chadwick, James (1891-   ), 103n, 160, 386
Charles II of England (1630-1685), 105
Claude, Georges (1870-1960), 209
Coblentz, 399
Copernicus, Nicolaus (1473-1543), 104
Coulomb, Charles A. (7136-1806), 114
Couper, A. S. (1831-1892), 378
Crookes, William (1832-1919), 125, 208
Curie, Marie Sklodowska (1867-1934), 138
Curie, Pierre (1859-1906), 138

Dalton, John (1766-1844), 10, 14-16, 27-30
Davisson, Clinton J., (1881-1958), 276, 386
Davy, Humphry (1778-1829), 118, 340
Democritus (460-370 B.C.), 14
Dempster, A. J., (1886-1950), 388
Descartes, René (1596-1650), 105
Dirac, P. A. M. (1902-   ), 295
Döbereiner, Johann (1780-1849), 61
Dorn, F. E. (1848-1916), 209
Du Fay, C. (1698-1739), 111

Dulong, Pierre (1785-1838), 42-46
Dumas, Jean-Baptiste (1800-1884), 36

Einstein, Albert (1879-1955), 220, 229
Empedocles (490-430 B.C.), 3
Euler, Leonhard (1707-1783), 239

Fahrenheit, Gabriel Daniel (1686-1736), 413
Faraday, Michael (1791-1867), 199, 125, 285n, 362, 378
Fermi, Enrico (1901-1954), 103n
Fizeau, A. H. L. (1819-1896), 173
Foucault, J. B. L. (1819-1869), 171, 173, 205
Franck, James (1882-1964), 248
Franklin, Benjamin (1706-1790), 112, 127n
Fraunhofer, Joseph (1787-1826), 183, 204, 397, 399
Fresnel, Augustin (1788-1827), 180
Frisch, Otto R. (1904-   ), 218n

Galilei, Galileo (1564-1642), 81, 84, 104, 167, 171-172, 189, 403
Galvani, Luigi (1737-1798), 115
Gay-Lussac, J. L., (1778-1850), 22
Geiger, Hans (1882-1947), 147
Geissler, Heinrich (1814-1879), 125
Germer, Lester H. (1896-   ), 276, 386
Gilbert, William (1540-1603), 110
Goldstein, Eugen (1850-1930), 132, 135n, 387
Grimaldi, Francesco (1618-1663), 176
Guericke, Otto von (1602-1686), 105, 115n

Harding, Warren G. (1865-1923), 139n
Harvey, William (1578-1657), 105
Heisenberg, Werner (1901-1965), 287, 295
Heraclitus (540-475 B.C.), 3
Herschel, John Frederick William (1792-1871), 195, 399
Herschel, William (Friedrich Wilhelm) (1738-1822), 194, 397
Hertz, Gustav (1887-   ), 248
Hertz, Heinrich (1857-1894), 196, 227
Hillebrand, W. F. (1853-1925), 208
Hitler, Adolf (1889-1945), 267n
Holmes, Sherlock, 267n
Hooke, Robert (1635-1703), 105, 177
Humphreys, C. J., 1898-   ), 214, 235
Huygens, Christian (1629-1695), 105, 177

Janssen, Pierre Jules Cesar (1824-1907), 207
Joliot, Frederic (1900-1958), 138n
Joliot-Curie, Irene (1897-1956), 138n
Joule, James P. (1818-1889), 101, 414
Julius, 399

Kekule, F. A., (1829-1896), 225, 378
Kelvin, Lord (William Thomson) (1824-1907), 143, 413

Subject Index

in atoms, 239, 255
in molecules, 345
Inverse-Square Law of Light Intensity, 170
Ion-dipole attraction, 372
Ionic bond, 336–340
Ionic crystals, properties of, 338, 340
Ionization energy, chart of, 258, 321
    definition of, 247, 249
    of helium, 250, 254
    of hydrogen, 247, 249
    of sodium, 336
    table of, 256
Ions, in canal rays, 132, 152–153
    charges on, 121, 264
    and electrolysis, 118–121, 264
    mass-to-charge ratios of, 123, 132, 152, 387
    in solutions and crystals, 259, 337
Isomers, 362–371
    classification of, 371
    definition of, 363
    numbers of, 366
Isotopes, 18, 48
    discovery of, 152–153, 388
    definition of, 152, 160
    origin of name, 154n
    uses of, 154

Joule (energy unit), 96, 414
Jupiter, moons of, 172

Kelvin scale, 413
Kilocalorie, 336, 414
Kilogram, 412
Kinetic energy, 97–98
    of orbiting electron, 243
    of a particle in a box, 301, 304
    of revolving bodies, 238n
Kinetic theory of gases, 26
Krypton, discovery of, 209

Lanthanides, 70, 323
Large calorie, 336, 414
Lattice defects, 18–19
Laws of nature (general), 82, 266
Lead isotopes, 152
Length, units of, 411–412, 415
Levorotatory compounds, 369
Light, aberration of, 270
    corpuscular theory of, 169–171, 233
    diffraction of, 176–178
    dispersion of, 168, 177, 186, 188
    dual nature of, 232–233, 292
    energy of, 193, 200, 220
    interference of, 275, 288
    momentum of, 275, 288
    monochromatic, 168, 188
    photons of, see Quantum theory of light
    polychromatic, 188
    pressure of, 288
    quantum theory of, 231–233
        quanta, 219, 228
    rectilinear propagation of, 170, 177

reflection of, 170
refraction of, 171, 173, 185, 187
speed of, 172
    in media, 171, 173, 175
    in vacuum, 175, 416
    visible, wavelengths of, 183, 199, 394
    wave theory of, 177–186
    see also Electromagnetic radiation; Electro-
        magnetic spectrum; Electromagnetic
        theory; and Electromagnetic waves
Light Intensity, Inverse-square Law of, 170
Light source in spectroscopes, 393
Light-year, 189
Logarithms, 105
Loschmidt's number, see Avogadro's number
Limiting reagent, 53
Liter, 412
Lyman series, 212, 214, 223, 246

Macromolecules, 348
Magic numbers, 263, 313, 322
Magnetic field, effect, on alpha and beta
        particles, 139
    on atoms, 312
    interaction with electric field, 124, 126, 196
Magnetic quantum number, 310
Malleability of metals, 350
Manometer, 105
Mars-Earth distance, 173n
Mass, conservation of, 6, 177
    of electron, 132, 416
    units of, 412
    versus weight, 83–85
Mass spectrometer, 49, 153, 387–389
Mass-to-charge ratio, of alpha particles, 140
    of electron, 129, 285
    of hydrogen atom, 123, 216n
    of ions, 129, 132, 387
Methematical procedures, 403–410
Matrix mechanics, 295; see also Wave
        mechanics
Matter waves, 276, 279
Mean free path, 145
Measurement limitation of, 287
Mechanical energy, 100
Mechanical equivalent of heat, 101, 414
Melting point, definition of, 339
Melting points, of fluorides, 351
    of ionic crystals, 339, 372
    of molecular crystals, 348, 372
Mercury vapor, 36
Metallic bond, 349–350
Metallic character, 71, 328
Metals, and nonmetals, 328
    number of, 65
    properties of, 65–66, 71, 350
Meter, 411
Method of Holmes, 267n
Metric system, 411–416
Michelson-Morley experiment, 175n
Microwaves, 199, 397
Mixtures, 12–13
Models, 35

Eastern College Library
St. Davids, PA

3 1405 001 59281 9

QD
461
.L5